Special Publications 73

SCIENTIFIC INTEGRITY AND ETHICS IN THE GEOSCIENCES

Edited by
Linda C. Gundersen

T0176632

This Work is a co-publication of
the American Geophysical Union and John Wiley and Sons, Inc.

WILEY

Published under the aegis of the AGU Publications Committee

Brooks Hanson, Senior Vice President, Publications

Lisa Tauxe, Chair, Publications Committee

For details about the American Geophysical Union visit us at www.agu.org.

Wiley Global Headquarters

111 River Street, Hoboken, NJ 07030, USA

For details of our global editorial offices, customer services, and more information about Wiley
products visit us at www.wiley.com.

Limit of Liability/Disclaimer of Warranty

Library of Congress Cataloging-in-Publication data is available.

ISBN: 978-1-119-06778-8

Cover image: Global view of the city lights of North and South America, composite (April and
October 2012) from the Suomi National Polar-orbiting Partnership satellite. Shows Hurricane
Sandy, illuminated by moonlight. NASA Earth Observatory image by Robert Simmon,
data provided courtesy of Chris Elvidge (NOAA National Geophysical Data Center).

Cover design: Wiley

Set in 10/12pt Times New Roman by SPi Global, Pondicherry, India

10 9 8 7 6 5 4 3 2 1

CONTENTS

CONTRIBUTOR LIST

David M. Abbott, Jr.
Chair of the Ethics Committee
American Institute of Professional
Geologists
Consulting Geologist
Denver, Colorado, USA

Melissa S. Anderson
Professor
Director of Graduate Studies
University of Minnesota
Minneapolis, Minnesota, USA

Thomas Arrison
Program Director
Policy and Global Affairs
National Academies of Sciences,
Engineering, and Medicine
Washington, D.C., USA

Peter Bobrowsky
Senior Scientist
Geological Survey of Canada
Sydney, British Columbia, Canada

Maeve A. Boland
Director of Geoscience Policy
American Geosciences Institute
Alexandria, Virginia, USA

Monica Z. Bruckner
Science Education and Evaluation
Associate
Science Education Resource Center
Carleton College
Northfield, Minnesota, USA

Richard A. Coleman
Science Integrity Coordinator
U.S. Geological Survey
Denver, Colorado, USA

Vincent S. Cronin
Professor
Department of Geosciences
Baylor University
Waco, Texas, USA

Giuseppe Di Capua
Research Geologist
Istituto Nazionale di
Geofisica e Vulcanologia
(Italian Institute of Geophysics
and Volcanology)
Rome, Lazio, Italy

John W. Geissman
Professor
Department of Geosciences
University of Texas at Dallas
Richardson, Texas, USA

Linda C. Gundersen
Geologist, Retired
U.S. Geological Survey
Ocean View, Delaware, USA

Brooks Hanson
Senior Vice President, Publications
American Geophysical Union
Washington, D.C., USA

Susan W. Kieffer
Professor Emeritus
Department of Geology
Emeritus Walgreen University Chair
University of Illinois
Champaign, Illinois, USA

Sabine Kleinert
Senior Executive Editor
The Lancet
London, United Kingdom

Vance S. Martin
Research Associate
National Center for Professional &
Research Ethics
University of Illinois at
Urbana-Champaign
Illinois, USA

Tony Mayer
European Representative and
Research Integrity Officer
Nanyang Technological University
Republic of Singapore

Michael McPhaden
Senior Scientist
NOAA/Pacific Marine
Environmental Laboratory
Seattle, Washington, USA

David W. Mogk
Professor
Department of Earth Sciences
Montana State University
Bozeman, Montana, USA

Robert M. Nerem
Professor and Parker H. Petit
Distinguished Chair Emeritus
Petit Institute for Bioengineering
and Bioscience
Georgia Institute of Technology
Atlanta, Georgia, USA

Silvia Peppoloni
Research Geologist
Secretary General IAPG,
Istituto Nazionale di Geofisica e
Vulcanologia (Italian Institute of
Geophysics and Volcanology)
Rome, Lazio, Italy

Nicholas H. Steneck
Professor Emeritus of History
University of Michigan
Research Integrity Consultant
Ann Arbor, Michigan, USA

Alan D. Thornhill
Director, Western Ecology Division
Environmental Protection Agency
Corvallis, Washington, USA

Donna C. Tonini
Research Associate
National Center for Professional &
Research Ethics
University of Illinois at
Urbana-Champaign
Illinois, USA

Nancy Tuana
DuPont/Class of 1949 Professor
of Ethics
Department of Philosophy
Director, Rock Ethics Institute
The Pennsylvania State University
University Park, Pennsylvania, USA

John W. Williams
Past President, National Association
of State Boards of Geology (ASBOG)
Professor Emeritus
Department of Geology
San José State University
San Jose, California, USA

PREFACE

Welcome to scientific integrity and ethics in the Anthropocene! We are now a densely populated and globally connected species, impacting every aspect of life on Earth, continually generating new technology, new substances, new data, and new challenges. We change the course of large river systems; destroy, replant, and harvest millions of acres of vegetation; modify the chemistry of water and soil on planetary scales; and move millions of tons of earth materials. From climate change to pandemics, from food and energy security to natural catastrophes, we are faced with high-stake dilemmas demanding solutions. Science and technology will need to help mitigate and solve these problems, especially the geosciences. However, without more careful attention to scientific integrity and ethics, we are headed into a dangerous future. Scientific integrity protects and upholds the framework of science itself, which is currently suffering under a barrage of research misconduct issues from data falsification and fabrication to systemic harassment and discrimination that is disrupting science and marginalizing women and minority students and scientists. Science skeptics have become even more outspoken, and some of that skepticism is being woven into public policy. Scientific misconduct fuels this skepticism and jeopardizes the trust the public has in the scientific enterprise.

Integrity and trust are the foundations of science. This is, in fact, the only way that science actually works. Every scientist trusts that the knowledge and data they use from other scientists is the truth and was produced honestly, objectively, and with integrity. Society is dependent on science and generally believes what scientists communicate and the way science is incorporated into their lives. Without trust and integrity, the system breaks down, and as a result advancement in science is hindered, time and research funds are wasted, society may be harmed and lose confidence in scientific institutions, there may be significant financial impacts to individuals and corporations, and funding for new science may diminish. There is a great deal at stake when a scientist chooses to forgo integrity. Science drives a significant portion of the wealth, health, safety, and well-being of our world, and here in the twenty-first century, science is also in the midst of significant change. The world is increasingly complex, making our integrity and ethical challenges more complex. The conduct of science is transitioning from individual-based, single-discipline research to large teams with multidisciplinary approaches. Scientific education, funding, and hiring are more competitive. The scientific community is connected and global with different cultural attitudes toward the scientific process and integrity. Data and communication are instantaneous, and technology and data accessibility are advancing at an unprecedented pace, well ahead of policy, standards, and our ability to adapt to them.

Generally, scientific or research integrity codes focus on the individual behavior of the scientist, standards of professional behavior and knowledge, and integrity in the scientific process and publications; they may contain guidance on ethical treatment of humans, animals, and the environment when conducting science. Some codes also include rules on bias, conflict of interest, privacy, confidentiality, and issues of quality that may affect the integrity of the data and interpretation. Ethics underpins scientific integrity but also needs to be a foundation for our decisions regarding how we undertake science and the application of the scientific advancements we make. Ethics in science includes our broader responsibilities to society, moral decisions on the subject and use of science, and our behavior and interactions with both the scientific community and the public. The development of professional or applied ethical codes is well established in the medical, biological, and engineering fields and more recently in the environmental, geographic, and geoscience fields. Geoethics is one such emerging field that has garnered significant attention in the last few years through the focused efforts of several organizations and scientists. Chief among them is the International Association for Promoting Geoethics, which recently released the Cape Town Statement on Geoethics, the first international set of applied ethical principles for the geosciences (http://www.geoethics.org/ctsg). Additionally, the past 10 years have seen the emergence of new global and national scientific integrity codes, the emergence of new applied ethics codes and ideas, and the growing awareness of unacceptable behaviors in the research and educational environment. This volume presents an overview of the current thinking on scientific integrity and ethics from academic, professional, and governmental perspectives, with particular attention to the geosciences. Much of this book is also applicable to all the sciences, addressing common issues such as publishing, data stewardship, and the need for scientific integrity and ethics education for students and early career scientists.

The first section of the book features new codes and reports that are having a strong influence on the landscape of scientific integrity. Chapter 1, on the Singapore and Montreal Statements, discusses the first international research integrity codes, created by the historical World Research Integrity Conferences, that speak beyond traditional fabrication, falsification, and plagiarism and include strong statements on professional behavior, collaboration, and values. Chapter 2 provides insight into the new National Academy of Science report on research integrity that breaks new ground, defining six core values that shape the norms of research, and goes beyond traditional research misconduct by examining detrimental research practices. Chapter 3 provides in detail the Department of Interior Scientific Integrity and Ethics Policy that set the standard for new policies in federal science agencies in the wake of the landmark Memorandum on Scientific Integrity from President Obama in March of 2009.

The second section of the book examines the latest codes of conduct from several major geoscience professional societies and the challenges they face in the current science environment in supporting research integrity and ethical values. Chapter 4 presents the new American Geosciences Institute Guidelines for Ethical and Professional conduct that has been adopted by most American geoscience societies. Chapter 5 presents a discussion by the American Geophysical Union's past president on the society's recent scientific integrity and ethics policy, the challenges faced implementing it, outreach efforts, and the latest update that encompasses discrimination, harassment, and bullying. Chapters 6 and 7 provide current and historical perspectives from geoscience industry groups, including the National Association of State Boards of Geology, the American Association of Petroleum Geologists, and the Society of Economic Geologists, with a particular emphasis on the ethical issues most valued by professional geologists and the importance of enforceable codes.

The third section of the book addresses two very critical subjects in science: publications and data stewardship. Chapter 8 discusses the past and present ethical issues in science publication, the industry-wide challenge to scientific journals related to reproducibility, and the new movement in publication to ensure reliability and provide the data that underpin published science. Chapter 9 walks the reader through the scientific process within the framework of the research data lifecycle, providing checklists of practical ethical questions for every step of the lifecycle that students and faculty can use to ensure the ethics and integrity of their science.

The fourth section of the book introduces the concept of value and ethics in conducting science and the emerging field of geoethics. Geoscientists have traditionally stayed out of policy and secondary applications of their work. Increasingly geoscientists are asked to estimate risks, map areas of vulnerability, and think about the impact of their work on the health, benefit, and welfare of society. Chapter 10 discusses the role of ethical values in scientific integrity using the example of climate change, and Chapter 11 provides an extensive overview of the new field of geoethics.

The last section provides resources for educators on best practices for teaching scientific integrity, ethics, and geoethics. Chapter 12 provides strong support for an experiential approach to teaching integrity and ethics. Chapter 13 presents important understanding and best practices for teaching geoethics within the geoscience curriculum. Chapter 14 is an impassioned appeal on the importance of science ethics education for undergraduates that includes practical examples for implementation.

The book closes with two appendices providing teaching and reference resources for classroom practice and further research and understanding. It is hoped that students, faculty, and professionals in sciences and ethics will be able to use this book to learn, share, and dialogue about scientific integrity and ethics in this changing world and incorporate those lessons into their professional work and teaching.

Linda C. Gundersen
Ocean View, Delaware

ACKNOWLEDGMENTS

The editor would like to thank the American Geophysical Union for their leadership and efforts in promoting scientific integrity and geoethics, and in fighting harassment, discrimination, and bullying in the science environment. Much appreciation goes to Brooks Hanson, who first suggested creating this book, especially for his advice and support throughout its assembly. Thank you to my husband Joseph Smoot for his unfailing support and good counsel. Thank you to the many reviewers and copy editors for their excellent comments and edits, but especially to Kathie Rankin who unselfishly provided specialty editing for several papers. Finally, thank you to Dr. Ritu Bose, editor at Wiley for her patience and guidance throughout.

Section I: Examples of Recently Developed International and National Codes and Policies

1. THE ORIGIN, OBJECTIVES, AND EVOLUTION OF THE WORLD CONFERENCES ON RESEARCH INTEGRITY

Nicholas H. Steneck[1], Tony Mayer[2],
Melissa S. Anderson[3], and Sabine Kleinert[4]

Abstract

The World Conferences on Research Integrity (WCRI) have grown over the past decade from a proposal to convene a joint U.S.–European conference on research integrity into a global effort to foster integrity in research through research, discussion, the harmonization of policies, and joint action. Over the course of the first four WCRIs, held in Lisbon, Portugal, in 2007; Singapore in 2010; Montreal, Canada, in 2013; and Rio de Janeiro, Brazil, in 2015, participation has grown from 275 participants from 47 countries in 2007 to 474 participants from 48 countries in 2015. The WCRIs have produced two global statements on research integrity: the Singapore Statement in 2010 and the Montreal Statement in 2013. In addition, three sets of proceedings and numerous papers and working reports archived on the WCRI website (www.researchintegrity.org) are available. The WCRI effort celebrated its tenth anniversary at the Fifth WCRI in Amsterdam, May 28–31, 2017. A total of 836 participants from 52 countries attended.

[1] University of Michigan, Ann Arbor, Michigan, USA
[2] Nanyang Technological University, Republic of Singapore
[3] University of Minnesota, Minneapolis, Minnesota, USA
[4] The Lancet, London, United Kingdom

Scientific Integrity and Ethics in the Geosciences, Special Publications 73,
First Edition. Edited by Linda C. Gundersen.
© 2018 American Geophysical Union. Published 2018 by John Wiley & Sons, Inc.

1.1. Introduction

In an ideal world, integrity should be a regular element of all aspects of research. In practice, it is too often a topic that gets attention when there is a crisis and then is put on the shelf until the next crisis arises. Thus, over the 40 or so years that research integrity has been a topic of public discussion, universities, professional societies, and governments have responded to crises, issued reports, and then, too often, moved on to other issues, hoping that no further crises would arise.

The World Conferences on Research Integrity have evolved into an ongoing forum for the study and discussion of ways to promote responsible behavior in research. This was not, however, the goal of the initial and somewhat audaciously titled "World Conference on Research Integrity" held in Portugal in 2007. The aim of the initial conference was more modest.

The World Conferences began as an experimental extension of the U.S. Office of Research Integrity's (ORI) conference program to Europe. In 2000, ORI's authority was "changed to focus more on preventing misconduct and promoting research integrity through expanded education programs" [*Federal Register*, 2000]. Under its new authority, ORI initiated programs aimed at improving researcher training and engaging researchers and professional organizations in the discussion of integrity in research. The conference program (small grants to organizations and institutions to organize conferences) was part of this effort. In 2006, a consultant working at ORI, Nicholas Steneck, University of Michigan, was heading to Europe for an academic meeting and suggested that he explore the possibility of holding a Europe–United States conference to discuss research integrity issues of common interest. The ORI Director, Chris Pascal, and the Director of the Division of Education and Integrity, Larry Rhoades, agreed to provide $25,000 for this effort, with the understanding that a European partner be found to match ORI funding.

In 2006, a number of European countries and groups of European researchers were engaged in efforts to develop misconduct policies and otherwise promote integrity in research. However, most did not have enough funding to support a collaborative U.S.–European conference. In a series of meetings, World Conference initiator Steneck was assured of European interest in promoting integrity but received no commitment of support until one final meeting in Strasbourg with European Science Foundation (ESF) Chief Executive, Bertil Andersson.

While some countries had responded to research misconduct incidents at the national level, ESF was the first European organization to formally engage the topic of research integrity in its 2000 Science Policy Briefing, *Good Scientific Practice in Research and Scholarship* [*ESF*, 2000]. Andersson was deeply committed to taking an active role in promoting integrity in research and quickly agreed to match ORI's funding. More importantly, he also agreed to take the lead in seeking additional support in Europe, starting with the European Commission, and appointed an ESF consultant, Tony Mayer, to co-organize and co-chair the proposed joint U.S.–European conference on research integrity. From this agreement on, Co-Chairs Steneck and Mayer assumed major responsibility for securing funding and organizing the first World Conference on Research Integrity.

1.2. The First World Conference on Research Integrity

With strong encouragement from Andersson and colleagues consulted during the early planning process, Co-Chairs Steneck and Mayer broadened the U.S.–European plan to an International Conference for Fostering Responsible Research, justifying the effort in their unpublished planning report to the ESF and ORI as follows:

> Research, which prides itself on its internal self-governance and its integrity, is now faced with a number of well publicized cases of misconduct, fraud and questionable research practices. The research community worldwide has to face this challenge in order to retain public confidence and establish clear best practice frameworks at an international level.

However, planning also included the need to address "questionable research methods and environments in which such methods are tolerated." With these broad objectives in mind, the overall purpose of the first World Conference was

> … to assemble an international group of researchers, research administrators from funding agencies and similar bodies, research organizations performing research, universities and policy makers for the purpose of discussing and making recommendations on ways to 1) improve, 2) harmonize, 3) publicize, and 4) make operationally effective international policies for the responsible conduct of research.

At roughly the same time that planning for the first WCRI began, two members of the Organization of Economic Cooperation and Development (OECD), Canada and Japan, proposed the development of a Working Group on research integrity, with the goal of producing recommendations for action by all OECD members [*OECD*, 2007]. Steneck and Mayer soon established a collaborative working relationship with this effort and also began working with the International Council of Science (ICSU), which was also interested in increasing attention to integrity by the global scientific community [*ICSU*, 2002]. And most importantly, through the efforts of Andersson and Mayer, the European Commission agreed to provide major support for the first WCRI and to encourage Portugal to host the Conference during its upcoming presidency of the European Union. Through these and other related developments, what became the founding WCRI was set for September 2007 in Lisbon, Portugal, at and with further support of the Gulbenkian Foundation headquarters.

Opening talks by the Portuguese Minister of Science, the late Jose-Mariano Gago, the European Commissioner of Research Janez Potocnik, and others, challenged participants to engage the issues through discussion and further action. As summarized in the final report [*Mayer and Steneck*, 2007], over the 2.5 days of meetings, the 275 participants from 47 countries participated in "a series of plenary sessions, three working groups, formal opening and closing sessions, and other events designed to promote discussion and begin a global exchange about ways to foster responsible research practices." More information on the first WCRI will be available on the World Conference for Research Integrity Foundation website: researchintegrity.org.

1.3. The Second World Conference on Research Integrity

One of the outcomes of the first WCRI was support for convening a second global conference, with some preference for a country in the rapidly expanding Asian research world. Given that by the time of the first WCRI, both Andersson (as Provost) and Mayer had moved to the Nanyang Technological University (NTU) in Singapore, Singapore quickly became the logical site for the Second WCRI. The NTU is one of the two highly ranked, research intensive universities in the city state, the other being the National University of Singapore (NUS).

Working within the Singaporean system, Andersson and Mayer were able to mobilize substantial funding for the Second WCRI through the two major universities (NTU and NUS), the Singapore Management University (SMU), and the Agency for Science, Technology and Research (A*STAR). All four institutions had high international research profiles and recognized the importance of carrying out research to the highest standards of integrity. In addition to these organizations, the Ministry of Education provided significant extra funding. The organizers also had the financial support of a number of other organizations, including the Committee on Publication Ethics (COPE), which has supported every WCRI held to date. The level of funding achieved enabled not only the support for the conference program but also provided the wherewithal for Co-Chairs Steneck and Mayer to offer modest travel grants to participants from disadvantaged countries. This was an important development in the transformation of research integrity into a global issue.

The Second WCRI was a truly global event with more than 340 participants from 51 countries attending. Building on the results of the first WCRI, the Second WCRI focused on national and international structures for promoting integrity and responding to misconduct, global codes of conduct and best practices for research, common curricula for training students and researchers in best practices, and uniform best practices for editors and publishers [*Mayer and Steneck*, 2012].

During planning for the Second WCRI, Steneck proposed developing some lasting legacy from the conference, such as a global code of conduct for research. With Planning Committee support, Steneck, Mayer, and Melissa Anderson, University of Minnesota, took the lead in drafting the *Singapore Statement on Research Integrity* (Box 1.1). A draft *Singapore Statement* was sent to all

Box 1.1   The Singapore Statement on Research Integrity.

Preamble. The value and benefits of research are vitally dependent on the integrity of research. While there can be and are national and disciplinary differences in the way research is organized and conducted, there are also principles and professional responsibilities that are fundamental to the integrity of research wherever it is undertaken.

Principles

Honesty in all aspects of research
Accountability in the conduct of research
Professional courtesy and fairness in working with others
Good stewardship of research on behalf of others

Responsibilities

1. *Integrity:* Researchers should take responsibility for the trustworthiness of their research.
2. *Adherence to Regulations:* Researchers should be aware of and adhere to regulations and policies related to research.
3. *Research Methods:* Researchers should employ appropriate research methods, base conclusions on critical analysis of the evidence, and report findings and interpretations fully and objectively.
4. *Research Records:* Researchers should keep clear, accurate records of all research in ways that will allow verification and replication of their work by others.
5. *Research Findings:* Researchers should share data and findings openly and promptly, as soon as they have had an opportunity to establish priority and ownership claims.
6. *Authorship:* Researchers should take responsibility for their contributions to all publications, funding applications, reports, and other representations of their research. Lists of authors should include all those and only those who meet applicable authorship criteria.
7. *Publication Acknowledgement:* Researchers should acknowledge in publications the names and roles of those who made significant contributions to the research but do not meet authorship criteria, including writers, funders, sponsors, and others.
8. *Peer Review:* Researchers should provide fair, prompt, and rigorous evaluations and respect confidentiality when reviewing others' work.

9. **Conflict of Interest:** Researchers should disclose financial and other conflicts of interest that could compromise the trustworthiness of their work in research proposals, publications, and public communications as well as in all review activities.

10. **Public Communication:** Researchers should limit professional comments to their recognized expertise when engaged in public discussions about the application and importance of research findings and clearly distinguish professional comments from opinions based on personal views.

11. **Reporting Irresponsible Research Practices:** Researchers should report to the appropriate authorities any suspected research misconduct, including fabrication, falsification, or plagiarism, and other irresponsible research practices that undermine the trustworthiness of research, such as carelessness, improperly listing authors, failing to report conflicting data, or the use of misleading analytical methods.

12. **Responding to Irresponsible Research Practices:** Research institutions, as well as journals, professional organizations and agencies that have commitments to research, should have procedures for responding to allegations of misconduct and other irresponsible research practices and for protecting those who report such behavior in good faith. When misconduct or other irresponsible research practice is confirmed, appropriate actions should be taken promptly, including correcting the research record.

13. **Research Environments:** Research institutions should create and sustain environments that encourage integrity through education, clear policies, and reasonable standards for advancement, while fostering work environments that support research integrity.

14. **Societal Considerations:** Researchers and research institutions should recognize that they have an ethical obligation to weigh societal benefits against risks inherent in their work.

participants prior to the conference and became an underlying theme for much of the discussion during the meeting. This effort paralleled the ESF/All European Academies (ALLEA) initiative to develop a *European Code of Conduct on Research Integrity* [*ESF/ALLEA*, 2011]. At the closing session, participants acting as individuals rather than as institutional representatives discussed the few areas where there were differences of opinion about coverage and/or wording. Finding proper wording for Responsibility 14, Social Considerations, took the most time. At the end of the session, those present broadly endorsed the code, pending a few minor revisions. These revisions were made after the Second WCRI and sent to all participants for comments and approval. The final 4 principles and 14 responsibilities set out in the *Singapore Statement* were then posted on the Web and have since been translated into 27 languages [*Singapore Statement*, 2010].

The Second WCRI achieved its objective to consolidate the work of the first WCRI and set the pattern for an ongoing series of World Conferences on Research Integrity. Having held meetings in Europe and Asia, consideration was given to other major regions. Steneck and Mayer also wanted to bring in new leadership and turned to Melissa Anderson to take on organizing and chairing responsibilities. She quickly brought in Sabine Kleinert, from *The Lancet*, to continue the practice of having conference co-organizers and co-chairs. Exchanges between the new conference Co-Chairs and the Conference Services Office of the National Research Council Canada confirmed a mutual interest in siting the conference in Montréal, Canada, in May 2013 and established a financial mechanism for support through the council's practice of funding conferences on a reimbursement basis.

1.4. The Third World Conference on Research Integrity

The Third WCRI continued the practice of previous conferences in engaging government officials, publishers, and leaders in policy and education, but it also intentionally recruited participants who were actively conducting research on or relating to the responsible conduct of research. A broad search through publications in the field yielded a list of hundreds of scholars who had recently published research on research integrity. To encourage their participation, Anderson and Kleinert issued a broad call for presentation proposals. The many presentation proposals received in response to this call led to the decision to expand the conference from 2.5 to 3 full days. Attendance at the Third WCRI grew to 366 participants from 44 countries.

Building on the success of the workshops that concluded the Second WCRI, the Third WCRI incorporated four tracks of focused discussions on the following topics: integrity in international research collaborations, cooperation between research institutions and journals in cases of suspected misconduct, education in the responsible conduct of research, and research integrity in relation to societal responsibility [*Steneck et al.*, 2015]. The track related to international research collaborations was devoted to discussion of a draft document that was eventually published as the *Montréal Statement on Research Integrity in Cross-Boundary Research Collaborations* [2013]. The *Montréal Statement* (Box 1.2) is intended to serve as a companion document to the *Singapore Statement*. The 20-point document focuses on aspects of research integrity that have particular relevance to collaborative research that crosses national, institutional, disciplinary, or sector boundaries (the last representing, for example, public-private or academy-business collaborations). It is now available in 14 different languages.

Box 1.2   The Montreal Statement on Research Integrity in Cross-Boundary Research Collaborations.

Preamble. Research collaborations that cross national, institutional, disciplinary, and sector boundaries are important to the advancement of knowledge worldwide. Such collaborations present special challenges for the responsible conduct of research because they may involve substantial differences in regulatory and legal systems, organizational and funding structures, research cultures, and approaches to training. It is critically important, therefore, that researchers be aware of and able to address such differences, as well as issues related to integrity that might arise in cross-boundary research collaborations. Researchers should adhere to the professional responsibilities set forth in the *Singapore Statement on Research Integrity*. In addition, the following responsibilities are particularly relevant to collaborating partners at the individual and institutional levels and fundamental to the integrity of collaborative research. Fostering the integrity of collaborative research is the responsibility of all individual and institutional partners.

Responsibilities of Individual and Institutional Partners in Cross-Boundary Research Collaborations

General Collaborative Responsibilities

1. *Integrity.* Collaborating partners should take collective responsibility for the trustworthiness of the overall collaborative research and individual responsibility for the trustworthiness of their own contributions.
2. *Trust.* The behavior of each collaborating partner should be worthy of the trust of all other partners. Responsibility for establishing and maintaining this level of trust lies with all collaborating partners.
3. *Purpose.* Collaborative research should be initiated and conducted for purposes that advance knowledge to the benefit of humankind.
4. *Goals.* Collaborating partners should agree at the outset on the goals of the research. Changes in goals should be negotiated and agreed to by all partners.

Responsibilities in Managing the Collaboration

5. *Communication.* Collaborating partners should communicate with each other as frequently and openly as necessary to foster full, mutual understanding of the research.
6. *Agreements.* Agreements that govern collaborative research should be understood and ratified by all collaborating partners. Agreements that

unduly or unnecessarily restrict dissemination of data, findings, or other research products should be avoided.

7. *Compliance with Laws, Policies, and Regulations.* The collaboration as a whole should be in compliance with all laws, policies, and regulations to which it is subject. Collaborating partners should promptly determine how to address conflicting laws, policies, or regulations that apply to the research.

8. *Costs and Rewards.* The costs and rewards of collaborative research should be distributed fairly among collaborating partners.

9. *Transparency.* Collaborative research should be conducted and its results disseminated transparently and honestly, with as much openness as possible under existing agreements. Sources of funding should be fully and openly declared.

10. *Resource Management.* Collaborating partners should use human, animal, financial, and other resources responsibly.

11. *Monitoring.* Collaborating partners should monitor the progress of research projects to foster the integrity and the timely completion and dissemination of the work.

Responsibilities in Collaborative Relationships

12. *Roles and Responsibilities.* Collaborating partners should come to mutual understandings about their roles and responsibilities in the planning, conduct, and dissemination of research. Such understandings should be renegotiated when roles or responsibilities change.

13. *Customary Practices and Assumptions.* Collaborating partners should openly discuss their customary practices and assumptions related to the research. Diversity of perspectives, expertise, and methods, and differences in customary practices, standards, and assumptions that could compromise the integrity of the research should be addressed openly.

14. *Conflict.* Collaborating partners should seek prompt resolution of conflicts, disagreements, and misunderstandings at the individual or institutional level.

15. *Authority of Representation.* Collaborating partners should come to agreement on who has authority to speak on behalf of the collaboration.

Responsibilities for Outcomes of Research

16. *Data, Intellectual Property, and Research Records.* Collaborating partners should come to agreement, at the outset and later as needed, on the use, management, sharing, and ownership of data, intellectual property, and research records.

17. **Publication.** Collaborating partners should come to agreement, at the outset and later as needed, on how publication and other dissemination decisions will be made.

18. **Authorship and Acknowledgement.** Collaborating partners should come to agreement, at the outset and later as needed, on standards for authorship and acknowledgement of joint research products. The contributions of all partners, especially junior partners, should receive full and appropriate recognition. Publications and other products should state the contributions of all contributing parties.

19. **Responding to Irresponsible Research Practices.** The collaboration as a whole should have procedures in place for responding to allegations of misconduct or other irresponsible research practice by any of its members. Collaborating partners should promptly take appropriate action when misconduct or other irresponsible research practice by any partner is suspected or confirmed.

20. **Accountability.** Collaborating partners should be accountable to each other, to funders, and to other stakeholders in the accomplishment of the research.

1.5. Recent and Future Conferences

During the Third WCRI, Steneck, Mayer, Anderson, and Kleinert agreed to work together as a steering committee to assure the continuity of the WCRI effort. Their first task was to review proposals from several countries that had responded to a call for bids to host the Fourth WCRI. Brazil was selected as the site for the next conference, under the local leadership of Sonia Vasconcelos, Edson Watanabe, and Martha Sorenson of the Federal University of Rio de Janeiro. The selection of Rio de Janeiro brought the World Conferences to South America, with the goal of encouraging participation from countries that had previously been underrepresented. Representatives from 48 countries participated, with total conference participation of 474.

The theme of the Fourth WCRI was "Research Rewards and Integrity: Improving Systems to Promote Responsible Research." It was expressed not only in the plenary sessions but also in focus tracks that addressed the relationships between research integrity and systems represented by funders, countries, and research institutions. The conference continued to attract decision makers, publishers, and researchers, in a somewhat greater spread in their experiential bases. Some countries had made considerable strides in policy development, oversight, and education in the responsible conduct of research. They brought to the Fourth WCRI relatively well-developed models of programs, documents, and instructional programs. Other countries represented at the Fourth WCRI were at earlier stages in their efforts to foster research integrity. In some cases, delegates from these latter countries illustrated ways in which integrity initiatives were developing along lines that diverged somewhat from earlier models, showing how important local context is to policy, instruction, and oversight related to research integrity. Selected papers from the Conference were published as: *Proceedings of the 4th World Conference on Research Integrity* [2016].

Continuing the tradition of naming the next site at each meeting, the bid submitted by a team organized by Lex Bouter, Vrije Universiteit Amsterdam, was warmly accepted, with the Fifth WCRI held in late May 2017 in Amsterdam. Information on the Fifth WCRI can be found at http://www.wcri2017.org The Conference was co-chaired by Steneck, Mayer, and Bouter and mark the 10th anniversary of the WCRI effort and the founding conference in Europe. In Rio, the Steering Committee also added Susan Zimmerman, Secretariat on Responsible Conduct of Research, Canada, and Sonia Vasconcelos, to its membership as representatives of the countries hosting the Third WCRI and the Fourth WCRI, respectively.

During the Fifth WCRI, the Steering Committee met and made the decision to establish the World Conferences on Research Integrity Foundation to coordinate future planning. The new Foundation is led by Board Chair Lex Bouter. The Steering Committee also accepted a bid for the 6th World Conference in 2019 to be hosted by Hong Kong and organised jointly by WCRIF, Hong Kong and Australia. Further information on these and other efforts will be available on the Foundation website, researchintegrity.org.

References

European Science Foundation (2000), Good Scientific Practice in Research and Scholarship. http://www.esf.org/fileadmin/Public_documents/Publications/ESPB10.pdf.

European Science Foundation and ALLEA (2011), The European Code of Conduct for Research Integrity. http://www.esf.org/fileadmin/Public_documents/Publications/Code_Conduct_ResearchIntegrity.pdf.

Federal Register (May 12, 2000), *Department of Health and Human Services*, 65(93), 30600–30601. DOCID:fr12my00-83.

International Council for Science (2002), Standards for Ethics and Responsibility in Science: An Empirical Study. http://www.icsu.org/publications/reports-and-reviews/standards-responsibility-science/.

Mayer, T., and N. Steneck (2007), Final Report to ESF and ORI First World Conference on Research Integrity: Fostering Responsible Research. http://www.esf.org/index.php?id=4479.

Mayer, T., and N. Steneck (2012), *Promoting Research Integrity in a Global Environment*. World Scientific Publishing, Singapore. doi:10.1142/9789814340984.

Montreal Statement (2023), Montreal Statement on Research Integrity in Cross-Boundary Research Collaborations. http://www.researchintegrity.org/Statements/Montreal%20Statement%20English.pdf.

Organisation for Economic Co-operation and Development (2007), Global Science Forum, Best Practices for Ensuring Scientific Integrity and Preventing Misconduct. http://www.oecd.org/sti/sci-tech/40188303.pdf.

Proceedings of the 4th World Congress on Research Integrity (2016), Research Integrity and Peer Review, 1 (Suppl 1), 9. doi:10.1186/s41073-016-0012-9

Singapore Statement (2010), Singapore Statement on Research Integrity. http://singapore statement.org.

Steneck, N. H., M. Anderson, S. Kleinert, and T. Mayer (2015). *Integrity in the Global Research Arena*. World Scientific Publishing, Singapore.

2. FOSTERING INTEGRITY IN RESEARCH: OVERVIEW OF THE NATIONAL ACADEMIES OF SCIENCES, ENGINEERING, AND MEDICINE REPORT

Thomas Arrison[1] and Robert M. Nerem[2]

Abstract

Research integrity is essential to the health of the research enterprise, providing the foundation for good science. The past decade has seen a steady flow of high-profile cases of data fabrication from around the world, a sharp increase in retractions of scientific articles, and an increase in the number of research misconduct allegations investigated by U.S. research institutions. Research misconduct and detrimental research practices can damage science and its reputation. Much still needs to be learned about why researchers engage in these behaviors. Future studies should focus not only on individual behavior but also on practices, incentives, and institutional environments. Mitigating hypercompetitive research environments, setting expectations of integrity and excellence at the highest levels of institutions and professional societies, and creating common standards for authorship, data and model accessibility, and reporting will greatly improve the current situation. Providing tools to institutions to aid in addressing responsible conduct of research education and for handling misconduct is strongly recommended, including establishing an independent Research Integrity Advisory Board to bring neutrality and focus to understanding and responding to research misconduct across all disciplines. This chapter summarizes the key themes, findings, and recommendations of the report *Fostering*

---

[1] *National Academies of Sciences, Engineering, and Medicine, Washington, D.C., USA*
[2] *Petit Institute for Bioengineering and Bioscience, Georgia Institute of Technology, Atlanta, Georgia, USA*

Scientific Integrity and Ethics in the Geosciences, Special Publications 73,
First Edition. Edited by Linda C. Gundersen.

Integrity in Research, released by the National Academies of Sciences, Engineering, and Medicine in 2017. The report contains broad guidance and specific recommendations for fostering integrity and addressing breaches in integrity directed to all participants in the research enterprise: researchers, research institutions, research sponsors, societies, and science, engineering, technology, and medical publishers.

2.1. Introduction

The National Academies of Sciences, Engineering, and Medicine (NASEM) released the report *Fostering Integrity in Research* in 2017 [*NASEM*, 2017]. The 13-member authoring committee included representation from a range of research disciplines and various career stages as well as experience in administrative and educational roles related to research integrity. The study was sponsored by several U.S. federal agencies, the National Academies, and other organizations. This article summarizes the report's key themes, findings, and recommendations. The full text of the findings and recommendations is provided at the end of the chapter.

In framing its treatment of research integrity, the committee draws on past National Academies' work. The 1992 report *Responsible Science: Ensuring the Integrity of the Research Process* was issued in the midst of major shifts in approaches to research misconduct and research integrity on the part of the U.S. government and research institutions in the wake of several highly publicized investigations of research misconduct allegations [*NAS-NAE-IOM*, 1992]. The 2002 report *Integrity in Scientific Research: Creating an Environment that Promotes Responsible Conduct* described what was known about how research environments may support (or not support) research integrity, and it outlined an approach to assessing research environments [*IOM-NRC*, 2002). In 2009, the National Academies released the third edition of the popular educational guide *On Being a Scientist: A Guide to Responsible Research Conduct* [*NAS-NAE-IOM*, 2009a]. Also in 2009, *Ensuring the Integrity, Accessibility and Stewardship of Research Data* was released [*NAS-NAE-IOM*, 2009b]. This report described the growing challenges and opportunities facing the research enterprise in the area of digital data and recommended principles for addressing those challenges.

The committee benefited from the presentations of numerous experts from academia, industry, and government using a wide range of sources from around the world. These include surveys aimed at shedding light on the incidence and causes of research misconduct and detrimental research practices, policy reports framing national approaches to addressing misconduct, explorations of research values and research best practices, responsible conduct of research educational materials, and institutional and media reports on notable cases.

Although much of the report's discussion of government and institutional policies focuses on the United States, the committee was informed by international developments and efforts to foster integrity underway in many countries.

2.2. The Integrity of Research

The research enterprise is a complex system in which the participants, stakeholders, systems, and processes continually interact, adapt, and change. Effective approaches to fostering research integrity in the contemporary environment need to be grounded in an understanding of this complexity. There is an extensive literature on the values and norms of research [*IOM-NRC*, 2002; *Merton*, 1973; *Popper*, 1999; *Council of Canadian Academies*, 2010; *European Science Foundation*, 2017; *InterAcademy Council*, 2012; *Irish Council for Bioethics*, 2010; *InterAcademy Partnership*, 2016]. The committee's synthesis yielded six core values that shape the norms of research and the practices that uphold integrity:

- **Objectivity** is fundamental to the scientific method. Researchers have a responsibility to design experiments so that it is possible for their hypotheses to be refuted. Researchers should seek to ensure that personal beliefs and motivations do not influence their work.
- **Honesty** requires that researchers completely and accurately report what they have done. Researchers who are honest not only refrain from out-and-out fabrication or falsification of data but also avoid misrepresentation, nonreporting of phenomena, and inappropriately enhancing digital images.
- **Openness** means being transparent and presenting all the information relevant to a decision or conclusion. It also means making the data and other information on which a result is based available to others so that they may reproduce and verify results or build on them.
- **Accountability** means that researchers are responsible for and stand behind their work, statements, actions, and roles in the conduct of their work. At its core, accountability implies an obligation to explain and/or justify one's behavior.
- **Fairness** comes into play in activities such as reviewing proposals for funding, reviewing articles for publication, and making hiring or promotion decisions. Being fair in these contexts means making professional judgments based on appropriate and announced criteria, including processes used to determine outcomes.
- **Stewardship** describes the researcher's responsibility to colleagues and to the broader research enterprise. This involves being aware of and working to sustain healthy relationships within the lab and across the research enterprise, as well as performing service activities at the institutional or disciplinary levels.

The research enterprise faces particular challenges in socializing and training individuals into responsible practices based on these values. Research differs from some other professions, such as medicine and law, in that research does not require a formal certification to practice, and there is no distinctive ethical code that researchers are bound to follow. Instead, the research enterprise has traditionally relied on informal approaches to professional training centered on mentorship.

Significant trends are affecting the conduct of research and efforts to promote research integrity. These include the movement toward larger, more interdisciplinary research teams and the continuously growing relevance of research results to industry and public policy. Perhaps the most significant of these trends are the ongoing globalization of research activity and the central role of information technologies in enabling researchers to tackle new questions through the collection and analysis of digital data.

Although research has always been an international endeavor, with researchers and knowledge crossing borders, the globalization trend has accelerated in recent decades. Research activity has grown rapidly in China and several other emerging economies, international coauthorship of scientific articles is on the rise, and large numbers of foreign born students and researchers come to the United States and likewise to Europe to study and work. International cooperation has many benefits and can speed the advance of knowledge by making it possible to do work that cannot be undertaken by one country working alone. At the same time, globalization can complicate efforts to foster integrity. Education in the responsible conduct of research is not universal, even in the United States, and requirements elsewhere in the world are uneven. Researchers from different countries and cultures might have different views on issues such as conflicts of interest, the deference to be accorded to supervisors and mentors, data handling, and authorship practices.

The growing power of information technology has also enabled new forms of research and new collaborations while creating new challenges. For example, in some fields intermediate complex analyses are undertaken between the collection of "raw" data from sensors and observations, meaning that detailed knowledge of the software used for analysis is required to recreate the steps from data to results. This creates opportunities to manipulate data and results. Digital technologies can also be used to manipulate images, and many journals now utilize tools that can detect such manipulations.

2.3. Research Misconduct and Detrimental Research Practices

Notable examples of fabrication and falsification have appeared throughout the history of research. Prompted by a series of high-profile cases, Congress, federal agencies, and research institutions in the United States adopted new policies and institutional arrangements in the late 1980s and early 1990s aimed at achieving greater consistency and accountability in how allegations of research misconduct were investigated and addressed. During this time, there were debates within the research community about how research misconduct should be defined. In 2000, the U.S federal government adopted a policy that defines research misconduct as fabrication of data, falsification of data, or plagiarism (FFP) [*Office of Science and Technology Policy*, 2000]. This definition excludes nonspecific formulations such as "other serious deviations from accepted practices" that had been in use by federal agencies up to that time.

The report explores the arguments for and against FFP, discusses alternative definitions that are used by other countries and by U.S. research institutions, and reviews the U.S. experience with the current federal definition. Using FFP provides a clear, limited set of behaviors that constitute research misconduct. Advocates for a broader definition or the inclusion of "other serious deviations from accepted research practices" argue that a broader set of behaviors should be targeted. FFP has become firmly established in U.S. policy, and change would be strongly opposed by many in the research community. Advocates of making a change would need to consider the potential benefits and weigh them against the difficulty in realizing such a change. The committee endorsed the current federal definition but also pointed out areas where greater consistency in implementation is needed, such as defining and addressing plagiarism. In the federal government's research misconduct policy, plagiarism is defined as "the appropriation of another person's ideas, processes, results, or words without giving appropriate credit" [*Office of Science and Technology Policy*, 2000]. Both the National Science Foundation Office of Inspector General (NSF-OIG) and the Office of Research Integrity (ORI) at the Department of Health and Human Services, two of the primary agencies that oversee research misconduct investigations by universities, say that they exclude "authorship disputes" as possible cases of misconduct. However, there are indications that they might take different approaches. The NSF-OIG is open to considering as potential plagiarism alleged intellectual theft by former collaborators or by noncollaborating researchers working at the same institution. The ORI appears to not be open to considering these sorts of allegations. For example, the report describes a case in which an institution did not report to ORI an allegation of intellectual theft by a researcher against a noncollaborating researcher at the same institution in the belief that ORI would not consider the alleged behavior to be plagiarism. Recognizing that some allegations of plagiarism involving intellectual theft between former collaborators might be difficult or impossible to prove or disprove, the committee believes that they should not be automatically dismissed. Certainly, an allegation of plagiarism against a researcher at the same institution with whom the accuser has never collaborated should not be automatically dismissed.

In addition to misconduct, the 1992 report *Responsible Science: Ensuring the Integrity of the Research Process* defined two other categories of damaging behavior within the research community: "questionable research practices" and "other misconduct" [*NAS-NAE-IOM*, 1992]. *Fostering Integrity in Research* revisits these categories. The report reaffirms the "other misconduct" category, which includes damaging behaviors that occur within the research environment but can also occur in other work or educational settings, including financial improprieties and sexual harassment. These behaviors should be addressed outside the research misconduct framework. Some behaviors in the "other misconduct" category, such as retaliation against good-faith whistleblowers in research misconduct cases, addressed in Recommendation Three further ahead in this chapter, are clearly relevant to addressing research misconduct, even if they do not constitute research misconduct in themselves. There are several options for strengthening protection of research misconduct whistleblowers discussed in the report that do not involve changing the definition of research misconduct, such as including whistleblower protection provisions directly in research funding legislation, as was done in the American Recovery and Reinvestment Act of 2009.

In the case of *questionable research practices*, the research community should rethink this category, which has included irresponsible authorship practices, misleading use of statistical methods and data manipulation short of falsification, and inattentive or abusive supervision. These behaviors should be clearly recognized as being detrimental to research, and labeled as *detrimental research practices*. The category should also be expanded from one that focuses solely on the behaviors of individual researchers to one that includes the detrimental practices committed by research institutions, journals, and other research participants [*NASEM*, 2017]. These include the failure by research institutions to maintain the policies and capabilities needed to effectively undertake research misconduct investigations, and irresponsible publishing practices such as pressuring authors to add citations from the journal to their work (the goal being to raise the journal's impact factor) as a condition of acceptance.

The main sources of information about the incidence of research misconduct and detrimental research practices are statistics from federal agencies about investigations and results, surveys of researchers about their own behavior and their knowledge of the behavior of colleagues, and article retractions [*DuBois et al.*, 2013; *Fanelli*, 2009; *Fang et al.*, 2012; *Grieneisen and Zhang*, 2012; *Martinson et al.*, 2005; *NSF-OIG*; *ORI*, 2012-2013; *Steen*, 2011]. All of these sources of information have limitations, may overstate or understate the actual incidence, and cannot serve as direct proxies for how often these behaviors are occurring or as trends over time. Since the number and rate of allegations and retractions have shown marked increases over the past decade, research misconduct and detrimental research practices have become more visible phenomena than in the past. In any case, research misconduct and detrimental research practices occur frequently enough to drive home the point that they are realities that need to be addressed by the research enterprise on an ongoing basis. The report's proposed Research Integrity Advisory Board, described in more detail below, could make a significant contribution over time.

The costs and consequences of research misconduct and detrimental research practices were examined, including resources spent on work that produces fabricated or falsified data, resources spent building on fabricated research, the cost of investigations, damaged or destroyed careers, the damage to reputations of institutions, the public health costs of acting on incorrect results, and the damage to public trust in the research enterprise. For some of these costs and consequences, it is possible to conceptualize how a dollar estimate might be generated, and some attempts have been made to generate estimates for certain categories of costs. However, the information needed to generate a reliable estimate of the total cost of these behaviors is lacking. Based on the available information, it is reasonable to assume that the costs incurred by the U.S. research enterprise due to research misconduct and detrimental research practices total at least several hundred million dollars annually, and may run as high as several billion dollars [*NASEM*, 2017]. This may be a small fraction of the total investment in research, but it still constitutes a significant problem.

Reproducibility has become a significant issue in research over the past several years. Reproducibility can be conceptualized or defined in several different ways, including the replication of experiments using the exact same methods and materials, and verifying or validating the conclusions of research where exact replication is impossible or impractical, such as observations of natural phenomena or clinical trials. High rates of irreproducibility of research results have been estimated in several fields and disciplines [*Academy of Medical Sciences*, 2015; *Freedman et al.*, 2015; *Global Biological Standards Institute*, 2015; *Ioannidis*, 2005; *Open Science Collaborative*, 2015]. There are a number of potential causes of failure to reproduce the results of research, including uncharacterized independent variables, errors in either the original experiment or the attempt to replicate, insufficient detail in methods sections of articles, and unavailability after publication of data, reagents, or materials. Some inherent level of irreproducibility due to unknown variables, false positives or negatives, or differences across labs in quality control is consistent with progress in a healthy research field. At the same time, fabrication, falsification, and detrimental research practices are responsible for some lack of reproducibility as well. Efforts are underway within several disciplines to better understand and address the reproducibility problem. The implementation of more rigorous standards and best practices in sharing data, methods, code, and materials should help to improve reproducibility and would help to uncover or deter some research misconduct and detrimental research practices. Several of these efforts are discussed in other chapters in this volume.

Efforts to understand the causes of research misconduct and build approaches to addressing it have often focused on the pathologies of a small number of individual researchers. Yet, as an example, shifts in research funding, institutional policies and practices, and publication practices help to shape the choices and incentives facing researchers at the laboratory level. Particularly in fields such as biomedicine, declining success rates in grant applications, an increasing ratio between the number of PhDs awarded and available full time faculty positions, and peer cultures that accept corner cutting are creating workforce imbalances and hypercompetitive research environments that may not be conducive to fostering integrity [*Alberts et al.*, 2014]. Much remains to be learned about why researchers commit research misconduct or engage in detrimental practices and how research environments influence this behavior. A key challenge will be designing and implementing steps to ensure that intense competition for funding and workforce imbalances do not exacerbate the problem. One approach would be to ensure that award criteria and recognition explicitly follow a defined set of best practices.

This chapter reviews the available evidence on the effectiveness of current approaches to addressing research misconduct and detrimental practices. These approaches include policies and practices of research sponsors and regulators, including federal agencies, research institutions, societies, and science, engineering, technology, and medical publishers. Approaches used in the United States are compared with those used in other countries. Based on this examination of policies and practices, current U.S. practices have achieved "stability and effectiveness in ensuring that misconduct allegations involving federally funded work are investigated," but "there are gaps and inconsistencies" [*NASEM*, 2017]. The unevenness among institutions in their policies and capacities for fostering integrity and addressing misconduct is a significant vulnerability. Institutions have sometimes been unable to effectively investigate and address misconduct, notably in cases where the results could affect their financial or reputational interests. In addition, regarding implementation of the federal research misconduct policy, approaches to publicly naming respondents on the part of federal agencies differ considerably. The ORI posts the names of those who have been found guilty of misconduct on its website, while NSF-OIG enters the names in a publicly accessible data base that requires one to enter a name in order to search. Consequently, those who have been found guilty of research misconduct while performing National Institutes of Health–funded research are well known, while those who have been found guilty while performing NSF-funded research may not become known at all. The approaches of federal agencies to naming respondents should be harmonized, but there is currently no consensus regarding the approach that should be used.

The discussion also identifies information gaps and questions. In particular, developing approaches to discourage and address some detrimental research practices involves a number of challenges that are distinct from the challenges of dealing with research misconduct. For example, the effort to ensure persistent access to data and code will require new standards from journals and disciplines, funding from sponsors, and changed attitudes from the community. This is an area where progress is being made, but much additional alignment and effort is needed, including finding sustained support for data repositories.

It is clear that federal and private spending on efforts to discourage research misconduct and detrimental research practices and to promote research integrity is quite low relative to conservative estimates of the costs and consequences.

Potential new approaches include the establishment of a new Research Integrity Advisory Board (RIAB) as an independent nonprofit organization and the recommendation that disciplines establish clear authorship standards based on a common framework.

The RIAB would "bring a unified focus to understanding and addressing challenges across all disciplines and sectors" [*NASEM*, 2017]. It would facilitate the exchange of information and diffusion of best practices in areas such as handling allegations of misconduct and investigations. It would also provide "advice, support, encouragement, and, where helpful, advocacy on what needs to be done by research institutions, science, engineering, technology, and medical journal and book publishers, and other stakeholders in the research enterprise to promote research integrity" [*NASEM*, 2017]. The RIAB would have no role in investigations or regulation but would serve as a neutral resource for the community.

Realizing improved authorship standards will require focused efforts across disciplines and sectors. It will also require changes in attitudes and culture to establish the norm that practices such as "guest" or honorary authorship, coercive authorship, and unacknowledged ghost authorship are always wrong. A recent workshop held by the NASEM and attended by scientific societies and publishers has resulted in a new compendium of authorship best practices [*McNutt et al.*, 2017].

2.4. Fostering Integrity in Research

The research enterprise and its participants can take several affirmative steps to foster integrity. High priority should be to examine and update their policies and practices. Progress will require participation and engagement by individual researchers, research institutions, research sponsors, journals and scholarly communicators, and societies but in different ways. For example, at the laboratory level, supervisors need to model appropriate behaviors, check the work of students and postdocs, and be skeptical of results that may be too good to be true. At the institutional level, university leaders need to understand that integrity is central to their research and educational missions and encourage faculty and administrators to uphold the highest standards of integrity. Research sponsors, scientific publishers, and scientific societies have similar tasks. The key themes underlying all of these practices are the values and norms of research, including honesty, openness, and accountability.

Responsible conduct of research (RCR) education has evolved from the early 1990s to today, but our understanding of what works is limited. Too many institutions approach RCR education as a regulatory compliance task. Efforts are needed to develop the evidence necessary to guide future improvements. For example, what mix of online and in-person RCR education is most effective for the most students? How much content should be embedded in disciplinary coursework? What are the appropriate targets for assessment?

2.5. Key Themes and Takeaways

Participants and stakeholders can foster integrity in research in support of a healthy research enterprise over the long term, including the following:

Maintain attention on individuals while broadening the focus to include institutions and environments.

Future progress toward realizing a research enterprise characterized by high levels of integrity and quality will depend on the attitudes and actions of research institutions and their leadership, research sponsors, policymakers, publishers, and societies. These attitudes and actions will be influenced by how they see and frame the issues and challenges. A primary focus on individual misbehavior that does not consider the role of institutions and environments lets those with significant power avoid responsibility. The research enterprise can achieve the highest levels of integrity only when all the participants recognize and fulfill their responsibilities.

Maintain attention on research misconduct while broadening the focus to include detrimental research practices.

Research misconduct is directly connected with detrimental research practices. Not only are detrimental practices harmful to research in themselves, tolerance for practices such as withholding data and other information necessary to replicate or validate work by disciplines, journals, and sponsors can enable misconduct to continue undiscovered. While improving approaches to discouraging and addressing research misconduct is a critical task, it will be necessary to raise standards and practices so that detrimental practices are discouraged and seen as unacceptable.

Develop new knowledge that can guide evidence-based approaches.

Many gaps in knowledge and information remain, including the connections between research environments and research misconduct, how institutions can improve these environments, and basic information such as the true incidence of research misconduct. Filling in these gaps will take significant time and effort in many cases. Using scientific methods and approaches will be necessary to better understand the research enterprise, as well as to develop and implement interventions that foster integrity. Assessing research environments at the institutional level can play an important role.

Develop new institutions and approaches that can help improve standards and practices on a continuing basis.

In the United States and many other countries, there is currently no institutional focus for efforts to foster integrity across sectors and disciplines on a continuing basis. The RIAB could fulfill a number of necessary tasks. Similarly, the effort to more clearly define authorship standards across disciplines using a common framework can improve transparency and discourage detrimental authorship practices such as honorary authorship and excluding those who have earned authorship. While the focus of the report is largely on the United States, effective long-term approaches to fostering integrity will require building on international efforts such as the World Conferences on Research Integrity, the Organization for Economic Cooperation and Development's Global Science Forum, the Global Research Council, the InterAcademy Partnership, and others.

2.6. Full Text of the Findings and Recommendations [*NASEM*, 2017]

FINDING A: Developing and implementing improved approaches to fostering research integrity and meeting the current threats to integrity posed by research misconduct and detrimental research practices are urgent tasks. These improved approaches should reflect an understanding of the complex interactions among the many components of the research enterprise and its multiple stakeholders.

RECOMMENDATION ONE: To better align the realities of research with its values and ideals, all stakeholders in the research enterprise—researchers, research institutions, research sponsors, journals, and societies—should significantly improve and update their practices and policies to respond to the threats to research integrity identified in this report.

RECOMMENDATION TWO: Since research institutions play a central role in fostering research integrity and addressing current threats, they should maintain the highest standards for research conduct, going beyond simple compliance with federal regulations.

RECOMMENDATION THREE: Research institutions and federal agencies should work to ensure that good faith whistleblowers are protected and their concerns assessed and addressed in a fair, thorough, and timely manner.

RECOMMENDATION FOUR: To provide a continuing organizational focus for fostering research integrity that cuts across disciplines and sectors, a Research Integrity Advisory Board should be established as an independent nonprofit organization. The RIAB will work with all stakeholders in the research enterprise—researchers, research institutions, research sponsors and regulators, journals, and scientific societies—to share expertise and approaches for addressing and minimizing research misconduct and detrimental research practices. RIAB will also foster research integrity by stimulating efforts to assess research environments and improve practices and standards.

FINDING B: Ensuring greater openness and accountability in science is essential to fostering research integrity and improving research quality. Establishing and agreeing on new standards and building the infrastructure needed to implement those standards will require collaborative, focused efforts on the part of the research enterprise and its stakeholders.

RECOMMENDATION FIVE: Societies and journals should develop clear disciplinary authorship standards. Standards should be based on the principle that those who have made a significant intellectual contribution are authors.

Significant intellectual contributions can be made in the design or conceptualization of a study, the conduct of research, the analysis or interpretation of data, or in drafting or revising the manuscript for intellectual content. Those who engage in these activities should be designated as authors of the reported work, and all authors should approve the final manuscript. In addition to specifying all authors, standards should (1) provide for the identification of one or more authors who assume responsibility for the entire work, (2) require disclosure of all author roles and contributions, and (3) specify that gift or honorary authorship, coercive authorship, ghost authorship, and omitting authors who have met the articulated standards are always unacceptable. Societies and journals should work expeditiously to develop such standards in disciplines that do not already have them.

RECOMMENDATION SIX: Through their policies and through the development of supporting infrastructure, research sponsors and science, engineering, technology, and medical journal and book publishers should ensure that information sufficient for a person knowledgeable about the field and its techniques to reproduce reported results is made available at the time of publication or as soon as possible after that.

RECOMMENDATION SEVEN: Federal funding agencies and other research sponsors should allocate sufficient funds to enable the long-term storage, archiving, and access of datasets and code necessary for the replication of published findings.

RECOMMENDATION EIGHT: To avoid unproductive duplication of research and to permit effective judgments on the statistical significance of findings, researchers should routinely disclose all statistical tests carried out, including negative findings. Research sponsors, research institutions, and journals should support and encourage this level of transparency.

FINDING C: Improved strategies for fostering research integrity and for addressing threats to integrity posed by research misconduct and detrimental research practices need to be based on knowledge and evidence that does not currently exist. Investments are needed in research that improves understanding of key issues such as the relationship between structural conditions in science and the tendency for individuals to practice research according to the values and norms of integrity or to deviate from those values and norms. Improving knowledge in this area is essential to the long-term health of the research enterprise itself.

RECOMMENDATION NINE: Government agencies and private foundations that support science, engineering, and medical research in the United States should fund research to quantify, and develop responses to, conditions in the research environment that may be linked to research misconduct and detrimental research practices. These research sponsors should use the data accumulated to monitor and modify existing policies and regulations.

RECOMMENDATION TEN: Researchers, research sponsors, and research institutions should continue to develop and assess more effective education and other programs that support the integrity of research. These improved programs should be widely adopted across disciplines and across national borders.

FINDING D: Working to ensure research integrity at the global level is essential to strengthening science both in the U.S. and internationally.

RECOMMENDATION ELEVEN: Researchers, research institutions, and research sponsors who participate in and support international collaborations should leverage these partnerships to foster research integrity through mutual learning and sharing of best practices, including collaborative international research on research integrity.

Acknowledgments

This study was sponsored by the Office of Research Integrity of the U.S. Department of Health and Human Services, the Office of Science of the U.S. Department of Energy, the U.S. Environmental Protection Agency, the U.S. Geological Survey of the U.S. Department of the Interior, the Office of Inspector General of the U.S. National Science Foundation, and the Burroughs Wellcome Fund, with additional support from the Society for Neuroscience, the American Chemical Society, the American Physical Society, the National Academy of Sciences Arthur L. Day Fund, and the W. E. Kellogg Foundation Fund.

References

Academy of Medical Sciences, BBSRC, MRC, Wellcome Trust (2015), *Reproducibility and reliability of biomedical research: Improving research practice*, Academy of Medical Sciences, London.

Alberts, B.M., M.W. Kirschner, S. Tilghman, and H. Varmus (2014), Rescuing US biomedical research from its systemic flaws, *Proceedings of the National Academy of Sciences*, 111(16): 5773–5777, doi:10.1073/pnas.1404402111.

Council of Canadian Academies (2010), *Honesty, accountability and trust: Fostering research integrity in Canada*, Ottawa, Ontario.

DuBois, J.M., E.E. Anderson, and J. Chibnall (2013), Assessing the need for a research ethics remediation program, *Clinical Translational Science*, 6(3): 209–213.

European Science Foundation and All European Academies (2017), *The European code of conduct for research integrity* (Rev. ed.), Berlin, Germany. Available at www.allea.0rg/wp-content/uploads/2017/03/ALLEA-European-Code-of-Conduct-for-Research-Integrity-2017-1.pdf.

Fanelli, D. (2009), How many scientists fabricate and falsify research? A systematic review and meta-analysis of survey data, *PLoS One*, 4(5): e5738.

Fang, F.C., R.G. Steen, and A. Casadevall (2012), Misconduct accounts for the majority of retracted scientific publications, *Proceedings of the National Academy of Sciences*, 109(42): 17028–17033. doi:10.1073/pnas.1212247109.

Freedman, L.P., I.M. Cockburn, and T.S. Simcoe (2015), The economics of reproducibility in preclinical research, *PLOS Biology*, 13(6): e1002165.

Global Biological Standards Institute (2015), *The case for standards in life science research*, Washington, D.C.

Grieneisen, M.L., and M. Zhang (2012), A comprehensive survey of retracted articles from the scholarly literature, *PLoS One*, 7(10), e44118.

Institute of Medicine and National Research Council (IOM-NRC) (2002), *Integrity in scientific research: Creating an environment that promotes responsible conduct*, National Academies Press, Washington, D.C.

InterAcademy Council and InterAcademy Panel (2012), *Responsible conduct in the global research enterprise: A policy report*, Amsterdam, Netherlands.

InterAcademy Partnership (2016), *Doing global science: A guide to responsible conduct in the global research enterprise*, Princeton University Press, Princeton, New Jersey.

Ioannidis, J.P.A. (2005), Why most published research findings are false, *PLOS Medicine*, 2(8): e124.

Irish Council for Bioethics (2010), *Recommendations for Promoting Research Integrity*, Irish Council for Bioethics, Dublin, Ireland.

Martinson, B.C., M.S. Anderson, and R. De Vries (2005), Scientists behaving badly, *Nature*, 435: 737–738.

McNutt M., M. Bradford, J. Drazen, B. Hanson, B. Howard, K. Hall Jamieson, V. Kiermer, M. Magoulias, E. Marcus, B. Kline Pope, R. Schekman, S. Swaminathan, P. Stang, and I. Verma (2017), Transparency in authors' contributions and responsibilities to promote integrity in scientific publication, *bioRxiv*, 140228, doi:10.1101/140228.

Merton, R.K. (1973), *The sociology of science: Theoretical and empirical investigations*, University of Chicago Press, Chicago, Illinois.

National Academies of Sciences, Engineering, and Medicine (NASEM) (2017), *Fostering integrity in research*, National Academies Press, Washington, D.C.

National Academy of Sciences, National Academy of Engineering, Institute of Medicine (NAS-NAE-IOM) (1992), *Responsible science: Ensuring the integrity of the research process*, National Academies Press, Washington, D.C.

National Academy of Sciences, National Academy of Engineering, Institute of Medicine (NAS-NAE-IOM) (2009a), *On being a scientist: A guide to responsible research conduct*, National Academies Press, Washington, D.C.

National Academy of Sciences, National Academy of Engineering, Institute of Medicine (NAS-NAE-IOM) (2009b), *Ensuring the integrity, accessibility and stewardship of research data*, National Academies Press, Washington, D.C.

National Science Foundation, Office of Inspector General (Various Years), Semi-Annual Report to Congress. Arlington, Virginia.

Office of Research Integrity, Department of Health and Human Services (2012–2013), *Office of Research Integrity annual report*. Washington, D.C.

Office of Science and Technology Policy, Executive Office of the President (2000), *Federal policy on research misconduct*, Federal Register 65: 76260–76264.

Open Science Collaborative (2015), Estimating the reproducibility of psychological science, *Science*, 349(6251), 28 August.

Popper, K. (1999), *All life is problem solving*, Routledge, New York.

Steen, R, G. (2011), Retractions in the scientific literature: Is the incidence of research fraud increasing? *Journal of Medical Ethics*, 37(4): 249–253. doi:10.1136/jme.2010.040923.

3. SCIENTIFIC INTEGRITY: RECENT DEPARTMENT OF THE INTERIOR POLICIES, CODES, AND THEIR IMPLEMENTATION

Alan D. Thornhill[1] and Richard A. Coleman[2]

Abstract

Established on January 28, 2011, the Department of Interior's (DOI's) Scientific and Scholarly Integrity Policy was the first federal agency policy to respond to the Presidential Memorandum on Scientific Integrity (March 9, 2009) and guidance issued by the Office of Science and Technology Policy Memorandum on Scientific Integrity (December 17, 2010). The increasingly important role of science in DOI decision making and heightened awareness of science integrity issues across the science enterprise provided impetus for making this policy a priority and for incorporating it into the DOI Departmental Manual (Part 305: Chapter 3). This chapter discusses the history of scientific integrity in the DOI, the key provisions of the first Department-wide policy and its implementation, and the subsequent revision of the policy. During the first 4 years of implementing the scientific integrity policy, the Department received 35 formal complaints. As of March 31, 2015, only two formal scientific integrity complaints resulted in "warranted determinations," while the other complaints were closed and dismissed as "not warranted." Based on the experience of the first three years of implementation (2011–2014), the Department policy was revised on December 16, 2014.

[1] Western Ecology Division, Environmental Protection Agency, Corvallis, Washington, USA
[2] U.S. Geological Survey, Denver, Colorado, USA

Scientific Integrity and Ethics in the Geosciences, Special Publications 73, First Edition. Edited by Linda C. Gundersen.

3.1. Introduction

3.1.1. Who Speaks for the Science?

"Science is at the heart of Interior's mission, so it's important that we continue to lead federal efforts to ensure robust scientific integrity policies."
Department of the Interior Secretary Sally Jewell (December 17, 2014)

Imbedded in the Department of the Interior's (DOI's) mission are the strong ethical values of stewardship, conservation, preservation, protection of wildlife and resources, and ensuring the welfare of the nation through management of public lands, energy, water, and other natural resources. Because of the broad and far reaching scope of these activities, it is critical that the underlying science used for decision making be of the highest integrity. The breadth and purpose of DOI activities are well described in the DOI Strategic Plan [2014]:

Today, DOI manages the Nation's public lands and minerals, including providing access to more than 500 million acres of public lands, 700 million acres of subsurface minerals, and 1.7 billion acres of the Outer Continental Shelf. The DOI is the steward of 20 percent of the Nation's lands, including national parks, national wildlife refuges, and public lands; manages resources that supply 23 percent of the Nation's energy; supplies and manages water in the 17 Western States and supplies 17 percent of the Nation's hydropower energy; and upholds Federal trust responsibilities to 566 federally recognized Indian tribes and Alaska Natives. The DOI is responsible for migratory bird and wildlife conservation; historic preservation; endangered species conservation; surface-mined lands protection and restoration; mapping, geological, hydrological, and biological science for the Nation; and financial and technical assistance for the insular areas.

In the wake of various allegations of political appointees tampering with science [*Union of Concerned Scientists*, 2010] during the second Bush Administration (2001–2009), DOI scientists welcomed President Obama's 2009 inaugural address message, "We will restore science to its rightful place...." Within two months, a Presidential Memorandum on Scientific Integrity (March 9, 2009) was issued to Executive Branch agencies, stating: "The public must be able to trust the science and scientific process informing public policy decisions. Political officials should not suppress or alter scientific or technological findings and conclusions." There was significant concern and frustration among scientists and scholars at DOI due to the perception, and in some cases the experience, of data and scientific results being manipulated by nonscientists in leadership. DOI science leadership was eager to respond to the President's directive and the subsequent guidance to federal agencies provided by the Office of Science and Technology Policy Memorandum on Scientific Integrity [2010].

One thing was clear when the writing team for the DOI scientific integrity policy began their work in the summer of 2010: although there are highly trained professionals designated to manage and ensure compliance with a variety of regulations and policies across DOI (e.g., Acquisitions and Grants, Human

Resources, Inspector General, various law enforcement entities, solicitors, union affairs), there was no designated group of highly trained professionals to speak for science or for the integrity of the scientific process and products within DOI. The writing team had found their guiding principle question: Who speaks for the science?

Up to this point, allegations of interference with science or scientific misconduct had been adjudicated by the Inspector General or management, but by and large, there is little scientific expertise in those groups. Thus, there was concern that those groups would not be aware of the acceptable professional standards for a specific scientific discipline, the importance of the integrity of the scientific process and products of science, or be able to detect a deviation or loss of the former or latter. The writing team considered that scientists are not sought out to adjudicate legal or management issues, so why would attorneys or managers be expected to adjudicate science issues? As a starting point, this was a powerful idea that the writing team decided was a first principle: Put scientists in charge of sorting out deviations from accepted scientific practices.

We had established *who* would do the assessment of any allegation of something going wrong with science; next we needed to determine *what* exactly we were trying to prevent, and exactly *how* a scientist would assess an allegation of science gone wrong. Finally, we had to determine under what circumstances one might have a positive finding of science gone wrong and what to do about it. Fortunately, the effort of the writing team to develop the DOI policy over 18 months stood on a foundation that had been laid for many years. The following sections review that legacy and history.

3.1.2. Federal Policy on Research Misconduct, 2000

The Office of Science and Technology Policy (OSTP) published the Federal Policy on Research Misconduct (65 FR 7620-76264) on December 6, 2000 (http://www.gpo.gov/fdsys/pkg/FR-2000-12-06/html/00-30852.htm). This policy established the "scope of Federal government's interest in the accuracy and reliability of the research record and the processes involved in its development." All federal agencies were required to conform to this policy, and accordingly it was fundamental to the later Department of Interior's policy: Integrity of Scientific and Scholarly Activities.

The Federal Policy on Research Misconduct [*OSTP*, 2000] contained a specific definition of *research misconduct*: "fabrication, falsification, or plagiarism in proposing, performing, or reviewing research, or in reporting research results." Further definitions included the following:
- "Fabrication is making up data or results and recording or reporting them."
- "Falsification is manipulating research materials, equipment, or processes, or changing or omitting data or results such that the research is not accurately represented in the research record."

- "Plagiarism is the appropriation of another person's ideas, processes, results, or words without giving appropriate credit."
- "Research misconduct does not include honest error or differences of opinion."

This Federal Policy on Research Misconduct [*OSTP*, 2000] also established the criteria for a finding of research misconduct, requiring the following:

- "There be a significant departure from accepted practices of the relevant research community; and
- The misconduct be committed intentionally, or knowingly, or recklessly; and
- The allegation be proven by a preponderance of evidence."

A multiphase framework to ensure a thorough response to an allegation of research misconduct was provided, including the following:

1. Inquiry—"the assessment of whether the allegation has substance and if an investigation is warranted."
2. Investigation—"the formal development of a factual record, and the examination of that record leading to dismissal of the case or to a recommendation for a finding of research misconduct or other appropriate remedies."
3. Adjudication—"recommendations are reviewed and appropriate corrective actions determined." and "Adjudication is separated organizationally from inquiry and investigation."

> When the agency has made a final determination, it will notify the subject of the allegation of the outcome and inform the institution regarding the disposition of the case. The agency finding of research misconduct and agency administrative actions can be appealed pursuant to the agency's applicable procedures...appeals are separated organizationally from inquiry and investigation.

Fair and timely procedures were also included in this rule "to provide safeguards for subjects of allegations as well as for informants":

- Informants—"protection from retaliation for informants who make good faith allegations, fair and objective procedures for the examination and resolution of allegations of research misconduct, and diligence in protecting positions and reputations of those persons who make allegations of research misconduct in good faith."
- Subjects—"their rights are protected and the mere filing of an allegation against them will not bring their research to a halt or be the basis for other disciplinary or adverse action absent of compelling reasons...timely written notification regarding substantive allegations made against them; a description of all such allegations; reasonable access to the data and other evidence supporting the allegations; and the opportunity to respond to allegations, the supporting evidence, and the proposed findings of research misconduct (if any)."
- Objectivity and Expertise—individuals who review allegations or conduct investigations have the "appropriate expertise and have no unresolved conflicts of interests to help ensure fairness throughout all phases of the process."

- Timeliness—"reasonable time limits" for all phases of the process and "allowances for extensions where appropriate."
- Confidentiality—"To the extent possible consistent with a fair and thorough investigation and as allowed by law, knowledge about the identity of subjects and informants is limited to those who need to know."

These fundamental definitions, processes, and safeguards formed the basis for subsequent DOI scientific integrity policies (presented below in chronological order) and those of many other federal agency policies.

3.2. Department of the Interior, First Steps in Scientific Integrity

3.2.1. Interior's Research and Development Council, 2005

In January 2005, DOI formed a Research and Development Council to provide a forum for discussion of research and development within DOI, coordinate strategic science priorities, and facilitate integration, coordination, and leveraging of science needs between bureaus and offices within DOI. This council also drafted the first policy on scientific integrity.

3.2.2. First Draft of DOI Policy on Scientific Integrity, 2007

In 2007, DOI developed a draft Department policy, "Integrity of Scientific Activities and Code of Scientific Conduct." However, this policy was never finalized. According to Interior officials, a decision was made to delay the adoption of the policy. This was due to several reasons, such as DOI bureaus' inability to reach consensus on the provisions of the policy and the impending administration change" [*Office of the Inspector General (OIG)*, 2010]. The bureaus of the DOI have common values related to managing and protecting the nation's natural resources; however, they each have individual missions that emphasize science in different ways. Because the U.S. Geological Survey (USGS) is predominantly a science agency, it moved forward first in adapting the draft policy for the bureau.

3.2.3. U.S. Geological Survey Policy: Scientific Integrity, 2007

In 2007, the USGS was the first bureau within DOI to establish a scientific integrity policy (Survey Manual 500.25 Scientific Integrity: http://www.usgs.gov/usgs-manual/500/500-25.html) "ensuring scientific integrity in the conduct of scientific activities and procedures for reporting, investigating, and adjudicating allegations of scientific misconduct by USGS employees and volunteers." This was the first application of the scientific integrity policy to volunteers in DOI and acknowledged the significant contribution that volunteers make in assisting and carrying out science.

This USGS policy closely followed the definitions, procedures, criteria, and safeguards of the Federal Policy on Research Misconduct (*OSTP*, 2000) and expanded the scope to include "scientific activities," defined as "activities involving inventorying, monitoring, experimentation, study, research, modeling, and scientific assessment. Scientific activities are conducted in a manner specified by standard protocols and procedures and include any of the physical, biological, or social sciences as well as engineering and mathematics that employ the scientific method." In this way, the USGS scientific integrity policy became tied to all other Survey Manual guidance already established for peer review and science protocols.

Per the USGS policy, allegations of misconduct were required to be filed with the servicing human resource office within 60 days of discovering the alleged misconduct (a timeliness requirement later included in the DOI policy).

- **Inquiry phase:** Upon receipt of an allegation of scientific misconduct, the servicing human resources office will contact the immediate supervisor of the subject (henceforth referred to as the Subject) of the inquiry to inform them that an allegation of scientific misconduct has been filed. The immediate supervisor of the Subject, working with the servicing human resources office, will review the allegation and immediately secure all the original research records and materials relevant to the allegation. The Subject of the allegation will then be notified in writing that an allegation of scientific misconduct has been filed against them.

- **Investigation phase:** If further investigation is required, the matter is referred to the Chairperson, Scientific Misconduct Review Panel (SMRP). The Chairperson, with the concurrence of the Director (USGS), selects four additional scientists and an alternate scientist to serve four-year terms on the SMRP. The Subject of the investigation is then notified by the Chairperson and advised of the investigation and his/her rights and responsibilities during the process. Following their further investigation, the SMRP prepares a summary report with findings.
  Note: the SMRP concept was later adapted in the subsequent DOI policy as the Scientific Integrity Review Panel (SIRP), sometimes used in further inquiry of an allegation.

- **Administrative Action phase:** The SMRP report is forwarded to the immediate supervisor and servicing human resources office, who work together to determine the appropriate action to be taken. At the conclusion of the process, the person who filed the allegation will be notified by the servicing human resources office that management has taken appropriate action and that the matter has been resolved.

The USGS policy also included the "USGS Code of Scientific Conduct," listing 14 "I will…" statements. This code of conduct was further developed in the DOI's "Code of Scientific and Scholarly Conduct."

The USGS policy includes a reporting requirement: "Servicing human resources offices will submit a report each calendar year to the Bureau Human

Resources Officer. The report will list (1) the number of allegations of scientific misconduct filed against employees and against volunteers, (2) whether or not scientific misconduct was found and (3) the disciplinary or other action taken, if any."

3.2.4. Secretary of the Interior Memo: "Ensuring Integrity in Scientific Activities," 2007

Department of the Interior Secretary Dirk Kempthorne's memorandum, "Ensuring Integrity in Scientific Activities" (December 21, 2007), affirmed the Department's commitment to the Office of Management and Budget's (OMB's) guidelines in their "Final Information Quality Bulletin for Peer Review" (2004), ensuring that "all information disseminated by the Department must comply with basic standards of quality to ensure and maximize its objectivity, utility, and integrity." The secretary expanded this understanding, saying:

> It is imperative that we ensure that the quality, objectivity, utility and *integrity of the scientific information* provided by the Department are above reproach. Therefore, Bureau Directors need to ensure that all employees are adhering to the guidelines and standards expected for *information development and management*. In addition, Bureau and Office Directors are responsible for determining whether their agency is engaged in *significant scientific data collection, processing, or analysis activities* where further clarification beyond the guidance provided by the OMB and Department is warranted. (emphasis added)

While acknowledging the USGS action to develop earlier supplementary policy, the secretary also called on the Research and Development Council to provide a "summary report, findings, and any further recommendations for ensuring scientific integrity within the Department." The council made significant progress in 2008, assisting each bureau with developing guidance for their employees on scientific integrity.

3.2.5. U.S. Fish and Wildlife Service Manual Chapter: "Scientific Code of Professional Conduct for the Service," 2008

On January 28, 2008, the U.S. Fish and Wildlife Service (FWS) established a new uniform "Scientific Code of Professional Conduct for the Service," Service Manual chapter 212 FW 7, for all employees who engage in scientific activities. "To the best of their abilities," these employees "must" follow conduct statements specifically listed under the following categories: (1) general scientific code, (2) when using or applying scientific information, (3) to maintain scientific professionalism, (4) treating uncertainty in their scientific activities, and (5) administrative requirements for scientific activities. The policy also stated, "We will treat any violation of this policy as a finding of scientific or other misconduct…if we verify misconduct we will take action in accordance with our human resource policy and where applicable, contract policy." Note: 212 FW 7 was revised on December 22, 2011, to comply with the new Department manual chapter on scientific integrity, 305 DM 3.

3.2.6. National Park Service Interim Guidance: Code of Conduct, Peer Review, and Information Quality, 2008

The National Park Service (NPS) adopted the "Interim Guidance Document Governing Code of Conduct, Peer Review, and Information Quality Correction for National Park Service Cultural and Natural Resource Disciplines" on January 31, 2008, to "ensure that NPS scientific and scholarly activities comply with the OMB Final Information Quality Bulletin for Peer Review, the Department of Interior draft 305 DM 4 Peer Review," and other directives listed. This policy provided specific NPS procedures for compliance with the Information Quality Act (P.L. 106-554 Section 515) and associated OMB guidelines, including OMB "Final Information Quality Bulletin for Peer Review" (2004). This policy also included a "Code of Scientific and Scholarly Conduct," stating, "To enhance their contribution to quality, objectivity, utility, and integrity of such information, all NPS employees working with scientific and scholarly information will in performing their duties:..." (followed by a listing of 12 conduct standards).

3.2.7. Office of Surface Mines Policy on Technical Studies, 2008

The Office of Surface Mines established the policy "Technical Studies" on September 25, 2008, directing "procedures for the conduct and administration of technical studies that are funded in whole or part by the Office of Surface Mining Reclamation and Enforcement (OSM)." This policy also stated, "All scientific information generated or disseminated by OSM will comply with basic standards of quality to ensure and maximize its objectivity, utility and integrity."

3.2.8. Minerals Management Service Interim Policy Document, 2009

The Minerals Management Service (MMS) established an interim policy, "Integrity and Code of Conduct for Science, Scientific Assessment and Other Similar Technical Activities" (December 28, 2009), which directed "all personnel (including decision-makers, employees, and external participants) who engage in or use the results of scientific activities or scientific assessments to comply with the requirements" of this policy. A "Code of Conduct," consisting of 14 "I will" statements, was included in the MMS policy and specifically affirmed by the subsequent policy statement: "Decision-makers, employees and external participants who engage in scientific activities or assessments must comply with the Code of Conduct specified above. Failure to do so may result in further review and/or disciplinary action" (IPD 2010-01, 6. C). The interim policy did not include any specific procedures for scientific integrity complaints or the review of any complaints other than standard personnel practices (370 DM 752).

3.2.9. Bureau of Indian Affairs Policy: Scientific Integrity and Scientific Conduct, 2009

The Bureau of Indian Affairs (BIA) established two complimentary policies in 2009: "Scientific Integrity" (10 IAM 3) and "Scientific Conduct" (10 IAM 4). "Scientific Integrity" held BIA personnel and its contractors "accountable for the integrity of the information they collect and analyze, and the conclusions they present," and to "understand their obligation to abide by 10 IAM 4 ("Scientific Conduct") and by the Federal Policy on Research Misconduct (*OSTP*, 2000)." BIA "will take appropriate action" for dissemination of misleading or inaccurate scientific information or from violations of research misconduct. The "Scientific Conduct" policy lists 14 conduct statements that all BIA employees working with scientific information will follow.

3.2.10. Bureau of Reclamation Interim Policy: Scientific Integrity, 2010

On January 11, 2010, the Bureau of Reclamation (BOR) issued temporary Directives and Standards (D&S), "Scientific Integrity," "to increase awareness of the importance of scientific information and science as a method of discovery to maintain and enhance our effectiveness in fulfilling our mission, program require-ments, and other federal mandates, and in establishing credibility and value with the public, both nationally and internationally." The BOR policy is similar to the FWS policy, 212 FW 7 (2008), including the following: "To the best of their abilities," BOR employees "must" follow conduct statements specifically listed under the categories (1) Scientific Code of Professional Conduct, (2) Uncertainty in Scientific Activities, (3) Maintaining Scientific Professionalism, (4) Use of Scientific Information, and (5) Accountability and Documentation. "Verified violations of the requirements in this D&S will be treated as a finding(s) of scientific or other misconduct."

3.2.11. U.S. Geological Survey: Office of Science Quality and Integrity, 2010

In a first for Interior bureaus, the USGS in October 2010 established the Office of Science Quality and Integrity (OSQI) to monitor and enhance the integrity, quality, and health of USGS science through executive oversight and the development of strong practices, policies, and supporting programs. OSQI is an enterprise-level function housed in the director's office and charged with bringing together ethics, implementing the Information Quality Act, overseeing fundamental science practices (including peer review), conducting scientist review (Research Grade Evaluation), promoting employee development through its youth and postdoctoral fellowship programs, and supporting education, outreach, and tribal relations in a consistent and rigorous way throughout this geographically dispersed and organizationally complex bureau.

3.3. Raising the Bar on Scientific Integrity

3.3.1. President Obama, Restoring Science and Public Trust

In his first inaugural address on January 20, 2009, President Barack Obama raised a new expectation, saying, "We'll restore science to its rightful place," and later declared, "Those of us who manage the public's dollars will be held to account…because only then can we restore the vital trust between a people and their government."

Within two months of his inauguration, President Obama issued a "Scientific Integrity" memorandum for the Heads of Executive Branch Departments and Agencies (March 9, 2009). This memorandum established the role of science, "Science and scientific process must inform and guide decisions of my Administration," and addressed the emerging public concern and challenge, that "the public must be able to trust the science and scientific process informing public policy decisions." The President also warned that "Political officials should not suppress or alter scientific or technological findings and conclusions." The President's goal as stated in this memorandum was "ensuring the highest level of integrity in all aspects of the executive branch's involvement with scientific and technology processes." The memorandum also directed the Director of the Office of Science and Technology to recommend a plan to achieve this goal throughout the executive branch.

3.3.2. Department of the Interior, OIG Evaluation Report on Scientific Integrity and Secretary Action, 2010

In April 2010, the Department of Interior's Office of the Inspector General (OIG) issued an evaluation report, "Interior Lacks a Scientific Integrity Policy" [*OIG*, 2010]. The report found,

> Interior has never had a comprehensive scientific integrity policy, or any requirement to track scientific misconduct allegations. A decade ago, the White House's Office of Science and Technology Policy (OSTP) required all executive office agencies to implement scientific integrity policies that would address scientific misconduct. Further, in a 2002 investigative report, OIG recommended that Interior develop a code of scientific ethics. Despite this, Interior has failed to implement a comprehensive scientific integrity policy. The lack of a comprehensive policy leaves not only Interior, but those who rely upon the scientific information, vulnerable to tainted data and misinformed decisions and, as a consequence, could have a negative effect on public trust.

This report recommended:

1. The DOI "develop and implement an Interior-wide comprehensive scientific integrity policy that addresses required elements of the White House Office

of Science and Technology Policy (OSTP) scientific misconduct policy." [*OSTP*, 2000]

2. The DOI "designate a responsible official to guide the development and implementation" of such policy, and "to oversee the bureaus' implementation and application of the policy."

The OIG report also recommended that development of Interior's scientific integrity policy consider the USGS scientific integrity policy as well as the Health and Human Services (HHS) policy, especially the HHS policy elements that require public posting of research misconduct determinations, training programs for internal and external research to prevent misconduct, and newsletters and annual reporting on research misconduct.

Responding to this OIG report, Secretary Salazar designated the USGS Director as the responsible official to oversee the development and implementation of the Department's policy on scientific integrity and pledged to complete this draft policy by October 1, 2010. A policy writing team was formed representing many bureaus. This team began with the earlier DOI draft policy (2007) and the scientific integrity policies from USGS (2007) and MMS (2009) (MMS was temporarily renamed the Bureau of Ocean Energy Management, Regulation and Enforcement at that time). The OSTP was consulted to ensure that this comprehensive Department draft policy would be consistent with their forthcoming guidance.

On August 31, 2010, the Department publicly posted a proposed draft policy on scientific integrity in the Federal Register, requesting public comments. More than 200 stakeholders were alerted to this posting. The Department received many excellent comments that helped guide their thinking and prioritization of the many issues they were considering including in the final policy.

DOI Secretary Salazar issued Order No. 3305, "Ensuring Scientific Integrity within the Department of the Interior," on September 29, 2010, directing the establishment of a Departmental Manual Chapter that sets forth principles of scientific and scholarly integrity and clarifying the roles and responsibilities of all Department employees and political appointees in upholding these principles, including the following:

- "DOI employees, political and career, must never suppress or alter, without new scientific or technological evidence, scientific or technological findings or conclusions."
- "Employees will not be coerced to alter or censure scientific findings."
- "Employees will be protected if they uncover and report scientific misconduct by career or political staff."
- "It shall be the duty of each employee, career and political, to report such misconduct."
- "This policy and a code of ethics on scientific integrity will guide the conduct of scientists and decision makers in a manner that is above reproach."

In addition, the secretary's memorandum specified eight principles consistent with the President's memorandum on scientific integrity and the subsequent OSTP memorandum on the same subject.

3.3.3. Office of Science and Technology Policy, Scientific Integrity Guidance, 2010

Per the President's assignment, John P. Holdren, director of the Office of Science and Technology Policy, issued a "Scientific Integrity" memorandum on December 17, 2010, to the heads of executive branch departments and agencies, providing further guidance on implementation of the administration's scientific integrity policy. Holdren declared, "Successful application of science in public policy depends on the integrity of the scientific process both to ensure the validity of the information itself and to engender public trust in Government."

The memorandum stated that agencies should develop policies that (1) "ensure a culture of scientific integrity," (2) "strengthen the actual and perceived credibility of Government research," (3) "facilitate the free flow of scientific and technological information, consistent with privacy and classification standards," and (4) "establish principles for conveying scientific and technological information to the public." The memorandum also encouraged agencies to develop policies for (1) public communication "that promote and maximize, to the extent practicable, openness and transparency with the media and American people," (2) "convening Federal advisory committees tasked with giving scientific advice," and (3) "promot[ing] the professional development of Government scientists and engineers."

3.4. New Scientific Integrity Policy at DOI

3.4.1. Department of Interior Policy: "Integrity of Scientific and Scholarly Activities" ("Version 1.0")

Following the Secretary's Order (October 2010) the policy writing team considered all the existing bureau policies for scientific integrity and completed the draft Department policy. Department bureaus' review and comments were incorporated during several rounds of review and the final draft was reviewed by numerous scientists and scholars throughout DOI. The final policy was approved by Secretary Salazar and released on January 28, 2011, as "Department Manual, Management, Science Efforts" (305), Chapter 3: "Integrity of Scientific and Scholarly Activities" (305 DM 3). DOI bureaus were directed to amend their bureau policies to conform to the new Department policy. The following sections review the major components of the policy.

3.4.2. Policy Purpose

It is essential that the Department establish and maintain integrity in its scientific and scholarly activities because information from such activities is a critical factor that informs decision making on public policies. Other factors that inform decision making may include economic, budget, institutional, social, cultural, legal, and environmental considerations. The Department is dedicated to preserving the integrity of the scientific and scholarly activities it conducts, and activities that are conducted on its behalf. It will not tolerate *loss of integrity* [emphasis added] in the performance of scientific and scholarly activities or in the application of science and scholarship in decision making. (305 DM 3.1)

The policy purpose intentionally focuses on information from scientific and scholarly activities and sets aside other factors (i.e., economic, budget, institutional, social, cultural, legal, and environmental) that may also be part of decision making in the Department. This distinction, focusing on the science used in decision making, was important in the later implementation of the policy, addressing the concern that the science might be altered to justify a predetermined decision. Another important element of the purpose statement is the introduction of the term *loss of integrity* in the policy terminology.

3.4.3. Key Provisions of the Policy

3.4.3.1. Scientific misconduct
The DOI policy adopted the Federal Policy on Research Misconduct [*OSTP*, 2000] definition of *research misconduct* and the criteria for determining findings of "scientific and scholarly misconduct" similar to the "scientific misconduct" defined in the USGS policy (2007). However, 305 DM 3 also added, "Misconduct also includes: (a) intentionally circumventing policy that ensures the integrity of science and scholarship, and (b) actions that compromise scientific and scholarly integrity."

This additional definition came to be known as a "loss of scientific integrity," especially "actions that compromise scientific and scholarly integrity." This term helps provide a thoughtful litmus test for all potential integrity issues related to science activities, going beyond the traditional plagiarism, fabrication, and falsification to include any actions that compromise scientific integrity. In practice, complainants often referred to excerpts from the Code of Scientific and Scholarly Conduct to discuss standards that were allegedly compromised, leading to a "loss of scientific integrity."

3.4.3.2. Scientific integrity defined
The condition resulting from adherence to professional values and practices, when conducting and applying the results of science and scholarship, that ensures objectivity, clarity, reproducibility, and utility and that provides insulation from bias, fabrication, falsification, plagiarism, outside interference, censorship, and inadequate procedural and information security. (305 DM 3.5L)

3.4.3.3. Criteria for a finding of scientific misconduct or loss of scientific integrity
Consistent with the Federal Policy on Research Misconduct [*OSTP*, 2000], the policy required that a finding of scientific and scholarly misconduct determine that
1. "There be a significant departure from accepted practices of the relevant scientific and scholarly community.

2. The misconduct be committed intentionally, knowingly, or recklessly.
3. The allegation be proven by a preponderance of evidence."

The policy also explained that "scientific and scholarly misconduct does not include honest error or differences of opinion."

Note: The criteria above did not include the word "and" after #2, although implementation of the policy assumed that all three elements of the criteria were required for a finding of misconduct, consistent with the Federal Policy on Research Misconduct [*OSTP*, 2000]; "and" was added to this criteria in version 2.0 of the policy in December 2014.

3.4.3.4. Code of Scientific and Scholarly Conduct

The policy includes a "Code of Scientific and Scholarly Conduct" (305 DM 3.7), a series of 19 "I will…" statements, divided into three sections: (1) All Departmental Employees, Volunteers, Contractors, Cooperators, Partners, Permittees, Leasees and Grantees; (2) Scientists and Scholars; and (3) Decision Makers. Although this section states that these groups "will abide by" this code, there is no other specific reference to this requirement in the definition of scientific misconduct or a loss of scientific integrity. During the first three years of implementation, many complaints included references to specific statements in the Code of Scientific and Scholarly Conduct in describing the nature of their complaint. The complainants were using segments of the Code of Conduct to describe a variance from this standard as an alleged "loss of scientific integrity" in their workplace, perhaps using the Code of Conduct as the "acceptable standard" for maintaining scientific integrity in the workplace. The later policy revision, version 2.0 (12/16/2014), incorporated this Code of Conduct in the definition of "loss of scientific integrity."

3.4.3.5. Scientific integrity officers

The policy created scientific integrity officers (SIO), each one appointed by the bureau head as bureau scientific integrity officer (BSIO), and a department scientific integrity officer (DSIO) appointed by the deputy secretary. The DSIO provided "department-wide leadership for the implementation" of this policy, "ensuring integrity and consistency across the Department" while also serving as the SIO for scientific integrity cases involving bureau heads or Department-level personnel.

Although not specified in the policy, the appointed SIOs typically had a strong scientific background and postgraduate science degrees. The BSIOs implemented this policy within their respective bureaus, conducting reviews or inquiries on complaints received regarding their bureau, and kept their bureau head "informed of the status of the implementation." In a few cases, the DSIO assigned another BSIO to conduct the review/inquiry of a complaint about a different bureau.

Note: In other departments and agencies, scientific integrity officers, if they exist, provide administrative oversight of the scientific integrity program, while their OIG conducts the formal inquiries of scientific integrity complaints. The Department of Interior's OIG receives scientific integrity complaints, directly or concurrently with the Office of the Executive Secretariat, and often coordinates their inquiry with the appropriate BSIO.

3.4.3.6. Responsible manager and servicing human resources officer

When processing a scientific integrity complaint that required a full inquiry, the BSIO worked with a "responsible manager" and a servicing human resources officer (SHRO), assigned by the bureau head, "to determine if the allegation is covered under the provisions" of this policy and "provide consistency, oversight, and guidance throughout the entire process." The responsible manager was usually the immediate supervisor, provided that there was no apparent conflict of interest. Often the responsible manager provided a good sounding board for the BSIO and enhanced the understanding of the context of the concern and the applicable standards of the "relevant scientific community." The responsible manager and SHRO were not part of the BSIO's final determination on the complaint. In cases involving bureau heads or Department-level personnel, the deputy secretary selected and assigned the responsible manager and SHRO to work with the DSIO.

3.4.3.7. Scientific Integrity Review Panel

Similar to the USGS policy, the DOI policy provides for a Scientific Integrity Review Panel (SIRP) to assist with "impartially examining allegations" as requested by the SIO and approved by the bureau head. The SIRP could be a standing panel established by the bureau or formed for a specific allegation, then disbanded. In some instances, an independent science services contractor may provide a panel of subject-matter experts to perform an impartial examination of an allegation.

Note: During the first three years of implementation, only one SIRP was appointed to address a specific FWS complaint, then disbanded.

3.4.3.8. Complaint review process

In accordance with the Federal Policy on Research Misconduct [*OSTP*, 2000], the complaint review process followed several phases of review or inquiry:

1. Complaints were officially received and logged in by the Department's Office of the Executive Secretariat (OES) and assigned to the appropriate SIO. OES assigned a tracking number to each case, assuring a permanent documentation and accounting of that matter. The DSIO determined assignments when the complaint involved more than one bureau or other considerations.

2. Initial review of the complaint by the SIO determined if the complaint included all of the elements as required by the policy. The SIO contacted the complainant to confirm that their complaint was received, discuss any missing elements needed, and provide the BSIO's contact information.

Incomplete complaints were rejected if all the required elements were not provided in a timely manner.

3. Preliminary review by the SIO of the complaint and submitted material, determined whether an inquiry was warranted. Allegations that were previously resolved were not reopened unless substantial new information was submitted. In those situations, the SIO determined if there was substantial new information to warrant a new inquiry. In some cases, this review determined that the same complaint was before a judicial hearing, an OIG investigation, or a matter in another jurisdiction, requiring the SIO to consult with the other offices and DSIO to determine the best venue for this concern. If the SIO determined that the complaint did not warrant further investigation and was without merit, the SIO dismissed the allegation and notified the complainant.
Note: The subject of the complaint was usually not notified of the allegations that were closed at this stage.

4. Inquiry phase required notification to the subject of the allegation and the opportunity for the subject to provide any relevant information. Working with a responsible manager and servicing human resource officer assigned to this inquiry, the SIO conducted a full inquiry of the allegation(s). In some cases, the SIO requested a panel of experts, SIRP or contracted panel (see above), to assist with the inquiry and provide a consensus panel report to the BSIO, DSIO and bureau head. In at least one case, the SIRP request was not approved by the bureau head, stating that the complaint was not a viable claim of scientific misconduct.

5. Report phase required the SIO to prepare and submit a final report on the fact finding and determination of the merit of the allegation(s). The SIO report was sent to the DSIO, bureau head, and the responsible manager. A synopsis of the allegation/finding was also posted on the Department's public website: Scientific Integrity/closed cases. (https://www.doi.gov/scientificintegrity/closed-scientific-integrity-cases)

3.4.3.9. Adjudication: Administrative and disciplinary action

If a report of finding determined that there was scientific misconduct or a loss of scientific integrity, appropriate actions (administrative and/or disciplinary) were determined by the responsible manager or other supervisors, following Department personnel procedures. The responsible manager also considered recommendations for restoring scientific integrity in the workplace, provided in some SIO reports and/or the SIRP report.

3.4.4. Informal/Formal Complaints

In some cases, employees contacted the BSIO to discuss a concern about scientific misconduct and/or a loss of scientific integrity. Often the BSIO performed an "ombudsman role" in attempting to resolve the matter by facilitating a common understanding among the parties, or providing an "informal" opinion about the merits of the concern. Sometimes an "informal" complaint became "formal" when an allegation was sent to the Office of the Executive Secretariat (OES). However, in many cases, the informal process, facilitated by the BSIO, was effective in resolving many concerns, perhaps far more than the number of formal complaints filed with OES. The "ombudsman" role of the SIO was recognized and authorized in the revised policy (version 2.0, December 2014).

3.4.5. Whistleblower Protection

Fear of reprisal and/or alleged acts of reprisal have been reported by several complainants or employees (informal) to the SIO. In these instances, the SIO immediately informed the complainant that they should contact the OIG Whistleblower Protection Program (OIG-WPP) to discuss their concerns. The SIO independently contacted the OIG-WPP to alert them of a concern and the contact from the employee. Allegations of reprisal were addressed by the OIG-WPP.

3.4.6. Implementation/Results

3.4.6.1. Complaints filed

During the first four years of implementing the scientific integrity policy, the Department received 37 formal complaints. Three of these formal complaints were dismissed because (1) the complaint was not submitted in complete form and the SIO was not able to obtain the additional information to complete the complaint, (2) the subject was not a DOI employee, or (3) the complaint was an Information Quality Act complaint and referred to that office.

Figure 3.1 provides an overview of the 37 formal scientific integrity allegations that were received and recorded by the Department's Office of the Executive Secretariat (OES) from February 2011 through April 2015. Some of these complaints involved two DOI bureaus, while most were a single-bureau matter. One complaint involved a bureau head and was handled as a Department complaint, per the policy. Several cases were received by the Office of Inspector General and referred to the OES as a scientific integrity compliant, with the resolution of that complaint referred to the OIG. The 37 formal complaints processed, by Fiscal Year (FY) and bureau (Geological Survey—USGS, Fish & Wildlife Service—FWS, Bureau of Ocean Energy Management—BOEM, National Park Service—NPS, Bureau of Land Management—BLM, Bureau of Reclamation—BOR, Department of the Interior—DOI), were as follows:

- FY 11: 9 cases (USGS-1, FWS/USGS-1, FWS-2, BOEM-2, NPS-1, BLM-1, FWS/BOR-1)
- FY 12: 9 cases (DOI-1, FWS-5, BLM-1, NPS-2)
- FY 13: 5 cases (FWS-2, BOR-1, USGS-1, NPS/USGS-1)
- FY 14: 6 cases (FWS-2, NPS-3, USGS-1)
- FY 15*: 8 cases (FWS-3, NPS-2, USGS-2, BOEM-1) *cases filed by April 30, 2015, were included in this report.

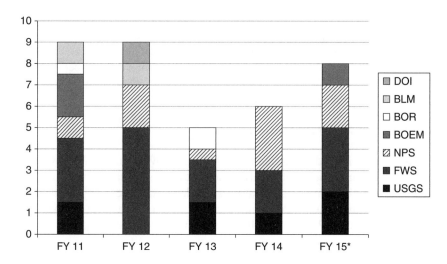

Figure 3.1 Summary of formal scientific integrity complaints submitted to the Department of Interior, Fiscal Years 2011–2015 (FY 2015 includes only October 2014–April 2015). Complaints regarding more than one DOI bureau are divided evenly between those bureaus.

During the first four years (2011 to early 2015), no formal scientific integrity complaints were filed regarding subjects in the Bureau of Indian Affairs, Bureau of Safety and Environmental Enforcement, Office of Surface Mining Reclamation and Enforcement, or the Office of the Secretary.

Figure 3.2 provides a graphical illustration of the kinds of complaints received. About one-third of these formal complaints alleged scientific misconduct (fabrication, falsification, plagiarism), but most often falsification. Another third of the complaints alleged a loss of scientific integrity in their workplace. The final third of complaints were incomplete and dismissed, and a few cases were referred to other processes such as Information Quality Act requests for information correction procedures.

3.4.6.2. Closed cases

As of March 31, 2015, only two formal scientific integrity complaints resulted in a "warranted" determination, while the other complaints were closed and dismissed as "not warranted." In many instances, the SIO did not find all three elements required for a "warranted" finding and therefore concluded the allegation as "not warranted" (see section 3.4.3.3. Criteria for a Finding of Scientific Misconduct or Loss of Scientific Integrity). For example, a "preponderance of evidence" was not found to support either that (1) the misconduct was "a significant departure from accepted practices of the relevant scientific and scholarly community," or (2) the misconduct was "committed intentionally, knowingly, or recklessly." A summary of each of these "closed" formal cases is posted on the Department's website, Closed Scientific Integrity Cases: https://www.doi.gov/scientificintegrity/closed-scientific-integrity-cases.

3.5. Revised Scientific Integrity Policy at DOI

3.5.1. Department Policy, Version 2.0

The Department manual chapter, 305 DM 3, was officially revised on December 16, 2014, following several years of experience in implementing the original version of the policy. This revision (version 2.0) included further clarification of the complaint requirement, the review process, and the roles and responsibilities of officials in this policy. Key revisions included the following:

1. The definition of "Loss of Scientific Integrity" was revised (<u>addition underlined</u>):

   "Loss of Scientific Integrity. Occurs when there is a significant departure from the accepted standards, professional values, and practices of the relevant scientific community, <u>including (for DOI employees and covered outside parties) the DOI Code of Scientific and Scholarly Conduct and Departmental standards for the performance of scientific or scholarly activities. Improperly using scientific information (including fabrication, falsification, or plagiarism of science) for decision making, policy formulation, or preparation of materials for public information activities, can constitute a loss of integrity</u>. Loss of scientific integrity negatively affects the quality or reliability of scientific information" (305 DM 3.5 B).

   <u>Reason for revision</u>: The intent was to provide more emphasis on and support for the Code of Scientific and Scholarly Conduct, and to establish the text of the Code as the expected standard for sustaining scientific integrity, making the Code "enforceable." Also, in the gradient between good professional behavior and scientific misconduct, we were trying to capture the middle part of that gradient and the potential to address concerns *before* misconduct occurred. The addition of the "using scientific information" sentence to the definition of loss of scientific integrity addressed concerns that scientific information was sometimes improperly described and used to explain a decision that may have be made for other non-scientific reasons.

2. Ombudsman role added to DSIO and BSIO:

   "Serves as the DOI ombudsman on matters of scientific integrity, which includes: fostering effective communication; acting as an intermediary between the complainant and the Department; acting as source of information, and provides advice and guidance on the Department's scientific integrity policy" (305 DM 3.6 C (8) and F (7)).

Scientific integrity complaint type

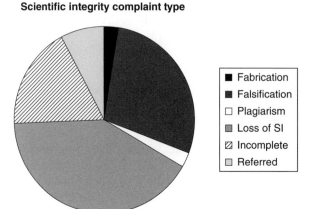

Figure 3.2 Summary of the type of formal scientific integrity complaints filed with the Department of the Interior, Fiscal Years 2011–2015 (FY 2015 includes only October 2014–April 2015). "Incomplete" complaints were not able to be completed and therefore dismissed. "Referred" complaints were not appropriate for the scientific integrity complaint process and were referred to more appropriate processes.

Reason for revision: We recognized the important ombudsman role that an SIO can and should perform in assisting with informal resolution of disputes over scientific work within the workplace.

3. "Coordinating Manager" replaces "responsible manager" and is defined (305 DM 3.5L) as "a supervisor or manager who is assigned to work with a scientific integrity officer on an inquiry. Supervisors or managers who have a real or potential conflict of interest should not be selected as the Coordinating Manager," with the responsibilities further specified.

    Reason for revision: We recognized the need for an independent yet knowledgeable manager to assist the SIO in understanding the context of the alleged misconduct in the workplace and the accepted practices of the relevant scientific and scholarly community in that bureau's workplace.

4. Complaint must also include (additions):

    a. The name of the person(s) or organization alleged to have committed the scientific misconduct or the loss of scientific integrity, if known (305 DM 3.8 A(3)(b)) Reason for revision: Experience indicated that sometimes the subject of the alleged misconduct could not be identified specifically by the complainant, especially if the complainant was outside of the agency.

    b. How and when the complainant learned of such facts of misconduct (305 DM 3.8 A (3)(c)). Reason for revision: We needed a way to more easily determine if the complaint is timely.

c. An explanation of how the criteria for scientific misconduct (as presented in Sections 305 DM 3.5.H, I, J) or loss of scientific integrity (as presented in section 3.5B) are met, including, for loss of scientific integrity: (1) citations or other information identifying the accepted practices of the relevant scientific community and (2) an explanation of how the complained-of actions constitute a significant departure from those practices (305 DM 3.8 A (3)(d)). <u>Reason for revision</u>: We recognized the importance of the complainant considering and describing how the alleged misconduct meets the criteria for a finding of misconduct: specifically, what are the accepted practices of the relevant scientific community and how the alleged misconduct is a significant departure from these practices.

d. Statement of whether this complaint was also submitted elsewhere (e.g., OIG, Office of Special Counsel, human resources office, Ethics Office, etc.) (305 DM 3.8 A (3)(f)). <u>Reason for revision</u>: Experience showed us that parallel submission does happen, and in order to coordinate effectively with other offices we need the complainant to disclose if the same complaint was filed elsewhere in order to avoid duplicate/competing review processes.

5. Scientific Integrity Review Panel (SIRP) (additions) (305 DM 3.8 D), a SIRP may be convened at any point in order to

a. determine the "accepted practices of the relevant scientific community,"

b. determine if the evidence gathered indicates a "significant departure" from those accepted practices, or

c. assist in further fact finding and review.

<u>Reason for revision</u>: We recognized the importance of more focus for the SIRP, as directed by the SIO.

6. Report of Inquiry (new addition section 305 DM 3.8 E) shall contain

a. record of all the evidence (documents, SIRP final report, signed or affirmed witness statements, etc.) relied upon,

b. findings of fact that reference the evidence of record, and

c. a determination as to whether scientific misconduct or loss of scientific integrity has occurred and an explanation of the reasons for the determination.

<u>Reason for revision</u>: Experience revealed the importance of providing final documentation of the full record, in compliance with the federal records requirements.

7. Reconsideration of a DSIO/BSIO finding (new addition section 305 DM 3.8 G):

a. A request for reconsideration may be filed by the subject(s) of the finding in order to present new relevant and material information about the findings of fact or determination.

Reason for revision: We corrected an omission in the first version of the policy and provided an "appeal" opportunity for the subject.

8. SI Council. The revised policy also added recognition of the Scientific Integrity Council, the assembly of the Department SIO (chairperson) and bureau SIOs who meet regularly for training, and further discussion on the consistent implementation of the policy.

Reason for revision: Experience helped us to appreciate the value of our peers and the need for the SIO to meet regularly for support and consistency in the policy implementation.

3.5.2. Scientific Integrity Procedures Handbook

The revised policy also contained a new important reference document, the Scientific Integrity Procedure (SIP) Handbook. (305 DM 3.11, new additional section). The SIP Handbook provides details on the scientific integrity policy, requirements, procedures, and processes. It contains the requirements and forms for filing and evaluating complaints of violations of the Department's scientific integrity policy, as well as standardized language for contracts, agreements, permits, or leases involving scientific activities and guidelines on how DOI employees may serve in their official or personal capacities with scientific and scholarly organizations. The SIP Handbook is available at http://elips.doi.gov/elips/0/doc/4054/Page1.aspx.

3.6. Scientific Integrity Website

Another aid to understanding and implementing the scientific integrity policy is the Department's website: Scientific Integrity (https://www.doi.gov/scientificintegrity). This site is updated regularly and contains links to the policy, the SIP Handbook, information on "closed cases," SIO's contact information, frequently asked questions/answers, and the Code of Scientific and Scholarly Integrity (in printable poster format).

3.7. Scientific Integrity Training

While the policy does not prescribe employee orientation procedures, training was developed to increase awareness and full implementation of the policy, and it provides a useful resource for new employees, enabling bureaus to effectively sustain compliance with the Scientific Integrity policy. Online training on scientific integrity was developed by the Federal Collaborative e-Learning Laboratory (Fed-CEL) and made available throughout the Department via DOI Learn. This one-hour training provides the knowledge to

*   Create conditions that support a culture of scientific integrity.
*   Recognize situations where scientific integrity may be at risk.

- Identify the roles and responsibilities of those who are obligated to uphold scientific integrity and ensure appropriate use and communication of scientific findings and data.
- Respond appropriately to potential violations of Departmental scientific integrity policy.

Recognizing the diversity of the bureau and office missions and employee responsibilities, each bureau/office director was provided latitude in defining their policy and identifying the employees who must take the scientific integrity training. It is expected that all bureau/office senior executives as well as employees in science, engineering, and technical positions and their supervisors will take this training. Some groups of employees may have a need to take the scientific integrity training on a recurring basis. Each bureau/office is required to develop scientific integrity training guidance that outlines which categories of employees are required to take the training on a onetime or recurring basis as well as the employees who are exempted from the training requirement.

Presentations were developed as well, for use by all the DOI bureaus, for their recurring training sessions or in conjunction with other training sessions, such as annual ethics training.

3.8. The Future Scientific Integrity Policy at DOI?

3.8.1. Ideas for Department Policy, Version 3.0

The Scientific Integrity Council continues to consider further revisions to the Department's policy, based on the continued experience of policy implementation and feedback from interested outside organizations. Suggested additional revisions include the following:
- More assertive support of zero tolerance for any retaliation against a complainant
- Extending "conflict of interest" to include the interests of any family member, the appearance of a conflict, cronyism, creating an unfair advantage for a favored associate, etc.
- Mandatory annual reporting of scientific integrity complaints/resolution
- More independence in the selection of the SIO and who oversees a scientific integrity inquiry
- Enabling scientific inquiry panelists/experts to recommend corrective measures, if needed
- Strengthening the definition of "loss of scientific integrity" to include political alteration of science or suppression of information
- Restore the earlier policy statement (version 1.0), which ensures that the selection and retention of employees in scientific or scholarly positions that rely on the results of scientific activities are based on the candidate's integrity,

knowledge, credentials, and experience relevant to the responsibility of the position
- Restore earlier policy responsibility (version 1.0) specifying that employees and volunteers report to the appropriate officials, as described in section 3.3.8 of this chapter, their knowledge of scientific misconduct that is planned, is imminent, or has occurred.
- Clarify the inquiry process jurisdiction of scientific integrity complaints that may be simultaneously filed in multiple complaint processes
- Assure OIG coordinates with SIOs in scientific integrity matters

3.9. Conclusion

The success of the DOI scientific integrity policy is dependent on creating, and sustaining, an environment where scientific integrity can flourish, where all employees believe in the scientific integrity principals embodied in the policy, where concerns or complaints about scientific integrity are promptly acted upon in a fair process, and where employees trust that there will not be reprisal when they make an allegation. If employees believe in the scientific integrity policy, then they will participate in its implementation, calling out science integrity concerns if they occur and collectively raising the bar of scientific integrity for practicing scientists and managers. Additionally, scientists, senior managers, and middle managers must share a common understanding of the underlying need for the scientific integrity policy and take responsibility for actively supporting the implementation of this policy.

Creating a culture that supports and expects scientific integrity is an ongoing process, and the mere existence of a policy will not ensure such a culture exists. However, by consistently making the expectations for protecting the integrity of the scientific process and products clear, we hope to see the entire DOI workforce embrace this concept that scientific information considered in DOI decision making must be robust, of the highest quality, and the result of as rigorous a set of scientific processes as can be achieved. Most importantly, the information must be trustworthy; integrity is a prerequisite to this.

Acknowledgments

Data supporting Figure 3.1 and 3.2 are available at the Department of Interior's website: scientific integrity, http://www.doi.gov/scientificintegrity/.

We thank Rama Kotra, Carolyn Reid, Nancy Baumgartner, Bruce Taggart, and Mike Diggles for their useful comments.

References

Department of the Interior (January 28, 2011), Policy: Integrity of Scientific and Scholarly Activities, (version 1.0), http://elips.doi.gov/elips/DocView.aspx?id=3045&searchid=5ec7a599-70e2-4115-a900-33ebb7f850ee&dbid=0.

Department of the Interior (December 16, 2014), Policy: Integrity of Scientific and Scholarly Activities, (version 2.0), http://elips.doi.gov/ELIPS/DocView.aspx?id=4056.

Department of the Interior (2014), Strategic Plan 2014–2018, http://www.doi.gov/bpp/upload/DOI-Strategic-Plan-for-FY-2014-2018-POSTED-ON-WEBSITE.pdf.

Office of the Inspector General (OIG), U.S. Department of the Interior (2010), Evaluation Report, Interior Lacks a Scientific Integrity Policy, Report No. WR-EV-MOA-0014-2009. http://www.doi.gov/oig/reports/upload/ScientificIntegrityPolicy.pdf.

Office of Science and Technology Policy (OSTP) (December 6, 2000), Federal Policy on Research Misconduct, 65 FR 76260-76264, https://www.federalregister.gov/articles/2000/12/06/00-30852/executive-office-of-the-president-federal-policy-on-research-misconduct-preamble-for-research#h-9.

Office of Science and Technology Policy (December 17, 2010), "Scientific Integrity" memorandum to heads of Executive Departments and Agencies, https://www.whitehouse.gov/sites/default/files/microsites/ostp/scientific-integrity-memo-12172010.pdf.

President of the United States (March 9 2009), Memorandum on Scientific Integrity, https://www.whitehouse.gov/the_press_office/Memorandum-for-the-Heads-of-Executive-Departments-and-Agencies-3-9-09/.

Secretary of the Interior (September 29, 2010), Order No. 3305, Ensuring Scientific Integrity within the Department of Interior, http://elips.doi.gov/ELIPS/0/doc/167/Page1.aspx

Union of Concerned Scientists (2010), Systematic interference with science at Interior Department exposed, http://www.ucsusa.org/our-work/center-science-and-democracy/promoting-scientific-integrity/political-interference-in.html.

Section II: The Role of Geoscience Professional Societies in Scientific Integrity and Ethics

4

4. THE AMERICAN GEOSCIENCES INSTITUTE GUIDELINES FOR ETHICAL PROFESSIONAL CONDUCT

Maeve A. Boland[1] and David W. Mogk[2]

Abstract

In response to a request from the geoscience community, the American Geosciences Institute coordinated the development of a geoscience-wide code of professional conduct in 1999. The Guidelines for Ethical Professional Conduct is an aspirational document that sets out values for the profession. The Guidelines were revised in 2015 to reflect the significant increase in geoethics scholarship and awareness since 1999. The Guidelines address expected personal and professional behaviors of geoscientists as members of a professional and scientific community. The revised Guidelines place more emphasis on the societal context of the geosciences, on ethical obligations to earn public trust and confidence, and on effective communication within and beyond the geoscience community, which all relate to the role of the geosciences at the intersection of Earth and human systems. The Guidelines are an overarching statement of principles for the geosciences that geoscience societies, organizations, and individuals are encouraged to use as the basis for further development.

Geoscientists study the Earth system, which is dynamic, heterogeneous, complex, and ancient. They deal with uncertain, ambiguous, and incomplete information. Geoscientists work at the intersection of the Earth and human

[1] American Geosciences Institute, Alexandria, Virginia, USA
[2] Department of Earth Sciences, Montana State University, Bozeman, Montana, USA

Scientific Integrity and Ethics in the Geosciences, Special Publications 73,
First Edition. Edited by Linda C. Gundersen.
© 2018 American Geophysical Union. Published 2018 by John Wiley & Sons, Inc.

systems, two systems that can have immense impacts on each other, from the local to the global scale. Geoscientists also work within the human system, with all the complexities of interpersonal interactions, legal and regulatory strictures, and moral expectations. The very nature of geoscience places its practitioners at the nexus of many potential ethical conflicts and dilemmas. Nevertheless, the geoscience community has lagged some other sciences in developing community-wide codes of ethics. A philosophical and practical framework addressing ethics and ethical conduct is needed to prepare geoscientists to recognize potential or emerging ethical issues, to provide them with the tools for ethical decision making, and to be prepared to act by preventing, mitigating, or responding to ethical dilemmas.

When introducing the first Guidelines for Ethical Professional Conduct in 1999, Stephen Stow posed a question: "The geosciences have grown and flourished for decades, if not for hundreds of years, with minimal visibility of published ethical standards. So what's the need to rush and develop something for the profession at this time?" Stow's main response was that public recognition of the geosciences was increasing and that geoscientists must be seen to adhere to the highest ethical standards, which required a statement of those standards [*Stowe*, 1999]. Building public trust in the geosciences is undoubtedly a strong motivation for developing a succinct statement of overarching ethical principles for the geosciences. Ethical guidelines also provide a framework for ethical decision making that can be used by individual geoscientists in their professional practice and that define standards of personal and professional behavior by members of the geoscience community.

Many geoscience organizations have codes of conduct or statements addressing ethical behavior. In this volume, David Abbott [2018], John Williams [2018], and Michael McPhaden [2018] review how some scientific and professional societies have responded to the challenge of crafting codes of ethics that are appropriate for their disciplinary interests. In each case, the organizations were considering the ethical issues related to their own members and activities. Few, if any, individual societies are in a position to develop a broad statement on ethics that would apply across all the geosciences or to define the ethical consequences of saying, "I'm a geoscientist." That role has been filled by the American Geosciences Institute (AGI).

AGI, which was founded in 1948, is a nonprofit federation of geoscientific and professional organizations. Although AGI itself has no individual members, its 50 member societies, which represent more than 250,000 geologists, geophysicists, and other earth scientists, span the geosciences from paleontology to seismology, from research scientists to industry, education professionals, and geoscience enthusiasts. One of AGI's roles is to serve as a voice for shared interests in the geoscience community; thus, AGI can foster collaboration and facilitate discussion among its member organizations on topics, such as ethics, that are of common or overarching interest.

This chapter reviews the American Geosciences Institute's role in developing the 1999 AGI Guidelines for Ethical Professional Conduct, a set of aspirational guidelines that are applicable across the geosciences, and their subsequent revision in 2015.

4.1. 1999 AGI Guidelines for Ethical Professional Conduct

The specific impetus for developing a geoscience-wide code of ethics originated in a five-day Geological Society of America Presidential Meeting on professional ethics, held in Welches, Oregon, in July 1997. The meeting was an "eclectic gathering" that included scientists from academia, industry, and government, plus philosophers, attorneys, a federal judge, and two members of the clergy [*Horten*, 1997]. Some of the impetus for the meeting appears to have been in response to growing public awareness of the geosciences, especially in relation to natural hazards, global change, and resources. This led to recognition that the profession should be seen to adhere to the highest ethical standards [*Stow*, 1999] and a realization that there was no broad-based standard of ethics for the geosciences.

Attendees at the conference concluded that "there is more to our profession than scientific knowledge and skills" and that "the geosciences as a whole should have a code of professional conduct—a commitment to protect the public well-being through the science" [*Stephenson et al.*, 1997]. One of the action items from the conference was to ask a sponsoring organization to develop guidelines for ethical behavior and to present them to the geoscience community [*Horten*, 1997; *Stephenson et al.*, 1997]. AGI took responsibility for developing the guidelines and a 10-person committee chaired by Stephen Stow from Oak Ridge National Laboratory developed the first Guidelines for Ethical Professional Conduct.

The Guidelines were designed to meet the need for "an expression of the highest common denominator of values for the profession" [*Stephenson et al.*, 1997]. Following an introductory statement, the Guidelines addressed the responsibilities of geoscientists under nine headings: the public and society, the environment, the geosciences, the profession, the employer, employees, clients, students, and colleagues and associates [*Stow*, 1999]. AGI gave credit to the American Chemical Society's *The Chemist's Code of Conduct*, from which many of the concepts in the 1999 Guidelines were drawn.

The Guidelines were aspirational, spelling out ideals and high levels of achievement for the profession. However, they had no provisions for disciplinary or enforcement action [*Stow*, 1999]. In an unpublished memo to AGI member society presidents and executive directors, dated April 2, 1999, AGI president David Stephenson explained the intended use of the Guidelines:

> You will see that the guidelines are rather generic and are designed to cover the breadth of interests encompassed by AGI's Member Societies. We realize that some of the Member

Societies already have codes of ethical conduct that are specific to their interests. In no way are the AGI guidelines intended to supplant these existing codes. Rather they are intended to represent the overarching fundamental values that should form the foundation of ethical conduct in the geosciences, regardless of the specific interests and needs of the Member Societies....I hope that you will use these guidelines in the fashion that is most appropriate to the needs of your society.

The 1999 Guidelines appear to have been broadly accepted as a high-level statement of the general principles of ethical professional behavior in the geosciences. AGI does not have detailed information on how the Guidelines have been used. We do know that SEPM (Society for Sedimentary Petrology), after careful review by their Council, adopted the Guidelines and they were added as SEPM's general ethics policy. Other organizations, such as the National Association of State Boards of Geology (ASBOG), endorsed the Guidelines and pledged to operate in accordance with them. The Guidelines have been used as an example of widely accepted standards for professional behavior.

4.2. Revision of the 1999 Guidelines for Ethical Professional Conduct

Every fall, AGI hosts a leadership forum for the presidents and executive directors of its member societies that focuses on a theme of broad interest to the geosciences. The September 2013 Leadership Forum addressed ethics and the geosciences. The emergence of geoethics as a distinct field of study provided a foundation for the forum. Also, as Stow [1999] had predicted, public recognition of the role of the geosciences in many important societal issues had grown, with a concomitant examination of the behavior of geoscientists and others working on geoscience-related issues. Accusations of bias and conflict of interest [e.g., *Augustine et al.*, 2012; *University at Buffalo*, 2012] and of scientific misconduct [e.g., *McPhaden*, 2018] forced geoscientists and institutions to develop standards and processes for ensuring ethical behavior. In response to these developments, AGI provided leaders in the geoscience community an opportunity to assess the profession's position on ethics and the geosciences.

Following a series of individual and panel presentations, Edmund Nickless from the Geological Society of London and Jack Hess from the Geological Society of America presented a draft Consensus Statement on Ethics in the Geosciences for discussion. Attendees developed an amended version of the statement that emphasized the importance of ethical behavior in the geosciences, with an agreement to work together to promote informed understanding of the role of ethics in the geosciences, to raise standards of professional practice among their memberships, and to develop and promote a common code of ethics for geoscientists. The statement was circulated to all of AGI's member societies and, by the end of January, 2014, twenty-seven member societies signed on to the consensus statement.

AGI again took on the role of coordinating and facilitating the process of developing a common code of ethics. All societies that signed the consensus statement were invited to nominate a member to an ad hoc AGI committee charged with developing a common code of ethics. Twelve nominations were received and the committee began work in May 2014. David Mogk from Montana State University chaired the committee, which included David Abbott, Robert Finkelman, Joe Gillman, Linda Gundersen, Denise Hills, Murray Hitzman, Edmund Nickless, Lee Phillips, Monica Ramirez, Robert Tepel, and Douglas Walker.

The committee reviewed the 1999 AGI Guidelines for Ethical Professional Conduct and ethics policies adopted by numerous geoscience scientific and professional societies in order to identify the common, overarching principles articulated in these policies. Following extensive debate and discussion, the committee presented a draft revision of the 1999 Guidelines to AGI member societies at the Joint AGI/GSA Societies Meeting in October 2014. AGI circulated the draft document to all member societies, inviting their comments and suggestions.

Twenty-five member societies submitted comments on the draft Guidelines and the committee reconvened in March 2015 to review this feedback. In addition to improving the language of the Guidelines, member societies raised important points about conflict of interest, the importance of workforce safety, responsible sampling, and data management. Several societies pointed out that the responsibility for earning the trust of the public lies with geoscientists themselves. Societies also discussed the role of geoscientists in relation to stewardship of the Earth. The committee gave careful consideration to identifying the fundamental ethical issues raised by the member societies and to addressing their concerns in a way that would be appropriate for all geoscientists. The Guidelines were expanded and some elements were rewritten to incorporate issues raised by the member societies.

The committee submitted the revised Guidelines for Ethical Professional Conduct to AGI's Executive Committee in early April 2015, and the Executive Committee unanimously approved and adopted the Guidelines on April 13, 2015. The text of the 2015 Guidelines is shown in Box 4.1 and is available for download, with the 1999 AGI Guidelines and the Consensus Statement on Ethics in the Geosciences, from the AGI website, which also provides links to member society codes of ethics and statements on ethics [*American Geosciences Institute*, 2015].

The 2015 Guidelines are structured around two themes: the ethical responsibilities of individual geoscientists and the ethical responsibilities of geoscientists as members of a professional and scientific community. The 2015 Guidelines are aspirational and voluntary; they are not binding on any AGI member society or its members, and they include no enforcement mechanisms. Nevertheless, they provide a succinct, strong statement on the professional responsibilities of geoscientists, both as individuals and as part of the geoscience community; they provide a framework within which other geoscience organizations may develop more focused codes of ethical conduct; and they give nongeoscientists an insight into the values of the geoscience profession.

Box 4.1   AGU Guidelines for Ethical Professional Conduct [2015]

These guidelines address common ethical topics across the geoscience community; the ethics statements of individual societies may expand beyond these guidelines.

Geoscientists play a critical role in ethical decision making about stewardship of the Earth, the use of its resources, and the interactions between humankind and the planet on which we live. Geoscientists must earn the public's trust and maintain confidence in the work of individual geoscientists and the geosciences as a profession. The American Geosciences Institute (AGI) expects those in the profession to adhere to the highest ethical standards in all professional activities. Geoscientists should engage responsibly in the conduct and reporting of their work, acknowledging the uncertainties and limits of current understanding inherent in studies of natural systems. Geoscientists should respect the work of colleagues and those who use and rely on the products of their work.

In day-to-day activities geoscientists should:

- Be honest.
- Act responsibly and with integrity, acknowledge limitations to knowledge and understanding, and be accountable for their errors.
- Present professional work and reports without falsification or fabrication of data, misleading statements, or omission of relevant facts.
- Distinguish facts and observations from interpretations.
- Accurately cite authorship, acknowledge the contributions of others, and not plagiarize.
- Disclose and act appropriately on real or perceived conflicts of interest.
- Continue professional development and growth.
- Encourage and assist in the development of a safe, diverse, and inclusive workforce.
- Treat colleagues, students, employees, and the public with respect.
- Keep privileged information confidential, except when doing so constitutes a threat to public health, safety, or welfare.

As members of a professional and scientific community, geoscientists should:

- Promote greater understanding of the geosciences by other technical groups, students, the general public, news media, and policy makers through effective communication and education.
- Conduct their work recognizing the complexities and uncertainties of the Earth system.

- Sample responsibly so that materials and sites are preserved for future study.
- Document and archive data and data products using best practices in data management, and share data promptly for use by the geoscience community.
- Use their technical knowledge and skills to protect public health, safety, and welfare, and to enhance the sustainability of society.
- Responsibly inform the public about natural resources, hazards, and other geoscience phenomena with clarity and accuracy.
- Support responsible stewardship through an improved understanding and interpretation of the Earth, and by communicating known and potential impacts of human activities and natural processes.

4.3. Comparison of the 1999 and 2015 Guidelines for Ethical Professional Conduct

The 1999 AGI Guidelines were based on the 1994 version of the American Chemical Society's (ACS) *Chemist's Code of Conduct*, which was first published as *The Chemist's Creed* in 1965. The ACS documents were subsequently updated and replaced by *The Chemical Professional's Code of Conduct*, one of a suite of ethical and professional guidelines prepared by the ACS to address different aspects of the profession [*American Chemical Society*, 2015]. Given that the 1999 AGI Guidelines were modeled on a code of conduct that addresses the responsibilities of scientists in a professional context, it is not surprising that the 1999 AGI Guidelines emphasize issues such as employer-employee relationships more than those associated with the scientist as citizen.

The committee crafting the 2015 Guidelines never questioned that there was adequate knowledge and expertise within the geoscience community to revise the Guidelines. The committee did consult important documents and codes of ethics from organizations outside the geosciences, but they were able to draw from an increasing depth of expertise and literature on geoethics to inform their deliberations. One of AGI's international associate societies, the International Association for Promoting Geoethics, provided expert feedback on the draft Guidelines.

There is considerable overlap between the 1999 and 2015 Guidelines, but there are also some significant differences that may reflect changes in the self-perception of the geoscience community, increased scholarship in and practice of geoethics, and public scrutiny of the geosciences. In comparison to the 1999 Guidelines, the 2015 Guidelines are more succinct and integrated. They show greater emphasis on the societal context of the geosciences, the importance of effective communication and education within and beyond the geosciences, and the uncertainties associated with our understanding of complex natural systems.

There is a change in tone in the 2015 Guidelines, which note that "geoscientists must earn the public's trust and maintain confidence in the work of individual geoscientists and the geosciences as a profession." This is a somewhat more modest phrasing than the 1999 Guidelines' statement that geoscientists should explicitly work to enhance the prestige of the geoscience profession. This change may reflect a growing awareness of the importance of public perception and trust based on events such as "Climategate," which involved the theft and release of documents and emails among climate scientists [*Jasanoff*, 2010]; the manslaughter conviction, subsequently overturned, of geophysicists in connection with the Italian L'Aquila earthquake [*Cartlidge*, 2014; M*ucciarelli*, 2015]; and the Bre-X Minerals scandal when a mineral exploration company reported fraudulent exploration results [*Danielson and Whyte*, 1997].

The 2015 Guidelines reaffirm the importance of ethical behavior in matters that may be regarded as essential to the health of the geosciences as a profession, for example, scientific and professional integrity, and professional growth and development. In a distinct change of emphasis, the 2015 Guidelines take a less hierarchical view of interpersonal relationships. The 2015 version promotes the development of a safe, diverse, and inclusive workforce while providing less detail on managing interactions between geoscientists and employers, employees, and clients. The revised Guidelines respond to the digital revolution by requiring the use of best practices in data management and prompt sharing of data. They reflect the growing awareness of geoheritage by calling for responsible sampling of geoscience materials and sites and archiving of data.

In drafting the 2015 Guidelines, the committee recognized the responsibilities that accompany geoscience professionals' specialized knowledge of the Earth system, including the known and potential impacts of human activities and natural processes on human and natural systems, while also acknowledging the limits of our understanding of these systems. The committee, being mindful of the appropriate limits of the Guidelines, concluded that the geosciences' unique contribution to responsible stewardship is through improving and communicating our understanding and interpretation of the Earth system. Scientific understanding has always been fundamental to the geosciences, but the obligation to promote greater understanding of the geosciences by individuals and groups outside the geoscience community is a significant addition to the expectations for ethical professional conduct.

4.4. Next Steps

The development of the 2015 Guidelines for Ethical Professional Conduct accomplishes one of the tasks set out in the Consensus Statement on Ethics and the Geosciences: the development of a common code of ethics for geoscientists.

AGI has published the Guidelines on its webpage and is in the process of inviting member societies to formally endorse the Guidelines. Within four months of publication, 12 societies and organizations have endorsed the revised Guidelines and other societies are actively considering endorsement. At least one society is using the Guidelines as a template for creating their own code or statement on ethics, and some societies that have their own, more comprehensive, ethics codes or statements expect to endorse the Guidelines as an adjunct ethics statement. The Guidelines are not intended to be a static document, and future revisions are anticipated.

The geoscience community still has much work to do in completing the other action items in the Consensus Statement, including promoting informed understanding of the role of ethics in the geosciences, raising standards of professional practice among geoscientists, and promoting the common code of ethics for geoscientists. Different parts of the community will approach these tasks in different ways, and other chapters in this book provide information on how several geoscience societies, organizations, and educators are working on these issues.

Publication of the revised Guidelines has provided impetus for special sessions at professional meetings and for talks about the importance of ethics. These sessions complement the increasing interest in geoethics in the academic geoscience community. AGI has an ongoing "I'm a Geoscientist" campaign that is designed to build a sense of community across the whole of the geosciences. The Guidelines for Ethical Professional Conduct are being used to stimulate conversations about the roles and responsibilities that accompany the statement "I'm a Geoscientist."

Educational institutions, individual societies, and organizations may choose to develop formal courses of study in geoethics for undergraduate and graduate students and for professionals, and to increase the emphasis on ethics in academic and professional development (see *Mogk et al.* [2018] and the related website *Teaching Geoethics across the Geoscience Curriculum* [n.d.]). Societies and organizations may raise the prominence of their ethics committees and procedures, either proactively or in response to internal or external events. The broad-based, cooperative, and collaborative process that led to the 2015 AGI Guidelines for Ethical Professional Conduct lays a firm foundation for all of these efforts.

References

Abbott Jr., D.M. (2018), Brief history and application of enforceable professional geoscience ethics codes, in *Scientific integrity and ethics in the geosciences*, edited by L.C. Gundersen. *AGU Special Publications*, John Wiley and Sons, Hoboken, NJ.

American Chemical Society (2015), Chemical Professional's Code of Conduct. http://www.acs.org/content/acs/en/careers/career-services/ethics/the-chemical-professionals-code-of-conduct.html. Accessed 19 June 2015.

American Geosciences Institute (2015), AGI Guidelines for Ethical Professional Conduct. http://www.americangeosciences.org/community. Accessed 19 June 2015.

Augustine, N.R., R.R. Colwell, and J.J. Duderstadt (2012), A review of the processes of preparation and distribution of the report "Fact-based regulation for environmental protection in shale gas development." Review prepared at the request of the University of Texas at Austin. 42p. http://www.utexas.edu/opa/wordpress/news/files/Review-of-report.pdf. Accessed 29 August 2015.

Cartlidge, E. (2014), Relief greets acquittals in Italy earthquake trial, *Science*, 346(6211), 794.

Danielson, V., and J. Whyte (1997), *Bre-X: Gold today, gone tomorrow, anatomy of the Busang swindle*, Northern Miner Press, Toronto.

Horten, H.A. (1997), Conferees tackle ethics questions, *Geology Today*, 7(10), 18.

Jasanoff, S. (2010), Testing time for climate science, *Science*, 328(5979), 695–696.

McPhaden M. (2018), American Geophysical Union adopts and implements a new scientific integrity and professional ethics policy, in *Scientific integrity and ethics in the geosciences*, edited by L.C. Gundersen. *AGU Special Publications*, John Wiley and Sons, Hoboken, NJ.

Mogk, D.W., J.W. Geissman, and M.Z. Bruckner (2018), Teaching geoethics across the geoscience curriculum: Why, when, what, how, and where? in *Scientific integrity and ethics in the geosciences*, edited by L.C. Gundersen. *AGU Special Publications*, John Wiley and Sons, Hoboken, NJ.

Mucciarelli, M. (2015), Some comments on the first degree sentence of the "L'Aquila trial," in *Geoethics and ethical challenges and case studies in Earth sciences*, edited by M. Wyss and S. Peppoloni, pp. 205–210, Elsevier, Amsterdam.

Stephenson, D., R. Grauch, T. Holtzer, J. Price, and P. Rose (1997), Ethics in the Geosciences: A conference summary, *Geotimes*, 42(11), 29–32.

Stow, S. (1999), Ethics: Creating guidelines for the geosciences, *Geotimes*, 44(8), 5, 13.

Teaching Geoethics across the Geoscience Curriculum (n.d.), http://serc.carleton.edu/geoethics/index.html. Accessed 26 August 2015.

University at Buffalo (2012), University at Buffalo president closes shale-gas institute. http://archive.pressconnects.com/article/20121119/NEWS11/311190051/University-Buffalo-president-closes-shale-gas-institute. Accessed 29 August 2015.

Williams, J. (2018), The National Association of State Boards of Geology (ASBOG) involvement in geoscience professional ethics, in *Scientific integrity and ethics in the geosciences*, edited by L.C. Gundersen. *AGU Special Publications*, John Wiley and Sons, Hoboken, NJ.

5. AMERICAN GEOPHYSICAL UNION ADOPTS AND IMPLEMENTS A NEW SCIENTIFIC INTEGRITY AND PROFESSIONAL ETHICS POLICY

Michael McPhaden

Abstract

The American Geophysical Union (AGU), an organization of 60,000 scientists from 140 countries, has a mission to promote discovery in Earth and space science for the benefit of humanity. It is the largest organization of its kind, with membership that spans atmospheric sciences, oceanography, hydrology, solid earth geophysics, space physics, planetary science, biogeosciences, climate, natural hazards, and related fields. In 2012, AGU adopted a new scientific integrity policy and code of conduct to raise awareness of ethical norms for professional behavior within its worldwide membership [*American Geophysical Union*, 2013]. This chapter provides the context for what motivated this new policy, an outline of the policy itself, and a discussion of how it is being communicated, applied, and evolved over time.

5.1. Background

AGU went through a major reorganization in 2010 to redefine its governance structure, which had been in place, with only slight modifications, since the organization was founded in 1919. This new governance structure [*Grove*, 2009] consisted of a scientific council, chaired by the president-elect, to provide scientific leadership; and a board of directors, chaired by the president, to provide oversight

NOAA/Pacific Marine Environmental Laboratory, Seattle, Washington, USA

Scientific Integrity and Ethics in the Geosciences, Special Publications 73,
First Edition. Edited by Linda C. Gundersen.

of organizational, financial, and legal matters, including issues related to scientific misconduct [*Gazley and Kissman*, 2015]. As part of the effort to transform itself, AGU also adopted a new strategic plan in 2010 to clearly articulate its mission, ideals, and specific goals that would guide the organization for the next decade [*McPhaden*, 2010]. This new plan was motivated not only to ensure that the AGU focused its resources on those areas deemed high priority for advancing the science, but also by the growing realization that geosciences and geoscientists have a critical role to play in promoting the health and welfare of the planet. The plan reflects the input from many stakeholders across a broad spectrum of the membership as well as input from external partners. It includes explicit 3–5-year objectives in four major high-priority areas: scientific leadership and collaboration, talent pool development, science and society, and organizational excellence.

In addition to goals and objectives, the AGU strategic plan also contains 11 core values to which the organization aspires. These values are not a code of conduct or scientific integrity policy per se, but they reflect the ideals that AGU stands for and that AGU members are expected to promote, namely,
- The scientific method
- The generation and dissemination of scientific knowledge
- Open exchange of ideas and information
- Diversity of backgrounds, scientific ideas, and approaches
- Benefit of science for a sustainable future
- International and interdisciplinary cooperation
- Equality and inclusiveness
- An active role in educating and nurturing the next generation of scientists
- An engaged membership
- Unselfish cooperation in research
- Excellence and integrity in everything we do

Many of these ideals were adopted from earlier versions of AGU strategic plans. They also echo values articulated in the codes of conduct of other related professional and scholarly societies, such as the Geological Society of America, the Ecological Society of America, the American Chemical Society, and the American Geosciences Institute (of which AGU is a member).

Soon after AGU's organizational transformation, a serious allegation of scientific misconduct came to the attention of the first new board of directors. Review of AGU's existing policy and procedures for dealing with misconduct, first formulated in 1994, revealed that they were out of date and in need of revision. The procedures were defined based on the old governance structure, which was no longer in existence. The policy was based on "Responsible Science: Ensuring the Integrity of the Research Process," published jointly by the National Academy of Sciences, the National Academy of Engineering, and the Institute of Medicine [*National Research Council*, 1992]. The National Science Foundation and the Department of Health and Human Services had, in the late 1980s and early 1990s, adopted regulations for dealing with allegations of misconduct in response to congressional legislation passed in 1985, but the fair and impartial application of those regulations required clarity on what constituted scientific misconduct. Thus, the National Academies commissioned a major study that led to publication of "Responsible Science," in which misconduct was defined primarily in terms of plagiarism and the falsification and fabrication of data. Many scientific organizations, like AGU, used this standard to develop their policies on misconduct.

One of AGU's most important services to the community is as a publisher of highly respected scholarly scientific journals. Publications are also the greatest source of revenue to the union, enabling other programs, products, and services that benefit the membership. Ensuring the integrity and safeguarding the reputation of AGU journals is therefore of paramount importance. The publication of "Responsible Science" provided the opportunity to develop a policy on scientific misconduct consistent with federal regulatory standards, addressing key aspects of the publication process. However, AGU's policy, like "Responsible Science," did not define the responsibilities of reviewers and editors. Nor did it provide guidelines for ethical behavior outside the publication process, for example, interactions with one's colleagues and the public.

Moreover, AGU's 1994 policy of scientific misconduct was developed at the dawn of the Internet age, before the explosion in access to vast volumes of digital information, the instant news cycle, the blogosphere, and the advent of social media. The period since 1994 has also witnessed ever-intensifying competition for limited research funding and immense pressure on scientists to regularly publish new research results, a situation that has tempted some scientists to adopt lax research standards in an effort to gain advantage (http://www.nytimes.com/2015/06/01/opinion/scientists-who-cheat.html). Some areas of research have likewise become increasingly politicized since AGU's original policy was first formulated, such as climate change and sustainable development of natural resources. "Climategate," for instance, ensued after hackers stole thousands of private emails from a server at the University of East Anglia's Climate Research Unit and made them publicly available on the Internet without authorization (http://www.skepticalscience.com/Climategate-CRU-emails-hacked.htm). This manufactured scandal, based in part on the exaggerated claim that critical data sets had been manipulated to fit researcher biases about the reality of global warming, was intended to discredit climate science and scientists just before the 2009 Copenhagen Climate Summit. Several independent investigations exonerated the scientists involved of any wrongdoing, but nonetheless, the credibility of the science suffered in the eyes of the public.

Climategate, which embroiled many AGU members as central figures, illustrated how scientific knowledge and the scientific process could be distorted for political purposes. But it also raised some legitimate issues concerning professional ethics. Among these were the open accessibility of critical data sets for independently assessing key scientific findings and the responsibilities of individual scientists for transparency in the conduct of research. As a result, Climategate brought greater public scrutiny of research methods and practices, especially with regard to high-profile research topics of societal relevance. It also highlighted the need for ethical guidelines in the AGU community beyond those only related to publishing.

Climate change science also made headlines as a contentious political issue in the early to mid-2000s in the wake of allegations that the Bush administration attempted to muzzle high-profile government climate scientists and water down official federal agency reports that emphasized potential adverse consequences of climate change. These events helped spur a newly inaugurated President Obama to state in March 2009, "Science and the scientific process must inform and guide decisions of my Administration on a wide range of issues, including improvement of public health, protection of the environment, increased efficiency in the use of energy and other resources, mitigation of the threat of climate change, and protection of national security" [*Obama*, 2009]. He then charged John Holdren, the director of the Office of Science and Technology Policy (OSTP), with developing specific recommendations to guide federal agencies in implementing the administration's policies on scientific integrity. OSTP issued these guidelines in December 2009 [*Holdren*, 2009], which were subsequently translated into agency-specific scientific integrity policies across the federal government.

5.2. A New Scientific Integrity and Professional Ethics Policy

It was against this backdrop that as President of AGU, I commissioned a task force in mid-2011 to review AGU's scientific ethical standards in the context of those from other related professional and scholarly societies; and, based on that review, to propose an updated AGU policy as well as procedures for dealing with cases of misconduct. It was also part of the committee's charge to expand the scope of the policy to include AGU meetings, other AGU activities, and interactions of scientists with their professional colleagues and the public.

Ironically, the task force got off to a very rocky start with the abrupt resignation of its first chair because of a serious ethical breach in early 2012 [*American Geophysical Union*, 2012]. This episode temporarily slowed down but did not derail the effort. I asked task force member Linda Gundersen of the U.S. Geological Survey Office of Science Quality and Integrity to step in as chairperson. She agreed and successfully guided the task force, consisting of 13 representatives from academia, government, and industry, through its deliberations. The AGU council and board approved the task force's policy recommendations in December 2012, following revision based on a period of open review and comment by the membership.

The new AGU Scientific Integrity and Professional Ethics Policy is described in *Gundersen and Townsend* [2015] and is available in full from the American Geophysical Union [2013]. The policy provides a definition of misconduct based the Federal Policy on Research Misconduct (http://www.aps.org/policy/statements/federalpolicy.cfm) and the Department of Interior Scientific and Scholarly Integrity Policy (https://www.doi.gov/scientificintegrity), but it is also intended to be positive and aspirational. It draws from a number of other existing documents and resources dealing with scientific integrity and professional ethics, most notably the "Singapore Statement" produced by the 2010 World Conference on Research Integrity [*Mayer*, 2015].

The new policy is based on the premise stated in the preamble, that "the value and benefits of research are dependent on the integrity of the research and researcher." It echoes many of the core values stated in the seven guiding principles of AGU's strategic plan, which include "excellence, integrity, and honesty in all aspects of research; personal accountability in the conduct of research and the dissemination of results; professional courtesy and fairness in working with others; [and] unselfish cooperation in research." It also spells out 13 separate personal responsibilities that all AGU scientists should adopt in the conduct of their professional activities. The first of these, for example, is, "Members will act in the interest of the advancement of science and take full responsibility for the trustworthiness of their research and its dissemination." Other responsibilities address adherence to laws and regulations; sharing data and new research findings; appropriately acknowledging the scientific contributions of others; keeping adequate records; abiding by AGU's guidelines to authors; maintaining fairness, impartiality, and confidentiality in peer review; disclosing conflicts of interest in the conduct of AGU business of any kind; and the obligation to report instances of irresponsible research practices. There are also responsibilities for how one communicates with the public, distinguishing when appropriate between what represents scientific evidence and what is personal opinion. There is a general obligation to promote a positive and welcoming work environment that "allows science and scientific careers to flourish." Finally, there is a responsibility for members "to weigh the societal benefits of their research against the costs and risks to human and animal welfare and impacts on the environment and society."

There is a separate section in the new policy that focuses on AGU volunteer leaders who are ambassadors for the organization and therefore have special responsibilities. Scientific publication is addressed in depth because publication is so central to the research enterprise and to the communication of scientific results. This part of the policy highlights the responsibilities of authors, editors, and reviewers and also includes a discussion of AGU's responsibilities to its editors.

5.3. Implementation

AGU has been very proactive in communicating its new policy to members since its adoption in 2012. For instance, acknowledging AGU's professional ethics policy is now a requirement for membership. AGU has also developed a code of conduct specifically for AGU-sponsored meetings, and registering to attend these meetings requires a commitment to abide by this code. The AGU board and council have adopted codes of conduct based on both good governance principals and ethical considerations. Even before the new policy was developed, AGU had required board and council members to provide a written statement every year declaring potential conflicts of interest and to flag any potential conflicts as they came up during the course of business meetings. This practice has now been extended to all volunteer leaders serving on AGU committees, and it has been particularly valuable for committees dealing with awards and honors.

AGU adopted a new data policy for publications, where authors are required to state in the acknowledgments section of articles where relevant data can be acquired [*Hanson*, 2014]. This open data policy has been established to ensure the integrity and reproducibility of published results and is consistent with open data policies and regulations adopted by federal funding agencies like the National Science Foundation. It also builds confidence and trust in the scientific method with the public and policy makers. AGU is enforcing its open data policy and some papers have been rejected for lack of compliance.

AGU has partnered with the American Geosciences Institute, representing 50 different scientific societies and organizations, to issue both a consensus statement on the importance of ethics in the geosciences and a commonly agreed upon set of ethical guidelines [*Boland and Mogk*, 2018, this volume]. AGU has also sponsored town halls, student and early career scientist workshops, and special sessions at its scientific meetings to highlight scientific integrity and professional ethics. The 2012 annual fall AGU meeting, for example, featured a panel discussion on the ethical dilemmas surrounding the conviction of six seismologists in the wake of the L'Aquila, Italy, earthquake in 2009 and what lessons could be learned. Town halls at the 2013 and 2014 annual fall AGU meetings have provided tutorials on AGU's new policy and where to find information about it. These town halls have also included interactive role-playing scenarios to simulate real-life situations. In 2014, for example, participants were challenged to put themselves in the position of the Italian seismologists and consider what they could or should have done differently to avoid allegations of misconduct.

Recently, the emerging field of "geoethics" has engaged a growing segment of the AGU membership. The International Association for Promoting Geoethics (IAPG) defines geoethics as "research and reflection on the values which underpin appropriate behaviors and practices, wherever human activities interact with the geosphere. Geoethics deals with the ethical, social and cultural implications of Earth Sciences education, research and practice, and with the social role and responsibility of geoscientists in conducting their activities" (http://www.geoethics.org/). Geoethics topics have been formally integrated into AGU member education initiatives consistent with the Science and Society theme of its strategic plan and with its professional ethics policy that urges members "to weigh the societal benefits of their research against the costs and risks to human and animal welfare and impacts on the environment and society." For example, the 2014 fall AGU meeting included a town hall spotlighting "Scientific Integrity and Geoethics" and education sessions on teaching geoethics. AGU also signed a partnership agreement with IAPG in November 2015 for promoting initiatives and events focusing on geoethics and for advancing the adoption of ethical standards in geoscience research and practice in order to better serve society.

The new policy outlines processes for dealing with allegations of misconduct ensuring confidentiality, fairness, and timely resolution. As a practical matter, the policy encourages resolution first through dialogue with the involved parties. In many cases, misunderstandings or honest mistakes can be resolved in this manner. The most serious cases, however, require convening an ad hoc Scientific Ethics Committee chaired by the past president of AGU. This committee does a full investigation and recommends appropriate sanctions, if necessary, to the board of directors.

AGU has instituted a system to track and record each allegation of misconduct as it arises. In the first three years after the policy was adopted, there were 18 formal allegations. Only two cases rose to a level where board action was required. The remaining 16 cases were resolved through the intercession of AGU editors, AGU staff, and/or the chair of the AGU Ethics Committee.

5.4. Addressing Sexual Harassment and Other Unacceptable Behaviors

In 2015, several high-profile cases of sexual harassment involving prominent scientists surfaced in the media. These well-publicized cases, as exemplified by that of the renowned University of California at Berkeley astronomer Geoffrey Marcy [*Witze*, 2015], have emphasized how damaging this abusive behavior is to the lives and careers of young women in science. Sexual harassment in science is a pervasive and longstanding problem that has too often been tolerated, condoned, or covered up to shield perpetrators who are in positions of influence and power. The recent spate of disclosures, however, has motivated the scientific community to be more outspoken in condemning harassment of any kind in research and education and in sparking public discourse over how to ensure a safe, welcoming, and productive workplace for all scientists at every stage of their careers.

Scientific societies have a role to play in setting standards of professional behavior for both their members and for the field as a whole. AGU, in partnership with other organizations such as the Association for Women Geoscientists and Earth Science Women's Network, has responded by assuming a leadership role in promoting a no-tolerance culture against sexual harassment [*Marín-Spiotta et al.*, 2016]. Specific actions that AGU has taken include hosting a town hall at the fall AGU meeting in December 2015 titled "Forward Focused Ethics—What is the

Role of Scientific Societies in Responding to Harassment and Other Workplace Climate Issues?" This town hall not only explored the issues, but also recommended steps that AGU could take to address the problem. AGU later convened a workshop with sponsorship from the National Science Foundation in September 2016 titled "Sexual Harassment in the Sciences: A Call to Respond" that was attended by more than 60 leaders from the scientific community, government agencies, and professional societies [*Wendel*, 2016]. This workshop led to a draft statement on organizational principles for addressing harassment, including the roles and responsibilities for professional societies, academic departments, government agencies, and institutions that fund research. It also recommended that the definition of research misconduct be expanded to include harassment.

In light of these developments, it became evident that AGU could significantly strengthen its code of conduct to emphasize how harassment, as well as other behaviors such as discrimination and bullying, is a violation of professional ethics; and likewise to make it clear that AGU will hold all members accountable for their behavior in the professional, research, and learning environments. Thus, in 2016 AGU established a task force to review and update its ethics policy, stressing that harassment defined in the broadest terms (based on factors such as ethnic origin, race, religion, citizenship, language, political persuasion, sex, gender identity, sexual orientation, disability, age, or economic class) is unacceptable under any circumstances. The task force recommended that the ethics policy should extend to member conduct that takes place outside of AGU functions and that professional conduct should be factored into AGU honors and awards. It went a step further and recommended that the definition of scientific misconduct be expanded to include harassment, bullying, and discrimination because of the harm such behaviors cause to the scientific enterprise and to the ability of AGU to carry out its mission. The task force also proposed strategies to educate members on the issue of harassment and that AGU provide additional resources to support members who may be targets of harassment or other hostile workplace conduct. AGU leadership made public presentations on evolving its Scientific Integrity and Professional Ethics policy, with specific emphasis on the topic of harassment, at the annual European Geosciences Union General Assembly in Vienna, Austria, in both April 2016 and April 2017, and at the International Geological Congress in Cape Town, South Africa, in August 2016. In addition, AGU has launched a web page, http://stopharassment.agu.org/, with links to relevant information and resources.

5.5. Conclusion

Scientific societies have a critical role to play in educating members about the essential importance of adhering to the highest standards of professional ethics when conducting research, interacting with peers, engaging with the public, and providing advice to decision makers. AGU's philosophy is to create a culture of awareness about the importance of scientific integrity in all aspects of professional behavior through education and outreach to its members. As Benjamin Franklin famously said, "an ounce of prevention is worth a pound of cure." Serious breaches of professional ethics by even one individual can cause severe damage to the trust that the public and policy makers have in the scientific enterprise. That damage can undermine the credibility of, and investment in, research that is pursued for the ultimate benefit society. AGU's vision, articulated in its strategic plan, is to "galvanize a community of Earth and space scientists that collaboratively advances and communicates science and its power to ensure a sustainable future." This vision can only be realized if its members act with integrity, honesty, openness, and fairness in all that they do.

Acknowledgments

The author would like to thank Arthur Nowell, Linda Gundersen, Billy Williams, Randy Townsend, and an anonymous reviewer for valuable comments on earlier versions of this manuscript. This is PMEL contribution no. 4333.

References

American Geophysical Union (2012), We must remain committed to scientific integrity. Available from http://about.agu.org/president/presidents-message-archive/remain-committed-scientific-integrity/

American Geophysical Union (2013), AGU scientific integrity and professional ethics policy. Available from http://ethics.agu.org/policy/.

Boland, M.A., and D.W. Mogk (2018), The American Geosciences Institute guidelines for ethical professional conduct, in *Scientific integrity and ethics: With application to the geosciences*, edited by L.C. Gundersen, John Wiley & Sons, Hoboken, NJ.

Gazley, B., and K. Kissman (2015), *Transformational governance: How boards achieve extraordinary change*. American Society of Association Executives, Washington, D.C., and Wiley Publishing, Hoboken NJ.

Grove, T.L. (2009), Looking to the future: AGU council takes important action. *Eos*, 90, 213.

Gundersen, L.C., and R. Townsend (2015), Formulating the American Geophysical Union's Scientific Integrity and Professional Ethics Policy: Challenges and lessons Learned, in *Geoethics: Ethical challenges and case studies in Earth science*, edited by M. Wyss and S. Peppoloni, Elsevier, Waltham, MA, pp. 84–93.

Hanson, B. (2014), AGU's data policy: History and context. *Eos*, 95, 337.

Holdren, J. P. (2009), Scientific integrity: Fueling innovation, building public trust. Available from: https://www.whitehouse.gov/blog/2010/12/17/scientific-integrity-fueling-innovation-building-public-trust.

Marín-Spiotta, E., B. Schneider, and M. A. Holmes (2016), Steps to building a no-tolerance culture for sexual harassment, *Eos*, 97, doi:10.1029/2016EO044859.

Mayer, T. (2015), Research integrity: The bedrock of geosciences, in *Geoethics: Ethical challenges and case studies in Earth science*, edited by M. Wyss and S. Peppoloni, Elsevier, Waltham, MA, pp. 71–81.

McPhaden, M.J. (2010), Strategic plan approved by council. *Eos*, 91, 265.

National Research Council (1992), *Responsible science. Ensuring the integrity of the research process*, vol. I, National Academies Press, Washington, D.C.

Obama, B. (2009), Memorandum for the heads of executive departments and agencies, Subject: Scientific integrity. Available from http://www.whitehouse.gov/the-press-office/memorandumheads-executive-departments-and-agencies-3-9-09.

Wendel, J. (2016), AGU-sponsored workshop targets sexual harassment in the sciences, *Eos*, 97, doi:10.1029/2016EO059651.

Witze, A. (2015), Berkeley releases report on astronomer sexual-harassment case. *Nature News*, doi:10.1038/nature.2015.19068. Available from http://www.nature.com/news/berkeley-releases-report-on-astronomer-sexual-harassment-case-1.19068.

6. THE NATIONAL ASSOCIATION OF STATE BOARDS OF GEOLOGY (ASBOG): INVOLVEMENT IN GEOSCIENCE PROFESSIONAL ETHICS

John W. Williams

Abstract

The National Association of State Boards of Geology (ASBOG®) was founded in 1988 to provide oversight and coordination for state licensing boards for the geosciences. Currently, 30 states and Puerto Rico are members of ASBOG, which has coordination responsibility for approximately 40,000 licensees. Twice a year, two examinations, Fundamentals of Geology and Practice of Geology, are administered by ASBOG to evaluate the qualification of candidates for licensing in the member states. The exam content is driven by surveys of the practicing licensed professionals and academicians [*Williams et al.*, 2012]. Professional ethics is regarded as extremely important by survey responses of all components of the geoscience community, including practicing licensed geologists in Canada and the United States, and members of the academic community. There is general agreement among these practitioners as to the relative importance of different elements of professional ethics and the frequency of these issues encountered in daily activities. Comparison of results of the surveys between administrations in 2010 [*Warner and Warner*, 2010] and 2015 [*Warner*, 2015] indicate little change in the attitudes of practicing geologists toward

Department of Geology, San José State University, San Jose, California, USA
National Association of State Boards of Geology (ASBOG), Douglasville, Georgia, USA

Scientific Integrity and Ethics in the Geosciences, Special Publications 73,
First Edition. Edited by Linda C. Gundersen.

professional ethics. Recognizing the importance of professional ethics in the geosciences, ASBOG is promoting the inclusion of professional ethics in the college geoscience curriculum in addition to encouraging the instruction of practicing professionals [*Williams*, 2013; *Williams*, 2014].

The National Association of State Boards of Geology (ASBOG®) is a professional organization founded in 1988. The purpose of the organization is to provide oversight of the state licensing organizations that existed in 1988 as well as those that have been added subsequently. At the time of its founding in 1988, ASBOG consisted of only six, Florida, South Carolina, North Carolina, Tennessee, Virginia, and Arkansas. Since its founding, the association has added a number of states and the territory of Puerto Rico; at the present time there are 30 member states of ASBOG (Figure 6.1). It is anticipated that the number of ASBOG member states will continue to grow. For example, the state of New York

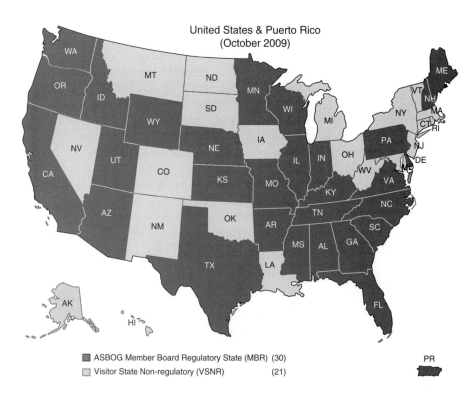

Figure 6.1 Map of the member states of ASBOG. Currently, 30 states and Puerto Rico require licensing of geologists doing work for the public. Source: *ASBOG* [2015]. Reproduced with permission of ASBOG.

passed a law in 2014 that requires the licensing of geologists within that state in order to practice for the general public. At the present time, member states of ASBOG oversee approximately 40,000 licensees within their combined jurisdictions.

The ASBOG organization has several specific goals, which include the following:

1. Provide oversight of state licensing operations,
2. Provide opportunities for states to develop common licensing procedures including educational and work experience requirements. Perhaps the most important activity of the organization is to develop national written licensing exams that can be used by each of the states that require licensing of geologists to practice for the public.

In 1991 ASBOG was incorporated in South Carolina. In 1992, the first ASBOG examinations were offered. The format of the written examinations includes two components:

1. The first component is the four-hour Fundamentals Examination that candidates in most states may take upon graduation with an undergraduate degree in geology.
2. The second component of the written examination, the four-hour Practice Examination, is taken after the candidate gains the appropriate amount of professional experience as mandated by the state in which the licensee wishes to obtain the license. The length of the work experience requirement varies from state to state but ranges between three and five years.

6.1. Administration of ASBOG Examinations

ASBOG currently administers the two examinations twice a year, generally in March and October. At present, approximately 1500 candidates take the examinations each year (Figure 6.2). Since the initial offering of the examinations, more candidates take the Fundamentals Examination compared to the Practice Examination. The number of candidates taking the Fundamentals Examination during the past five years has increased slightly. This is in contrast to the number of candidates taking the Practice Examination, which has shown a slight decrease over the same period of time. A number of reasons have been suggested for these contrasting trends; however, there is no consensus. One of the factors that may be contributing to the increasing number of individuals taking the Fundamentals Examination is that ASBOG currently provides or can provide data to academic departments of geology reflecting the performance of their students on the Fundamentals Examination. These data are helpful in evaluating course content of various geology courses offered. College accrediting organizations appreciate this type of information as they evaluate the overall progress of the college or university. At the present time, there is no accreditation process for individual

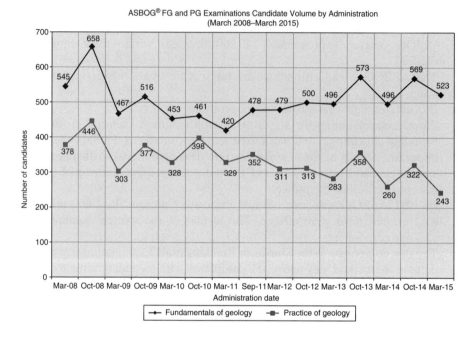

Figure 6.2 Number of candidates taking the Fundamentals and Practice Examinations March 2008 through March 2015. The trend for the Fundamentals Examination has shown a slight upward slope in recent years. The trend has been slightly downward for the Practice Examination. Source: *Warner* [2015]. Reproduced with permission of ASBOG.

geology departments in the United States. The overall performance of candidates on the two examinations is recorded in Figure 6.3.

In the past decade, ASBOG has developed stronger working relationships with Canadian provincial licensing groups to include cooperative testing using the fundamentals component of the written examinations. The opportunity for the mutual recognition of licenses across the international border has been enhanced by the closer working relationship.

6.2. ASBOG and Professional Ethics

What forces encouraged ASBOG to become actively involved with issues of professional ethics in the geosciences? It was during the time that ASBOG was developing the blueprints for the licensing examinations that the association documented the great importance of professional ethics for practicing members of the profession. In order to develop the examinations, including the number of questions from specific areas of expertise, surveys are conducted of licensed

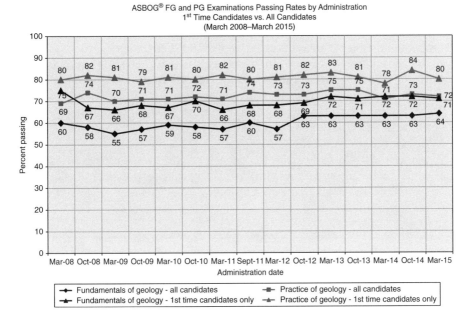

Figure 6.3 Candidate passing rates by examination dates March 2008 through March 2015: first-time test takers and all candidates. Source: *Warner* [2015]. Reproduced with permission of ASBOG.

practicing professional geologists throughout the United States. Their opinions are solicited to determine the relative importance of subject areas in the preparation of geologists to practice satisfactorily for the general public. As an example, in 2015, 7690 surveys were sent to licensed practicing geologists (USA total 5800; Canada total 1890) and to 2000 randomly selected academicians from the AGI Directory. A total of 2322 were returned from practicing geologists and 194 from academicians in the USA, while 399 were completed in Canada [*Warner*, 2015]. With regard to geological tasks considered, 43 are rated in terms of importance to public well-being by licensed geoscience practioners and members of geoscience academia. In the 2005 survey, participants rated the importance of the ethical issues. In the 2010 and 2015 surveys, 13 issues were rated with regard to frequency of encounter and seriousness. An explanation of the survey terms and questions is provided in Box 6.1, and results are presented in Figure 6.4.

Starting in 2005, ASBOG began to include professional ethics in many of its activities. Two activities were to include professional ethics questions on the written examinations that ASBOG administers and to provide instructional materials that might be used by faculty members and other instructors wishing to offer coursework in professional ethics (www.ASBOG.org). ASBOG has moved

Box 6.1

Professional ethics issues addressed on the 2005, 2010, and 2015 surveys of the geoscience profession. The respondents: U.S. practicing geologists, Canadian geoscientists, and academic geoscientists were asked two questions regarding each of the 13 ethics issues:

1. "How frequently have you encountered breaches of ethical behavior in these areas of the geological/geoscience profession?" (0 = Never; 1 = Seldom; 2 = Occasionally; 3 = Often)
2. "How serious do you believe a breach of ethical behavior is in each situation in terms of influencing the geological/geoscience profession?" (0 = Not serious; 1 = Somewhat serious; 2 = Very serious; 3 = Extremely serious)

Professional Ethics Issues

1. Conflict of interest
2. Failure to disclose regulatory violations
3. Failure to maintain confidentiality
4. Gifts—getting and giving
5. Inappropriate advocate for client
6. Insufficient "scope of work"
7. Invoicing
8. Misrepresentation of professional qualifications
9. Plagiarism
10. Practicing outside area of competence
11. Practicing without license
12. Selective data acquisition, analysis, or disclosure
13. Retaliation against whistle blowers

Source: *Warner* [2015]. Reproduced with permission of ASBOG.

aggressively to promote professional ethics instruction in the universities and colleges as well as to offer short courses on professional ethics in addition to ethics sessions at various scientific meetings, including organizations such as the Geological Society of America, the Association of Engineering and Environmental Geologists, etc.

Stimulated by the results of these surveys, ASBOG decided to include questions regarding professional ethics on both the Fundamentals and the Practice Examinations. Two questions are included on the Fundamentals Examination, and three questions are included on the Practice Examination. The average success rates on ethics questions tend to be higher than the success rate on more traditional geological questions (approximately 90% vs. approximately 65%).

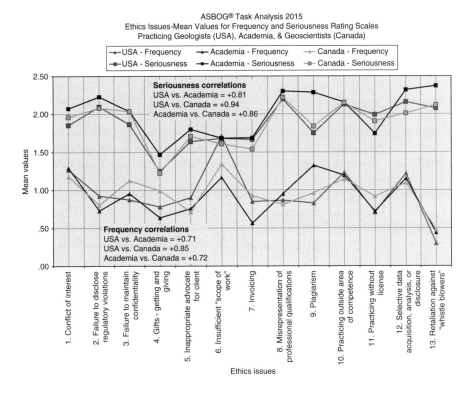

Figure 6.4 Mean values for frequency and seriousness of ethics issues as rated by U.S. geologists, Canadian geoscientists, and U.S. academic geoscientists. Task Survey Analysis (TAS) 2015 summary data for all three groups on both rating scales. In general, the correlations between groups are very high, with somewhat higher correlations between groups on the Seriousness rating scale:

Frequency Correlations:
USA vs. Academia = +0.71; USA vs. Canada = +0.85; Academia vs. Canada = +0.72
Seriousness Correlations:
USA vs. Academia = +0.81; USA vs. Canada = +0.94; Academia vs. Canada = +0.86
Source: *Warner* [2015]. Reproduced with permission of ASBOG.

The success rates on the fundamentals ethics questions tend to be higher than the success rates on the questions regarding professional ethics on the Practice Examinations (approximately 90% compared to approximately 70%).

6.3. Patterns of Data on Ethical Issues

Many individuals have sought data regarding the differences that might exist among practitioners in different subject areas as well as from different geographic regions with regard to what they view as the more frequently encountered ethical issues and their relative importance. The data collected during the surveys in 2010 and 2015 address this issue. A comparison of the data of respondents with regard to the frequency of a particular ethical issue being encountered and its relative importance, as gathered from practicing geologists in the United States, practicing geoscientists in Canada, and academic geologists, suggests that there are striking similarities. A detailed evaluation of the data provided from these three areas of practice indicate there are some minor differences. As an example, plagiarism is considered by members of the academic community to be one of the most critical and serious ethical issues encountered. The responses from practicing geologists in the United States and practicing geoscientists in Canada indicate that their collective opinion is that plagiarism is a less serious item. A graphical comparison of the responses by the three areas of practice across the 13 ethical issues is shown in Figure 6.4. The same ethical issues were addressed in the 2010 and 2015 surveys. A graphical comparison demonstrates that fundamentally there was almost no change in the expressed opinions of the respondents from 2010 to 2015 (Figures 6.5, 6.6, and 6.7).

6.4. State Enforcement of Ethics for Professional Geologists

State enforcement is important with regard to ethical issues and their violation by licensed members of the profession. There is general similarity among the ASBOG member states' codes of ethics. The penalties that a state may implement for code violations are also similar among states. A matrix of these penalties that states may impose is shown in Table 6.1. These range from admonition for violations of elements of the codes of ethics to revocation of the license of the individual found guilty of a violation.

In addition to having the tools for the enforcement of code violations, it is appropriate to evaluate how rigorous the various states are in enforcing these codes. ASBOG surveyed its member states' activity during the period of 2009–2011 to determine their rigor in investigating and taking action on complaints filed by the public with regard to the violations of elements of the code of ethics by its licensees. Twenty-six of the 30 ASBOG member states responded with data as to their level of enforcement. One hundred thirty-five cases of ethical violations were reported during the period of 2009–2011. Of those reported cases, approximately 90% were acted upon and resolved. This is encouraging, as one of the major criticisms that was made against many geoscience licensing boards in the early years is that these boards were negligent in terms of aggressively enforcing and acting upon complaints that were filed against their licensees. This is one of the issues that state legislatures consider as important during their legislative sunset reviews of the activities of the licensing board. As an example, the California licensing board for geologists and geophysicists was challenged during the legislative evaluation of the need for their continued existence; they were

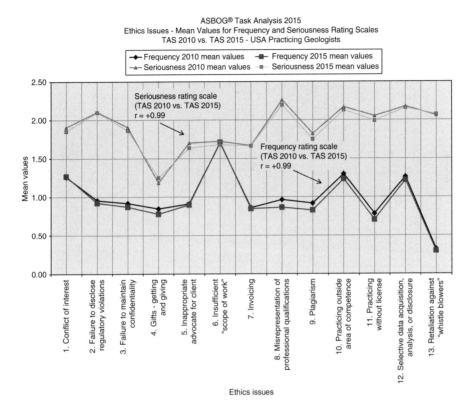

Figure 6.5 Comparison of mean values for seriousness and frequency of ethical issues for 2010 and 2015 surveys of U.S. geologists. Practicing Geologists: Survey 2010 vs. Survey 2015 Frequency Correlation = +0.99; Seriousness Correlation = +0.99. Source: *Warner* [2015]. Reproduced with permission of ASBOG.

accused of not demonstrating aggressive enough responses to complaints filed against California licensees. Currently, states seem to be much more responsive to complaints filed by members of the public against their licensees.

6.5. Implementing Professional Ethics Education for Geologists

In recent years ASBOG has enhanced opportunities for professional ethics education for both practicing licensed professionals and geoscience students. Many professions, including medical, legal, and engineering, require that their students complete formal instruction in professional ethics before they receive their undergraduate degree. There are many colleges of arts and sciences that require ethics courses and a few geology departments that do. With the

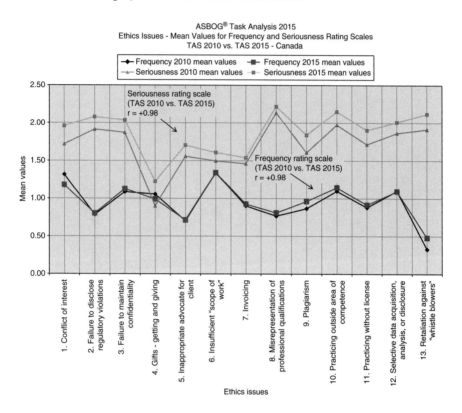

Figure 6.6 Comparison of mean values for seriousness and frequency of ethical issues for 2010 and 2015 surveys of Canadian geoscientists. Canadian Geoscientists: Survey 2010 vs. Survey 2015 Frequency Correlation = +0.98; Seriousness Correlation = +0.98. Source: *Warner* [2015]. Reproduced with permission of ASBOG.

increasing interest in professional ethics in the geosciences, it is apparent that formal instruction is appropriate for inclusion in the undergraduate and graduate geoscience curriculum. The challenge is to determine the most effective and/or practical way for implementation. With the increasing pressure to limit the number of courses and hours of instruction that can be required of a graduating student, there is less and less opportunity to insert more instructional requirements. Fundamentally, there are three ways in which instruction may be provided:

1. Formal classes in professional ethics offered within the department or required in departments outside of the geology department, such as within the philosophy department

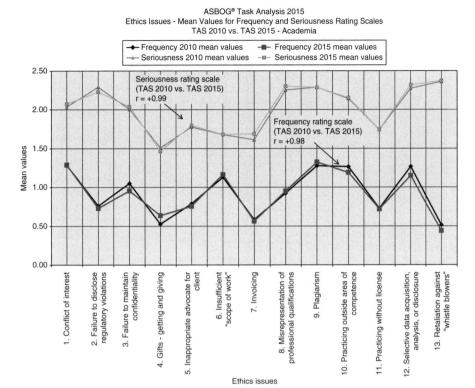

Figure 6.7 Comparison of mean values for seriousness and frequency of ethical issues for 2010 and 2015 surveys of academic geoscientists. Academia: Survey 2010 vs. Survey 2015 Frequency Correlation = +0.99; Seriousness Correlation = +0.98. Source: *Warner* [2015]. Reproduced with permission of ASBOG.

2. Modification of current required undergraduate geoscience curriculum to include a significant amount of professional ethics instruction
3. Allocations of minor amounts of time in existing courses to provide an introduction to professional ethics

Aside from the problem of finding time for the instruction of professional ethics, another significant challenge is finding qualified faculty with an appropriate background to do an effective job. ASBOG is working to provide effective ethics instruction. The association is making available ethics teaching materials that can serve as resources for those faculty willing to provide instruction. Having these ethics resources available minimizes the amount of time required for faculty to develop ethics curriculum. Currently, access to these resources is available via the ASBOG website, http://asbog.org/organization.html.

Table 6.1 Penalties that may be imposed by ASBOG member states for violation of their codes of ethics.

	State	Ethics Provisions	Stand-Alone Code	Within Enable Leg.	Monetary Fine	Required Remed. Ed.	License Suspension	Publication of Action	License Revocation	Forward Crim. Pros.	Other	Explanation of "Other"
1	Alabama	yes	no	yes			x	x	x		x	Public and/or private reprimand
2	Arizona	yes	no	yes	x	x	x	x	x	x	x	Probation, peer review, etc.
3	Arkansas	yes	no	yes	x		x	x	x	x		
4	California	yes	yes	no	x	x	x	x	x	x		
5	Delaware	yes	no	yes	x	x	x	x	x	x		
6	Florida	yes	no	yes					x			Data incomplete
7	Georgia											
8	Idaho	yes	yes	no						x	x	Not provided
9	Illinois	yes	no	yes	x		x	x	x	x	x	Practice with conditions
10	Indiana	yes	no	yes			x	x	x	x		
11	Kansas	yes	no	yes	x	x	x	x	x	x		
12	Kentucky	yes	no	yes			x	x	x	x	x	Probation
13	Maine											
14	Minnesota	yes	no	yes	x	x	x	x	x			
15	Mississippi	yes	no	yes	x	x	x	x	x	x		
16	Missouri	yes	no	yes		x	x	x	x			
17	Nebraska	yes	no	yes	x		x	x	x	x	x	Limitations on extent, scope of work
18	New Hampshire	yes	no	yes	x	x	x	x	x		x	Direct supervision

#	State											Refer to superior court
19	North Carolina	yes	no	yes	x	x	x	x	x	x	x	Not provided
20	Oregon	yes	no	yes	x	x	x	x	x	x	x	
21	Pennsylvania											
22	Puerto Rico											
23	South Carolina	yes	no	yes	x	x	x	x	x	x		
24	Texas	yes	no	yes	x	x	x	x	x	x		
25	Utah											
26	Virginia											
27	Washington	yes	no	yes	x	x	x	x	x	x	x	Local prosecution possible
28	Wisconsin											
29	Wyoming	yes	no	yes	x	x	x	x	x	x	x	Probation

No data received, repeat request issued.

Source: *Williams* [2013]. Reproduced with permission of ASBOG.

To meet the demands for instructional resources for those practicing professional geoscientists, ASBOG is doing the following:

1. Presenting formal short courses in professional ethics at geoscience society meetings such as the Geological Society of America, the American Institute of Professional Geologists, and the Association of Environmental and Engineering Geologists,
2. Generating professional papers resulting from professional presentations that can be included in geoscience publications regarding professional ethics, and
3. Encouraging member state boards to generate and provide resources to their licensees.

6.6. Conclusions

General conclusions from ASBOG's involvement with professional ethics are the following:

1. Regardless of specialty, practicing professional geologists consider professional ethics extremely important in their careers.
2. ASBOG as an umbrella organization for practicing licensed professional geosciences has been active on many fronts by promoting awareness of the importance of professional ethics and by providing professional ethics resources, including ethics curriculum, for the professional and academic community.

References

ASBOG (2015), Member map, http://asbog.org/map/asbogmap2017.pdf.

Warner, J.L. (2015), Task Analysis Survey 2015: A Study of the Practice of Geology in the United States and Canada. National Association of State Boards of Geology, www.asbog.org.

Warner, J.L., and S. Warner (2010), Task Analysis Survey 2010: A Study of the Practice of Geology in the United States and Canada (draft). National Association of State Boards of Geology, www.asbog.org.

Williams, J.W. (2014), Professional Ethics and the National Association of State Boards of Geology (ASBOG®): An Update. Geological Society of America Meeting (Cordilleran and Rocky Mountain Sections), Bozeman, Montana.

Williams, J.W. (2013), Ethical Issues and Licensing, the ASBOG® Perspective. American Geological Institute Leadership Forum, Washington D.C.

Williams, J.W., R. Spruill, J. Randall, and J. Warner (2012), National Association of State Boards of Geology (ASBOG): History, Exams, Assessment and Professional Ethics. 4th International Professional Geology Conference, Vancouver, Canada.

7. BRIEF HISTORY AND APPLICATION OF ENFORCEABLE PROFESSIONAL GEOSCIENCE ETHICS CODES

David M. Abbott, Jr.

Abstract

Fraudulent promotion of mining and petroleum ventures has existed as long as these ventures have. The American Association of Petroleum Geologists, founded in 1917, adopted an enforceable *Code of Ethics* in 1924 to address such frauds. Other professional societies have considered whether to adopt an enforceable ethics code and have decided not to for a variety of reasons. More recently, other professional societies have concluded that at least an aspirational guideline for professional geoscience practice was needed, and others have adopted enforceable ethics codes. What is included in such codes has evolved over the years. Those professional societies and organizations that adopt an enforceable ethics code should also adopt disciplinary procedures that describe how allegations of unethical or incompetent practice are to be brought and investigated and, should unethical or incompetent practice be formally charged, how the respondent can respond to those charges, appear at a hearing on the charges, and appeal the initial decision; further, they should describe how to address potential conflicts of interest. The possible sanctions against a respondent must also be specified. A potential consequence of adopting an enforceable ethics code is that a respondent may sue the organization, something most professional society budgets can ill afford.

American Institute of Professional Geologists, Denver, Colorado, USA

Scientific Integrity and Ethics in the Geosciences, Special Publications 73,
First Edition. Edited by Linda C. Gundersen.

7.1. Historical Overview

This article reviews some of the history of professional geoscience ethics codes by several American Geosciences Institute (AGI) member societies. Mining, the extraction and use of rocks and minerals for human use, is as old as our history of human activity. Stone and flint tools and clay potsherds are sought by archeologists in determining human cultures. It didn't take long for ancient cultures to recognize that the extraction of many of these minerals required a great deal of effort; they were, in modern terms, capital-intensive operations. Agricola's 1556 *De Re Metallica* is the earliest, commonly known book focused on the mining industry. Agricola observed, "Moreover, a prudent owner, before he buys shares, ought to go to the mine and carefully examine the nature of the vein, for it is very important that he should be on his guard lest fraudulent sellers of shares should deceive him." This statement notes that the selling of shares was a common means of raising the capital required by mining operations and that fraud, the making of unwarranted claims about a particular operation, sometimes occurred.

Professional geoscience societies generally have membership requirements that include some stated amount of professional education and perhaps experience. These requirements are, at least in part, intended to provide assurance to the public that the society's members can be relied on to act competently and, generally, honestly. The American Institute of Mining Engineers, the original incarnation of the Society for Mining, Metallurgy, and Exploration (SME), was founded in 1871, the Geological Society of America (GSA) in 1888, the American Association of Petroleum Geologists (AAPG) in 1917, and the Society of Economic Geologists (SEG) in 1920. However, professional ethics codes were not initially part of these societies' formal documents.

AAPG was the first of this group to address the issue of ethics. During AAPG's Fifth Annual Meeting in 1920, E.G. Woodruff introduced a resolution that stated in part, "Whereas, we feel that regularly qualified geologists are not now sufficiently protected from unqualified and unscrupulous men practicing geology, therefore be it resolved that efforts be made in every state in which petroleum is produced to obtain laws requiring practicing geologists to be licensed; that a condition of such licensing be that the applicant be a graduate of a school of recognized standing; that he be required to have two years or its equivalent of field work under a licensed geologist; and that he satisfy a qualifying board of his qualifications and ability before such license is issued." This resolution was referred to a committee that concluded, "It is both unwise and impractical to secure a discriminative classification of petroleum geologists and engineers by legislative means. It is also the view of the committee that the object can best be obtained by careful attention to standards of membership in this Association, and the enforcement of them, as well as the proper enforcement of discipline among its members" [*AAPG*, 1920]. Sidney Powers [1929] noted that at AAPG's Fifth Annual Meeting a recommendation was made that a list of requirements for regular members to be included in Article III, Section 1 and bylaw, Section 4, allowing for the expulsion by unanimous vote of the executive committee of any member "who shall be found guilty of flagrant violation of the established principles of professional ethics," provided that the accused member have a hearing prior to his suspension or expulsion. These were favorably recommended for adoption during the 1921 annual meeting. Powers went on to note that at AAPG's Seventh Annual Meeting in 1922, the executive committee was granted the power to expel by unanimous vote anyone "guilty of flagrant violation of the established principles of professional ethics." The question of a code of ethics was discussed at length. At AAPG's Ninth Annual Meeting in 1924, a detailed code of ethics prepared by Alexander Deussen and Max W. Ball was adopted [*AAPG*, 1924; *Powers*, 1929].

The expanded search for minerals after World War II and the uranium boom of the early 1950s, that included the issuance of shares by many new, small public companies, "combined to pose problems of ethical conduct on a scale not hereto encountered in the realm of Economic Geology" [*Davidson*, 1957]. Donald M. Davidson, the president of the SEG in 1954, appointed an SEG Committee on Ethics and Conduct "for the purpose of studying professional ethics and conduct to determine the desirability of a permanent committee on ethics, as well as a written code for the society" [*Davidson*, 1957]. In 1955, that committee reported the following:

1. That because of the nature and complexity of professional ethics the promulgation of a written code of ethics would not be an effective means of maintaining high standards;

2. That Article 2, Section 4 of the present By-Laws of the Society makes ample provision for dealing with breaches in ethical conduct on the part of members;

3. That inasmuch as the Council already has the power to deal with misconduct, either by direct action or with the aid of special committees, a standing committee on ethics and conduct is not necessary. In this respect the Committee believes that, other things being equal, the number of cases requiring investigation will continue to be few and far between, and that, inasmuch as acceptable standards of conduct vary to some degree with the time and local conditions, specially selected *ad hoc* committees would be better suited to the investigation of alleged misconduct than a standing committee whose membership might have no initial appreciation of the local circumstances.

In addition, a majority of the members of the committee was of the opinion that Article 2, Section 4, of the then-present By-Laws of the society not only made clear that the society expected its members "to conform with established standards of professional ethics," but that because of the very nature of ethics the admittedly vague but all-inclusive phrase "established standards of professional ethics" was adequate to serve its purpose.

However, one committee member wondered if the phrase "established standards of professional ethics" was sufficiently explicit and suggested that reference be made "to code of ethics of the AAPG or some other professional society as an example of what SEG has in mind." The committee noted that SEG's Admissions Committee and Council worked to ensure that "only men of integrity and proven capacity should be admitted" to SEG membership [*SEG Committee on Ethics and Conduct*, 1955].

In 1999, SEG appointed another ad hoc committee to examine whether SEG should adopt an enforced code of ethics. The ad hoc committee concluded [*Melrose et al.*, 1999],

The 1997 Conference on Ethics in the Geosciences combined with the Bre-X and several lesser scandals and resulting proposed regulatory changes, including the proposed Canadian requirements for becoming a Qualified Person, prompted the Society of Economic Geologists (SEG) to re-examine its commitment to high professional and ethical standards. The SEG participated in the American Geological Institute (AGI) committee that developed the *AGI Guidelines for Ethical Professional Conduct*. Half of SEG's members live outside the U.S. in 70 countries; 10-15% live outside North America. Diverse member career affiliations and the customs and practices create significant challenges in codifying professional ethics. Because an adverse ethical finding can result in loss of professional licensing or certification, professional societies having such programs can expect significant legal challenges by respondents in ethics cases. The potential financial liabilities to the society cannot be ignored. The *ad hoc* committee is examining alternative measures which will further ethical conduct and assist members who require an enforced code of ethics for professional certification. These alternatives include a voluntary SEG certification program, cooperative certification with organizations having acceptable codes of ethics, or limiting SEG action to ethics education and support of multi-organization ethics education programs, are measures currently being considered. Whatever actions are taken will reflect SEG's commitment to foster the highest level of professional and ethical conduct.

SEG has been aware of professional ethics issues since its founding. It has twice appointed committees to examine the need for a formal code of ethics, both of which decided that a formal code was unneeded. These conclusions were not a rejection of the importance of professional ethics but rather recognition that other professional organizations were better suited to the task of establishing and enforcing a code of geoscience ethics. SEG's position also recognizes that not every geoscience society needs to have its own code of ethics. Indeed, having only a few such codes, all of which differ in detail, makes it easier for geoscientists not to run afoul of some minor provision of a particular code. SEG is interested in professional ethics and an SEG representative was a member of both the 1999 and 2014 AGI committees that wrote and then revised the *AGI Guidelines for Ethical Professional Conduct* issued in 1999 and 2015.

When the American Institute of Professional Geologists (AIPG) was formed in 1963, it adopted the then-current version of the *AAPG Code of Ethics*. Various changes to both AAPG's and AIPG's *Code of Ethics* since 1963 have resulted in divergent language. These and other geoscience professional society codes of ethics were modified following a decision by the Federal Trade Commission in the 1980s that some statements in the codes could be considered as potential restraints of trade (unfair competition statements) and prohibited or greatly restricted a professional's ability to advertise offered professional services.

The American Geological Institute (AGI) formed a committee to develop the *AGI Guidelines for Professional Ethical Conduct* following the GSA Presidential Meeting on Professional Ethics held in 1997 in Welches, Oregon. (AGI was founded in 1948 as the American Geological Institute and did not change its name to the American Geosciences Institute until October 1, 2011.) That meeting was attended by representatives of the broad geoscience industry, government agencies (largely USGS), and academia. The committee issued *AGI Guidelines for Professional Ethical Conduct* in 1999 [*AGI*, 1999; *Stow et al.*, 1999]. A similar committee was formed following American Geosciences Institute's Leadership Forum in 2013 that focused on professional ethics. This committee, chaired by David Mogk of the University of Montana, was composed of individuals nominated by those AGI member societies that chose to nominate a member. The initial draft revision of the *AGI Guidelines for Professional Ethical Conduct* was released for AGI member society review and comment at the GSA Annual Meeting in October 2014 [*Mogk and Nickless*, 2014]. Following the comment period and further review by the committee, AGI formally adopted the revision in 2015. The *AGI Guidelines for Professional Ethical Conduct* is explicitly an aspirational document that does not require a professional organization that chooses to recognize the guidelines to adopt disciplinary procedures.

The great majority of professional geoscience society members do practice competently and honestly. As reflected by the concerns with fraudulent practices in the promotion of oil and gas and mining ventures discussed above by the AAPG and the SEG, some professional societies adopted codes of ethics that included the ability to discipline members who were alleged to have violated these codes. SME did not adopt an enforceable code of ethics until 2005, and that code applied only to those SME members seeking Registered Member status for the purposes of meeting the requirements of Canadian National Instrument 43-101 [2011]. Many other professional societies have either not adopted a code of ethics or have a code that does not subject members to specified discipline for ethical violations. Most professional societies do not feel that an enforceable code of ethics is needed [*Melrose et al.*, 1999, 2000]. Allegations of unethical conduct do occur but are relatively rare. I was aware of only a few occasions when a professional society member was an active perpetrator of fraud during my 21 years as a geologist with the U.S. Securities and Exchange Commission, where my duties included the investigation of oil and gas and mining frauds. A review of the AIPG, Canadian, Australasian Institute of Mining and Metallurgy (AusIMM), and American Institute of Mineral Appraisers (AIMA) disciplinary actions found that the number of allegations of ethical misconduct averaged around 0.5% of institutes' membership during the period 1995 through 2008, and over half of these allegations were dismissed due to lack of sufficient evidence for the allegation or the evidence demonstrated that no violation occurred [*Abbott*, 2005; *Abbott, et al.*, 2008].

In summary, although professional societies generally expect their members to act ethically and competently, many do not have a specified code of professional ethics, or if they do, it is strictly aspirational in that no disciplinary actions can be taken. The oil and gas and mining industries and related professional organizations have been more affected by unethical practices over time and therefore, starting with AAPG in 1924, adopted enforceable codes of ethics. As unethical practices in other areas of geoscience practice have emerged over the years, more professional organizations have adopted either their own ethics codes or a more generalized one like the *AGI Guidelines for Ethical Professional Conduct*. While the various codes of geoscience ethics have common features, there are individual variations reflecting (1) a particular organization's professional focus, (2) whether discipline for ethical violations is part of the code (these codes tend to be longer and more detailed), and (3) emerging ethical issues, for example, occupational health and safety, social licensing (a 21st century issue), and workforce diversity.

7.2. Comparison of the 1924 *AAPG Code of Ethics* and the 2015 *AGI Guidelines for Professional Conduct*

There are two types of professional ethics statements: aspirational and rules. Aspirational ethics statements encourage but do not require members to engage in or avoid particular behaviors. Ethical rules are statements that must be followed by members at all times. Statements concerning continuing professional development (CPD) or similar descriptors provide an example of either aspirational or rules statements. Some professional societies encourage members to engage in activities that maintain and expand their professional credentials by one of several means (credit courses, attending meetings, informal learning such as field trips, self-directed professional reading, writing papers and giving talks, and being a member of a society's committees) but do not require that a particular amount of CPD be done within a specified time period. These are aspirational statements. Increasing numbers of professional societies are now requiring their members, or some class of members, to engage in at least some minimum specified amount of CPD. In some cases, the requirements specify that at least a minimum number of hours must be spent in particular activities such as professional ethics, or that more than one type of CPD activity must be recorded within the specified time period. When CPD is required, some sort of log of CPD must be kept, usually on the society's website.

Careful comparison of the 1924 *AAPG Code of Ethics* and the 2015 *AGI Guidelines for Ethical Professional Conduct* shows that there are many differences between them, including organization, length, and topics addressed. A very important difference is that the AAPG code contains ethical rules, the violation of which can be the basis for disciplinary action against AAPG members. For example, Article II, Section 1 of the 1924 AAPG code states, "A geologist should avoid and discourage sensational, exaggerated, and unwarranted statements, especially those that might induce participation in unsound enterprises." The types of statements prohibited by this ethical rule are typical of fraudulent oil and gas venture promotions. The AGI guidelines are explicitly aspirational only, with no disciplinary action required. An example is the statement, "Encourage and assist in the development of a safe, diverse, and inclusive workforce." The current *AAPG Code of Ethics* has changed from the 1924 version. However, the organization and wording of 1924 *AAPG Code of Ethics* formed the basis for codes adopted by other professional societies, for example, the *AIPG Code of Ethics* adopted with AIPG's founding in 1963. Like the *AAPG Code of Ethics*, the *AIPG Code of Ethics* has been modified over the years, but it continues to use the same organizational structure and many of the same topics.

The texts of the 1924 AAPG code and 2015 AGI guidelines follow.

AAPG Code of Ethics

(as originally adopted March 26, 1924; drafted by Alexander Deussen of Houston and revised by Max W. Ball of Denver)

OBJECT

The object of this Association is to promote the science of geology, especially as it relates to petroleum and natural gas; to promote the technology of petroleum and natural gas and improvements in the methods of winning these materials from the earth; to foster the spirit of scientific research amongst its members; to disseminate facts relating to the geology and technology of petroleum and natural gas; to maintain a high standard of professional conduct on the part of its members; and to protect the public from the work of inadequately trained and unscrupulous men posing as petroleum geologists. (Constitution, Article II.)

ARTICLE I. GENERAL PRINCIPLES

SECTION 1. The practice of petroleum geology is a profession. It is the duty of those engaged in it to be guided by the highest standards of professional conduct and to subordinate reward and financial gain thereto.

SECTION 2. The confidence of the public and of the oil industry can be won and held only by the practice of the highest ethical principles.

SECTION 3. Honesty, integrity, fairness, candor, fidelity to trust, inviolability of confidence, and conduct becoming a gentleman are incumbent upon every member of this Association.

ARTICLE II. RELATION OF GEOLOGIST TO PUBLIC AND PROFESSION

SECTION 1. A geologist should avoid and discourage sensational, exaggerated, and unwarranted statements, especially those that might induce participation in unsound enterprises.

SECTION 2. A geologist should not knowingly permit the publication of his report or maps for the purpose of raising funds without legitimate and sound development in view.

SECTION 3. A geologist may accept for his services in the making of a report an interest in the property reported on, but it is desirable that the report state the fact of the existence of the interest.

SECTION 4. A geologist should not give an opinion or make a report without being as fully informed as might reasonably be expected, considering the purpose for which the information is desired. The opinion or report should make clear the conditions under which it is made.

SECTION 5. A geologist may publish simple and dignified business, professional, or announcement cards, but should not solicit business by other advertisements, or through agents, or by furnishing or inspiring exaggerated newspaper or magazine comment. The most worthy advertisement is a well-merited reputation for professional ability and fidelity. This cannot be forced, but must be the outcome of character and conduct.

ARTICLE III. RELATION OF GEOLOGIST TO EMPLOYER

SECTION 1. A geologist should protect, to the fullest extent possible, the interests of his employer so far as consistent with the public welfare and his professional obligations.

SECTION 2. A geologist who finds that his obligations to his employer conflict with his professional obligations should notify his employer of that fact. If the objectionable condition persists, the geologist should sever his connection with his employer.

SECTION 3. A geologist should not allow himself to become or remain identified with any enterprise of questionable character.

SECTION 4. A geologist should make known to his prospective employer any oil or gas interests which he holds in the region of his prospective employment.

SECTION 5. A geologist, while employed, should not directly or indirectly acquire any present or prospective oil or gas interest without the express consent of his employer.

SECTION 6. A geologist retained by one client should, before accepting engagement by others, notify them of this affiliation, if in his opinion the interests might conflict.

SECTION 7. A geologist who has made an investigation for a client should not, without the client's consent, seek to profit from the economic information thus gained, or report on the same subject for another client, until the original client has had full opportunity to act on the report.

SECTION 8. A geologist should not accept direct or indirect compensation from both buyer and seller without the consent of both parties; or from parties dealing with his employer without the employer's consent.

SECTION 9. A geologist should observe scrupulously the rules, customs, and traditions of his employer as to the use or giving out of information or the acquisition of interests, both while employed and thereafter; and, except as permitted by such rules, customs, and traditions or by the consent of the employer he should not seek to profit directly or indirectly from the economic information gained while so employed.

SECTION 10. A geologist employed by a state geological survey should not permit private professional work or the holding of private mineral interests in the state to interfere with his duty to the public or to lessen the confidence of the public in the survey. The preferable course is to avoid such private work and interests.

SECTION 11. A geologist should not divulge information given him in confidence.

ARTICLE IV. RELATION OF GEOLOGIST TO OTHER GEOLOGISTS

SECTION 1. A geologist should not falsely or maliciously attempt to injure the reputation or business of a fellow geologist.

SECTION 2. A geologist should not knowingly compete with a fellow geologist for employment by reducing his customary charges.

SECTION 3. A geologist should give credit for work done to those, including his assistants, to whom the credit is due.

ARTICLE V. DUTY TO THIS ASSOCIATION

SECTION 1. Every member of the Association should aid in preventing the election to membership of those who lack moral character or the required education and experience.

SECTION 2. A member of this Association who has definite evidence of the violation of the established principles of professional ethics by another should report the facts to the executive committee.

American Geosciences Institute Guidelines for Ethical Professional Conduct

Approved April 13, 2015

These guidelines address common ethical topics across the geoscience community; the ethics statements of individual societies may expand beyond these guidelines.

Geoscientists play a critical role in ethical decision making about stewardship of the Earth, the use of its resources, and the interactions between humankind and the planet on which we live. Geoscientists must earn the public's trust and maintain confidence in the work of individual geoscientists and the geosciences as a profession. The American Geosciences Institute (AGI) expects those in the profession to adhere to the highest ethical standards in all professional activities. Geoscientists should engage responsibly in the conduct and reporting of their work, acknowledging the uncertainties and limits of current understanding inherent in studies of natural systems. Geoscientists should respect the work of colleagues and those who use and rely upon the products of their work.

In day-to-day activities geoscientists should:
- Be honest.
- Act responsibly and with integrity, acknowledge limitations to knowledge and understanding, and be accountable for their errors.
- Present professional work and reports without falsification or fabrication of data, misleading statements, or omission of relevant facts.
- Distinguish facts and observations from interpretations.
- Accurately cite authorship, acknowledge the contributions of others, and not plagiarize.
- Disclose and act appropriately on real or perceived conflicts of interest.
- Continue professional development and growth.
- Encourage and assist in the development of a safe, diverse, and inclusive workforce.
- Treat colleagues, students, employees, and the public with respect.
- Keep privileged information confidential, except when doing so constitutes a threat to public health, safety, or welfare.

As members of a professional and scientific community, geoscientists should:
- Promote greater understanding of the geosciences by other technical groups, students, the general public, news media, and policy makers through effective communication and education.
- Conduct their work recognizing the complexities and uncertainties of the Earth system.
- Sample responsibly so that materials and sites are preserved for future study.
- Document and archive data and data products using best practices in data management, and share data promptly for use by the geoscience community.

- Use their technical knowledge and skills to protect public health, safety, and welfare, and enhance the sustainability of society.
- Responsibly inform the public about natural resources, hazards, and other geoscience phenomena with clarity and accuracy.
- Support responsible stewardship through an improved understanding and interpretation of the Earth, and by communicating known and potential impacts of human activities and natural processes.

The first point of the AGI guidelines, "Be honest," clearly implies that dishonesty is unethical. This implication bothered some members of the AGI committee that drafted the 1999 edition of the AGI guidelines because it therefore implied the need for discipline. However, all committee members agreed that honesty was a fundamental principle of professional ethics [*Abbott*, 2000]. The fact that dishonesty, or any other action that is contrary to the AGI guidelines, could be viewed as unethical behavior does not obligate AGI-member societies to develop or implement disciplinary procedures.

The issues addressed in two sections of the 1924 *AAPG Code of Ethics*, Article II, Section 5 on advertising and Article IV, Section 2's injunction against competing with fellow geologists by lowering one's billing rate, were deleted after similar provisions used by other professional societies were judged to be illegal. The statement on advertising was dropped after a similar provision in legal codes of conduct was found to violate freedom of speech. The prohibition on competition by cutting rates was found to violate the Sherman Antitrust Act.

7.3. Enforceable Codes of Professional Ethics

As noted above, AAPG developed an enforceable code of ethics within seven years of its founding, in part to combat "unqualified and unscrupulous men practicing geology" [*AAPG*, 1920]. AAPG's code served as a basis for codes adopted by some other societies. The 1997 Bre-X fraud in Canada and earlier problems with "penny dreadful" stocks issued in Australia led to the development of Canadian National Instrument 43-101 (NI 43-101) [2011] and the JORC Code [2012] in Australia, respectively. (JORC is the commonly used acronym for the Joint Ore Reserves Committee of the Australasian Institute of Mining and Metallurgy, Australian Institute of Geoscientists, and Minerals Council of Australia.) NI 43-101 specifies the filing requirements for and content of reports on mining properties listed on a Canadian stock exchange. The JORC Code does the same for listings on the Australian Stock Exchange and the New Zealand Stock Exchange. These two regulatory initiatives led to the development of the Competent Person and Qualified Person standards in Australia and Canada. While not identical, the definitions of *competent person* and *qualified person* are similar. Similar standards have been adopted by other regulatory bodies, including the Hong Kong Stock Exchange.

Canadian National Instrument 43-101 defines "professional association" and "qualified person" as the following:

"Professional association" means a self-regulatory organization of engineers, geoscientists, or both engineers and geoscientists that
a. is
   i) given authority or recognition by statute in a jurisdiction of Canada, or
   ii) a foreign association that is generally accepted within the international mining community as a reputable professional association;
b. admits individuals on the basis of their academic qualifications, experience, and ethical fitness;
c. requires compliance with the professional standards of competence and ethics established by the organization;
d. requires or encourages continuing professional development; and
e. has and applies disciplinary powers, including the power to suspend or expel a member regardless of where the member practices or resides.

"Qualified person" means an individual who
a. is an engineer or geoscientist with a university degree, or equivalent accreditation, in an area of geoscience, or engineering, relating to mineral exploration or mining;
b. has at least five years of experience in mineral exploration, mine development or operation, or mineral project assessment, or any combination of these, that is relevant to his or her professional degree or area of practice;
c. has experience relevant to the subject matter of the mineral project and the technical report;
d. is in good standing with a professional association; and
e. in the case of a professional association in a foreign jurisdiction, has a membership designation that
   i) requires attainment of a position of responsibility in their profession that requires the exercise of independent judgment; and
   ii) requires
      A. a favourable confidential peer evaluation of the individual's character, professional judgement, experience, and ethical fitness; or
      B. a recommendation for membership by at least two peers, and demonstrated prominence or expertise in the field of mineral exploration or mining;

Mining properties and the mining industry professionals who work on them occur throughout the world. As a consequence, these regulations identified "recognized overseas professional organizations" (ROPOs) [*NI 43-101*, 2011] or "recognized professional organizations" (RPOs) [*JORC*, 2012] whose members, or specified membership classes, were recognized as meeting the Competent or Qualified Person standards. The ROPOs required that the recognized members have basic educational and experience requirements and be subject to discipline for failure to comply with the ROPO's code of ethics regardless of where the member in question resides or where the property reported on was located [*Abbott*, 1999a, 1999b]. Because many mining engineers and mineral processing engineers never bothered to obtain professional engineer licenses that granted Competent or Qualified Person status, the Mining and Metallurgical Society of America (MMSA) [2005] adopted a code of ethics and a Qualified Professional category in 2005 that covered geologists, mining engineers, processing engineers, mineral resource and mineral reserve estimators, and environmental specialists. In 2006, the SME adopted a code of ethics for Registered Members (a new SME membership class) in order that SME Registered Members, particularly the mining and mineral processing engineers, could become recognized as Competent or Qualified Persons in those jurisdictions that have adopted such regulations.

7.4. Enforcing an Ethics Code

Once a professional organization has decided that it will have an enforceable code of ethics, it must determine which provisions of its code of ethics are ethical rules that cannot be violated and which are aspirational statements. Most professional societies have codes of ethics with both types of provisions, while state licensing boards and others will have codes of conduct containing only ethical rules.

Adoption of an enforceable code of ethics requires that disciplinary procedures be adopted. These disciplinary procedures should describe all of the following processes:

1. Review of allegations
2. Initiation of an investigation
3. Filing of formal charges
4. Adjudication process
5. Appellate process
6. Describe the disciplinary sanctions that can be imposed
7. Describe the rights of respondents
8. Avoidance of conflicts of interest within the procedures
9. Confidentiality of the proceedings
10. Reporting of violations to other organizations and regulators
11. Publication of annual summaries of disciplinary activity

7.4.1. Review of Allegations

Most organizations allow anyone to file allegations that the organization's ethics or conduct codes have been violated. The allegations should be in writing and be accompanied by all available supporting documents and a list of other persons who may have relevant information. A permanent organizational contact should be provided for receiving these allegations. The organization's headquarters office and executive director are usually the recipient of allegations. The procedures should make clear that all allegations received are passed on to the ethics committee chairman (or other named person or body) without review; the permanent contact is simply a "mailbox." Allegations can also be made directly to the ethics committee if the chairman's address is known. The ethics committee chairman reviews the allegations and, depending on the information received, can (1) determine that no violation occurred, (2) request additional information from the party making the allegations, or (3) initiate an investigation of the matter. Note that findings against a proposed respondent by another organization, a licensing board, or a legal proceeding may be grounds for a disciplinary action that is focused on whether the findings in that previous action demonstrate violation of the organization's code of ethics.

7.4.2. Initiation of an Investigation

Depending on the evidence received supporting the allegation of ethical misconduct, the ethics committee chairman can appoint an investigator having an appropriate technical background to make further inquiry about the allegations and to seek information from others who might have relevant information about the matter. The respondent (the member against whom allegations have been made) should be informed of the initiation of the investigation and the identity of the investigator. Relevant information and documents can be requested from the respondent. Failure of the respondent to respond in a reasonable amount of time can be used as evidence against the respondent. Following completion of the investigation, the investigator and ethics committee chairman determine whether the evidence collected demonstrates that no violation occurred, that there is not enough evidence to conclude whether or not a violation occurred (the allegation is unproven), or that formal charges should be made against the respondent.

While it is true that in U.S. criminal justice a person is presumed innocent until proven guilty, this principle is not true of scientific statements, which must be proven to be true by the claimant. ("Innocent until proven guilty" and the right against self-incrimination apply in criminal law. Civil law, which includes a professional organization's disciplinary actions, allows adverse implications to be drawn. This is also the basis for imposing sanctions in those cases where the respondent resigns his/her membership rather than challenging formal charges of ethical violation.) For example, mining and petroleum industry frauds usually involve claims that reserves of one or more minerals or of petroleum exist on the property being promoted. *Reserves* is an industry-defined term and the definitions require that specified varieties of supporting data must exist prior to claiming that reserves exist. If the supporting data are lacking, the statement that reserves exist cannot be proved and is scientifically false.

Differences of scientific opinion when each of the differing opinions is scientifically supported are not unethical. However, slandering another professional because he/she holds a differing opinion is unethical. It has been found that both parties in a difference of scientific opinion case warranted private letters of admonition.

The foregoing examples are brief. They serve to emphasize the importance of carefully weighing the evidence presented in an allegation of ethical misconduct. Experience has shown that only about 50% of the allegations received result in disciplinary action, largely because of lack of supporting evidence [*Abbott et al.*, 2008].

7.4.3. Filing of Formal Charges

Assuming that sufficient evidence, preferably documentary evidence, exists to support formal charges, a statement of the specific sections of the code of ethics that the respondent is accused of having violated is prepared, along with a memorandum supporting the charges and setting out in detail the evidence for each violation charged. The statement of formal charges should also include a statement of the sanctions being sought for the charged violations. The statement of formal charges and the memorandum supporting the charges are sent to the respondent, who is provided adequate time to respond to the charges.

A respondent's reply to the formal charges can include information and supporting documents that refute some or all of the formal charges. The response may result in modification of or dismissal of some or all of the charges.

7.4.4. Adjudication Process

The respondent has the right to a hearing before an adjudicatory committee or board whose composition is described in an organization's Disciplinary Procedures. The hearing is similar to, but is not a court of law, and therefore strict legal procedures need not be adhered to. This is a hearing before the respondent's professional peers. At this hearing, the evidence supporting the charges is presented by the ethics committee chairman, who acts as the prosecutor, and the respondent presents his/her defense. Witnesses against and for the respondent may be called and can be cross-examined. At the conclusion of the hearing, the adjudicatory body should prepare a report of findings of fact and conclusions regarding the ethical violations charged and determine what sanction, if any, should be imposed.

7.4.5. Appellate Process

Following receipt of the adjudicatory body's report, either the ethics committee chairman or the respondent may appeal the findings of fact, the conclusions regarding the ethical charges, and the sanctions recommended by the adjudicatory body. The organization's executive committee is normally the appellate body. Following completion of the appeals process, the name of the respondent and the findings of fact and conclusion of ethical violation may be made public.

7.4.6. Describe the Disciplinary Sanctions that Can Be Imposed

The potential disciplinary sanctions that can be imposed should be set out in the organization's bylaws. Sanctions can include penalties such as suspension of membership, reduction in membership grade, required continuing education, and permanent termination of membership. If the respondent chooses to resign his/her membership in the organization prior to the issuance of the adjudicatory body's report, this can be deemed a termination of membership with prejudice, effectively a permanent termination of membership. Such resignation normally terminates the disciplinary process prior to the adjudicatory body's report of formal findings of fact and conclusions of ethical violations.

7.4.7. Describe the Rights of Respondents

The respondent has the right to respond to the formal charges, to request an adjudicatory hearing, and the right to appeal the findings of fact and conclusions of ethical violations made by the adjudicatory body. The respondent has the right to present additional information and documents that seek to demonstrate that the charges brought should be modified or dismissed or that the recommended disciplinary sanction should be modified. The respondent may seek the advice of counsel at the respondent's sole expense during the disciplinary process. Failure to provide for due process can result in the respondent suing the organization.

7.4.8. Avoidance of Conflicts of Interest within the Procedures

The ethics committee chairman and anyone else involved in the investigation leading to the filing of the formal charges, including witnesses, the members of the adjudicatory body, and the members of the appellate body, must be independent groups. If, for example, a member of the organization's executive committee, which is normally the appellate body, is a witness in a case, that member must recuse himself or herself from hearing and passing on the appeal. Similarly, the members of the organization's executive committee should not be involved in screening allegations of ethical misconduct or informed of the specifics of a disciplinary investigation, including the names of those involved. This is often a problem with disciplinary procedures that have not been carefully thought out.

7.4.9. Confidentiality of the Proceedings

During the pendency of a disciplinary proceeding, the name of the respondent and the potential charges should be kept as confidential as possible. The mere act of asking potential witnesses the questions necessary for the inquiry into allegations can reveal confidential information, the name of the potential respondent in particular. Potential witnesses should be cautioned that no conclusions about the allegations have been reached (when this is true) and advised that the fact of the inquiry should not be made public. The investigation of allegations may result in a conclusion that no ethical violation occurred. Allegations of ethical misconduct, if made public, can severely impact the professional reputation of a respondent in such cases, and thus confidentiality is needed during the entire disciplinary process to protect against the possibility of a finding of no ethical violation or that the allegations made are unproven.

7.4.10. Reporting of Violations to Other Organizations and Regulators

The organization should retain the right to inform other professional organizations and appropriate regulators of the final results of a disciplinary action. Such notifications can be made in cases where the public's health, safety, or financial welfare would be harmed by the respondent's continued professional practice. A respondent's resignation prior to the preparation of formal findings of fact and conclusions of ethical violations can warrant notification to other professional organizations and appropriate regulators as long as the notification includes a statement that the disciplinary proceedings were terminated by the respondent's resignation and that no finding of fact or conclusions of ethical violation were formally reached.

7.4.11. Publication of Annual Summaries of Disciplinary Activity

An organization should prepare and publish on its website a summary of disciplinary activity during the year. This summary should generically provide the nature of the allegations and the results of the disciplinary proceeding without naming the respondent unless formal findings of fact and conclusions of ethical misconduct have been named. Such summaries provide other organizations and the public with a record of the organization's disciplinary activities. Experience has shown that several actions may occur in one year, while no allegations are received during a couple of subsequent years. A no-actions statement is made for such years. Where a disciplinary action extends from one year to the next, this should be stated until a conclusion is reached. An example of AIPG's "Disciplinary Summary" can be found at http://www.aipg.org/About/disciplinary.htm.

The AIPG's "Disciplinary Procedures" [2015] provides an example of such procedures that have been tested by use over many years and should be examined by organizations wishing to adopt or to evaluate their own disciplinary procedures.

7.5. Consequences of Adopting an Enforced Code of Ethics

An organization wishing to adopt an enforceable code of ethics should be aware of the potentially adverse consequences of doing so. Because a respondent to a disciplinary proceeding may seek the advice of counsel, there may be an attack on the disciplinary process as a means of deflecting attention from the respondent's activities to the organization's disciplinary process. Litigation is expensive, and many organizations' budgets do not allow for such costs. Indeed, the possibility of such litigation may well result in an organization's general legal counsel advising against adoption of an enforced code of ethics. Because an organization's executive committee has a fiduciary duty to maintain the assets of the organization, such advice is warranted. However, ethical integrity requires that the right thing be done regardless of the consequences, including the cost of litigation.

Recognizing the potential for litigation, it behooves an organization to ensure that its disciplinary procedures are fair to both the organization and the respondent, that conflicts of interest within the administration of the disciplinary procedures are avoided, and that the ethics committee chairman and all those involved in the investigation, the adjudicatory hearing (including members of the adjudicatory body), and appeal (executive committee) are familiar with the organization's disciplinary procedures and the consequences of deviating therefrom. Consideration of the potential adverse consequences suggests that experience with the legal system, including arbitration proceedings, is desirable for the ethics committee chairman and the head of the adjudicatory board. Recognition that litigation may occur despite an organization's best efforts and the consequent costs to the organization should be carefully considered by the organization's executive committee, including how litigation expenses can be met.

AIPG's experience with its disciplinary proceedings from 1989 through 2014 has been that only one case went to an adjudicatory hearing. In that case, the respondent did not appear and the case was decided based on the evidence presented by the ethics committee chairman. In all other cases, where it appeared that ethical violations had occurred, the respondent either resigned when the proceeding was initiated or when formal charges were brought. The advantage to the respondent in such cases is that there are no findings of fact or conclusions regarding the ethical violation. If such a formal report of ethical violations is prepared, then the formal report can be used against the respondent by other organizations or regulatory bodies (licensing boards) to which the respondent belonged. The 2015 amendments to AIPG's Disciplinary Procedures permit the chairman of AIPG's ethics committee to call for an adjudicatory hearing in egregious cases, even when a respondent has resigned. The adjudicatory hearing can then lead to the issuance of a formal report of ethical violations, regardless of whether the respondent chooses to appear at the hearing.

References

Abbott, D.M., Jr. (1999a), Who is a competent person and who cares? *The Professional Geologist*, January, 6-7.

Abbott, D.M., Jr. (1999b), The qualified person for resource and reserve estimation, *Mining Engineering* Rock in the Box column, September, 64 and 72.

Abbott, D.M., Jr. (2000), Honesty, the principal geoscience ethical principle, Geological Society of America Abstracts with Programs, 32(7), 293.

Abbott, D.M., Jr. (2005), A review of ethical enforcement proceedings against competent and qualified persons, *The Professional Geologist*, September/October, 35–38.

Abbott, D.M., Jr., O. Bonham, D. Larkin, and J.B. Gustavson (2008), Review of international disciplinary proceedings (abs), in *Proceedings of the AIPG/AHS/3rd IPGC Symposium, Flagstaff, AZ*, V. McLemore (ed.), 339–340.

Agricola, G. (1556), *De Re Metallica*, trans. H.C. Hoover, L.H. Hoover, 1909, Dover Publications, 1910, 1950 ed.

American Association of Petroleum Geologists (AAPG) (1920), Proceeding of the Fifth Annual Meeting, AAPG Bulletin, 4, 317–323.

American Association of Petroleum Geologists (AAPG) (1924), AAPG Code of Ethics, March 26, 1924.

American Geological Institute (AGI) (1999), Guidelines for ethical professional conduct, *Geotimes*, August, 13.

American Geosciences Institute (AGI) (2015), Guidelines for ethical professional conduct, http://www.americangeosciences.org/community/agi-guidelines-ethical-professional-conduct.

American Institute of Professional Geologists (AIPG) (2015), Disciplinary Procedures: http://www.aipg.org/Membership/Policies.htm#Disciplinary%20Procedures.

Canadian National Instrument 43-101 (2011), National Instrument 43-101 Standards of disclosure for mineral projects, http://www.osc.gov.on.ca/en/SecuritiesLaw_ni_20110624_43-101_mineral-projects.htm. Accessed 24 March 2015; previous versions in 1999 and 2005.

Davidson, D.M. (1957), Professional ethics and conduct for economic geologists, Economic Geology, 52, 312–316.

Joint Ore Reserves Committee (2012), JORC Code [various editions beginning in 1989; the most recent is 2012], Australasian Code for Reporting of Exploration Results, Mineral Resources and Ore Reserves: Joint Ore Reserves Committee of the Australasian Institute of Mining and Metallurgy, Australian Institute of Geoscientists, and Minerals Council of Australia, http://www.jorc.org/docs/JORC_code_2012.pdf, 44 p. Accessed 24 March 2015.

Melrose, T.G., D.M. Abbott Jr., H.M. Parker, and J.G. Price (1999), Considerations and consequences of adopting an enforced code of ethics (abs), Geological Society of America Abstracts with Programs, 31(7), A-136.

Melrose, T.G., D.M. Abbott Jr., H.M. Parker, and J.G. Price (2000), Considerations and consequences of adopting an enforced code of ethics, European Geologist, 10, 85–87.

Mining and Metallurgical Society of America (2005) Qualified Person category, http://www.mmsa.net/pdfs/MMSAHistFINAL.pdf, p. 35. Accessed 23 May 2015.

Mogk, D., and E. Nickless (2014), Development of the American Geosciences Institute common code of ethics for the geosciences, Presentation to the AGI Member Societies at the GSA Annual Meeting, Vancouver, BC, October 20.

Powers, Sidney (1929), The history of the American Association of Petroleum Geologists, AAPG Bulletin, 13, 153–170.

SEG Committee on Ethics and Conduct (1955), Report of the Committee on Ethics and Conduct, Economic Geology, 50, 549–552.

Stow, S., D. Abbott, V. Cronin, D. Dunn, R. Hatcher, T. Holzer, and T. Melrose (1999), Guidelines for ethical professional conduct issued by the American Geological Institute (abs), Geological Society of America Abstracts with Programs, 31(7), A-136.

Section III: Scientific Integrity and Ethics in Publications and Data

8

8. THE NEW LANDSCAPE OF ETHICS AND INTEGRITY IN SCHOLARLY PUBLISHING

Brooks Hanson

Abstract

Scholarly peer-reviewed publications serve five major functions: They (i) have served as the primary, useful archive of scientific progress for hundreds of years; (ii) have been one principal way that scientists, and more recently departments and institutions, are evaluated; (iii) trigger and are the source of much communication about science to the public; (iv) have been primary revenue sources for scientific societies and companies; and (v) more recently play a critical and codified role in legal, regulatory, and key societal decisions. Recent dynamics in science as well as in society, including the growth of online communication and new revenue sources, are influencing and altering particularly the first four core functions greatly. The changes in turn are posing important new challenges to the ethics and integrity of scholarly publishing and thus science in ways that are not widely or fully appreciated. Solutions, including for meeting the growing societal uses, pose new responsibilities for scientists, publishers, and scientific societies.

8.1. Introduction

Scholarly peer-reviewed publications serve five major functions, four of which have been core for some time and one of which developed and expanded in the last decades of the 20th century: (i) First, publications serve as the primary useful, citable archive of scientific progress, hypotheses, experimentation, results, and knowledge.

American Geophysical Union, Washington, D.C., USA

Scientific Integrity and Ethics in the Geosciences, Special Publications 73,
First Edition. Edited by Linda C. Gundersen.

(ii) Second, publications and their impact, variously measured, are among the principal ways that scientists are evaluated and recognized for career advancement. More recently, departments, institutions, and universities have also been formally evaluated based on the collective research output of their faculty, in large part measured by their publications. (iii) Third, research that has been peer reviewed and vetted by journals is a main source for much of the daily coverage of scientific results in the mass media and thus for public understanding of science. Journalists have come to rely on a journal's review, selection process, and brand when determining which results to communicate to the public. (iv) Fourth, publishing is the primary revenue source for many scientific societies. Scholarly publication is big business, and it is responsible for about $10 billion in revenues annually to publishers representing both scientific societies and companies [*Ware and Mabe*, 2012].

The important fifth core function is the codified use of the peer-reviewed literature in the legal and regulatory systems, and in groups advising governments on scientific matters, such as the Intergovernmental Panel on Climate Change (IPCC), a fact that underscores the importance of scientific integrity as a part of the implicit contract between the scientific community and society. Key steps leading to this function in the United States included the 1993 U.S. Supreme Court decision *Daubert v. Merrell Dow Pharmaceuticals, Inc.*, the Shelby Amendment of 1999, and a few years later, the Data Quality Act.

In *Daubert v. Merrell Dow* [1993], the majority set out five criteria for the evaluation of expert testimony by judges. One was "whether the theory or technique has been subjected to peer review and publication." This case has had a large influence in subsequent decisions. For example, a critical part of the original court decision overturning Proposition 8 (a California ballot measure overturning same-sex marriage), and thus, subsequent support of that decision, was the inclusion of peer-reviewed social science research supporting the plaintiffs, and rejection by the judge of studies submitted by the defendants that were not peer-reviewed and thus deemed unreliable or unscientific. Other recent cases, such as *Miller v. Alabama*, overturning the life incarceration of minors, were based heavily on peer-reviewed research in neuroscience and psychology. Other examples abound.

The Shelby Amendment and its interpretation by the Office of Management and Budget, as well as subsequent legislation, elevated the use of the peer-reviewed literature in justifying regulations. Despite fears at the time by leading scientists that this would hinder regulations, several recent verdicts have concluded that the overall effect has been neutral or positive [*Fisher*, 2013], and regulations, such as the recent EPA proposals for regulating atmospheric CO_2, are now heavily grounded in peer-reviewed publications.

With regards to the IPCC, its Principles and Procedures [2015] mandates use of the peer-reviewed literature, and it assigns deadlines for submissions to journals for inclusion in reports to allow enough time for publication and some vetting. Indeed, the IPCC was criticized heavily for basing one inference on non-peer-reviewed content [*Pearce*, 2010].

Through these and other developments, society has mandated that peer-reviewed research is an important and necessary resource for critical decision making, and has codified that it be weighted heavily relative to other resources. Many scientists are unaware of the importance society has assigned to peer-reviewed literature and don't appreciate the role it plays in guiding judicial and governmental processes. Because society is relying on the integrity of scholarly publications, securing that integrity is critical.

Beginning in the 1980s, major changes were introduced in the creation and dissemination of scholarly works that have impacted the aforementioned core functions ever since. Typewriters, manually drawn figures, and distribution by first class and express mail have been replaced by personal computers, digitally produced art, and electronic distribution through the Internet. The introduction of new technologies has greatly decreased the time, effort, and cost of producing and submitting a paper to a journal. Computers and advances in instruments have led to increases in the volume of data that can be collected and processed. Large global collaborations have become easier to accomplish; group authorship and team science have become more prevalent. Along with the advances in technology, competition for research funds has increased. Societies and commercial publishers have introduced new journals to publish the increasing research output, and to increase revenues.

No longer constrained by the need for ink and paper and curation by libraries, peer-reviewed research has become available to a much larger audience. With supplemental material, it now includes more background details and data. Discoverability has increased as more content has become digitized and available through scientific search engines and databases like PubMed and Web of Science. Google, Google Scholar, and other search engines have extended this reach beyond the scientific community to the general public. Postpublication scrutiny and commentary are no longer confined to subject-matter experts but are open to everyone.

With this increased access has come the ability for more people to question results and uncover problems in papers, ranging from sloppy lab practices to deliberate misconduct. Cases have come to light involving data and image manipulation, biased data presentation, plagiarism, lack of reproducibility, and more [e.g., see *Marcus and Oransky*, 2014]. In response, scientific societies, journal editors, and professional associations have expanded policies and systems to reinforce ethical practices and asked authors and funders to declare real or perceived conflicts of interest.

Discussions of ethics and integrity of scholarly publishing have usually focused on the various roles involved in writing, presenting, reviewing, editing, and publishing research. Problems addressed included authorship standards, personal and financial conflict of interest for authors and reviewers, and proper reviewing and editing. Current best practices are collected in the resources section of the AGU ethics site (http://ethics.agu.org/resources/); the Council of Science Editors' *White paper on publication ethics* [2012]; the National Academies', *On*

Being a Scientist: A Guide to Responsible Conduct in Research; and on the website of the Committee on Publication Ethics (COPE), among others.

While these established policies and guides continue to be relevant and important, they are not sufficient in light of changes affecting the core functions of scholarly publications. The focus of the remainder of this chapter is on the rapidly changing landscape for scholarly publishing and the emergence of new ethical responsibilities for scientists, editors, journals, and publishers. Although the focus will be on the Earth and space sciences, the issues apply across all scientific disciplines.

Changes involving each of the first four functions will be discussed in separate sections, but the expansion of the fifth function in particular places great importance on ensuring the integrity of research and how that integrity is conveyed. Arguments that peer-review is no longer useful or necessary are belied by the growing dependence of society on the peer-reviewed scholarly literature, and are reckless in this light. Because support for science stems from society, and society is increasingly depending on published science, the peer-review and publication process must be strengthened. Addressing the new emerging ethical and integrity issues is essential in that process.

Finally, discussions of ethics and integrity must be framed by the primary scientific goal of scholarly publishing of advancing science and our understanding of nature. Publications serve both to present tangible new results or ideas and to form a foundation for additional work. Hypotheses, speculation, inference, and tentative and even incorrect conclusions all play important roles in the process of science, as do comments and criticisms of prior results. Guided by peer review, editors of journals make publication decisions based on the level of contribution a particular article makes to these larger goals. They do not ensure that all claims are or will be irrefutable. Certain levels of confidence may be desirable but are not absolute. Ethics and integrity foster this primary scientific goal and provide authority for applying the integrated understanding that results. Many attacks on integrity in publishing derive from incorrect assumptions regarding this goal and the support for it provided by the peer-review process. Indeed, the occasional poorly reviewed or fraudulent paper is increasingly less important to the overall integrity of publishing. A greater concern is the ethical responsibilities that support the institution of scholarly publishing in this new environment. These need much more attention than they are receiving.

8.2. Integrity in Publications in Their Role as a Useful Archive

The expansion of electronic publishing has raised a number of new challenges for publishers, and the scientific community more broadly, with respect to their responsibility for curating scientific knowledge and even preserving the basic integrity of a manuscript. Some of these include determining or identifying the version of record, documenting errata and, more critically, retractions across multiple versions, not all of which are housed by a publisher; providing an audit trail; converting content to support evolving platforms and applications; and providing for a persistent archive to replace distributed print copies. Several organizations, some formed by publishers, have emerged to guide standards and best practices and to organize, curate, and identify content, such as NISO, CrossRef, Portico, LOCKSS, ORCID, and many more. While helpful, the need for these allied organizations illustrates the increasing complexity and challenges associated with maintaining integrity in today's scholarly publishing environment and the expanded responsibilities that publishers are acquiring. These responsibilities have also been generally poorly assumed by government agencies or institutional repositories that are also now typically hosting the accepted versions of papers. For example, retractions, corrections, and linked data sets are often missing from even the best-established repositories [*Davis*, 2012b].

Amid this increasingly complicated environment for traditional publishing, a new emerging trend is raising a fundamental challenge to integrity: the rise of data. The growing ability to collect, process, and analyze data is enabling much new science, but it is also altering scholarly work in a basic way. Publications in journals, while representing larger studies, contain relatively less of the overall science. They increasingly include pointers to other resources, such as data sets, methods, and code that are a growing foundation for advancing further science and for fostering the integrity of a result. These resources must therefore be preserved, discoverable, and usable, and the pointers must be persistent.

In the era of print journals, data for the most part were either included in papers or were absent; statements such as "data not shown" were common for data or observations that were not the primary reported results. In many fields, much of the relevant analytical results could be presented, and communities had expectations about which data merited inclusion. As more and more primary data were collected than could be included in a print format, and as the value of such data were appreciated (or the lack of availability of data in publicly relevant research criticized), research facilities, journals, and publishers developed a number of solutions to foster integrity and enhance reuse. Data inclusion and availability statements were strengthened, and "data not shown" statements and "unpublished" or "in press" references were banned to allow publications to be complete when first available, not at some unspecified future time or never [e.g., *Hanson et al.*, 2011]. Most leading journals now require authors to agree as a condition of submission that underlying data will be included or provided. Indeed, publication is the key leverage point in exposing much scholarly data. To facilitate availability and transparency, many journals allow and promote supplements to papers, encourage use of reliable repositories, and/or require authors to agree to host the data. All of these are proving problematic [*Vine et al.*, 2014].

Supplements have not all been reliably preserved by publishers over time (early versions were on microfiche or ftp servers that have not been maintained), but this situation is improving, and publishers recently helped develop standards for supplemental material [*NISO*, 2013]. Still, most supplements lack any metadata, are in a PDF format, are not easily discoverable, and provide no functionality. These problems inhibit broader use and meta-analyses. Many publishers limit their size, both to control their storage costs and to limit the burden on reviewers. While these supplements are provided to reviewers in most cases, there is broad concern that they are often not evaluated as part of the review or critically. These problems led the Society of Neuroscience to ban supplements from its journals [*Maunsell*, 2010] and require authors to host supporting data.

Authors, labs, and institutions, however, have a spotty record in hosting data related to publications. Several recent surveys [e.g., *Ferguson*, 2014; *Tenopir et al.*, 2011] show that most scientists are storing data on their own computers and servers; formal institutional repositories are in many cases just being developed. One recent study found that after just a few years, 20% of links to websites in publications are broken [*Klein et al.*, 2014].

Scientific communities are thus increasingly looking to data repositories as a preferred solution. These can provide permanent homes, enhance or in some cases ensure quality, and provide functionality. Repositories such as Genbank, the Protein Data Base, and in the Earth sciences, IRIS, have proven particularly valuable. While many such repositories have been started across scientific disciplines, too many have not shown long-term stability. For example, in 2005, a survey in *Nature* found that nearly half of 89 databases had no funding model and 7 had ceased operation [*Merali and Giles*, 2005], and funding challenges remain at the forefront today [e.g., see *ICPSR*, 2013]. Publishers have thus been shy in promoting repositories that do not have stable funding or are not supported by major institutions, a large chicken-and-egg problem. Institutions, in several cases led by their librarians, are starting to develop repositories, and these will be important where there are not established domain repositories.

Societies and publishers have been important in legitimizing use of repositories and in changing the culture toward increased availability and sharing of data. For example, a joint commitment by *Science*, *Nature*, and other journals led to the requirement that protein crystal structures be deposited in the Protein Data Bank at publication, eliminating a delay of 1 year in practice at the time [*Bloom*, 1988]. Publishers have worked with repositories such as the Protein Data Bank to allow confidential access to data during peer review. Societies and associated publishers in ecology and evolution similarly helped in the development of the Dryad digital repository and common data availability standards in 2011 (http://datadryad.org/pages/jdap). A coalition of publishers and data repositories in the Earth and space sciences has recently entered a statement of commitment to promote data standards, best practices, and especially availability in domain repositories [see http://COPDESS.org and *Hanson et al.*, 2015] and have developed with the Center for Open Science overall guidelines for transparency, openness, and reproducibility in publishing [see https://cos.io/top/ and *Nosek et al.*, 2015; *Mcnutt et al.*, 2016].

Another important challenge is that in many subject areas, more data are being collected than can be stored or transmitted. Domain data repositories, working in collaboration with societies, publishers, funders, and researchers, can lead in defining standards and priorities for retention to foster integrity and promote further research. Involving the repositories can enhance data quality, ease the burden on reviewing, and develop functionality that will facilitate new types of review and linkages between the underlying data and their representation in a publication. The best domain repositories include some quality control and can mandate standardization.

In biology and biochemistry, samples such as mice, bacterial strains, and reagents have been required to be shared after publication, again to foster integrity and advance further work. This has not been the culture or requirements in much of the Earth sciences. A notable exception is rare or valuable samples such as fossils and planetary materials including meteorites. Publishers and societies have required that these must be curated as a public trust (e.g., http://vertpaleo. org/Membership/Member-Ethics/Member-Bylaw-on-Ethics-Statement.aspx). In biology, community standards have emerged around acceptable "materials transfer agreements" for samples and originally proprietary data that specifically allow unfettered reuse in the scholarly literature (http://www.autm.net/ Technology_Transfer_Resources/8395.htm; *Bubela et al.*, 2015). In the geosciences, a standard sample identification and registry is emerging (IGSNs, or international geo sample number: http://www.geosamples.org/igsnabout), and publishers and repositories are aligning around requiring these [see *McNutt et al.*, 2016]. Although several major Earth and space science publishers require code related to important data processing and generating results to be available, others do not, although standards are emerging [*Stodden et al.*, 2016]. Privacy concerns, common in medical and social science research, are just entering Earth science research as geospatial, land-use, and geo-health research expands.

8.3. Integrity in Using Publications for Evaluations

Changes in how science is being produced and consumed are altering how publications are and should be used for evaluating scientists, institutions, journals, and publishers. Because these evaluations in turn have large financial implications, there are strong conflicts of interest that are endangering integrity throughout scholarly publishing.

8.3.1. Authorship and Fraud

One trend is the growth in the number of authors on publications, often from multiple institutions. The average number of authors per publication is now about six [*Ware and Mabe*, 2012], but some of the most important publications describing major results may have hundreds or even several thousand authors [*Castelvecchi*, 2015]. This is complicating credit and responsibility. In the past, the norm was that all authors were jointly responsible for the work in a paper, but this is no longer a common expectation. This growth in authorship is proving particularly problematic for promoting students and early career scientists, who have been traditionally evaluated on the basis of their primary publications. More authors dilute an individual's contribution, especially if some of the authors are honorary ones who do not meet the normal authorship requirements [*Zen*, 1988]. A larger question is: How does the scientific community steer top students to the most important challenges, which often requires larger collaborations, while rewarding their work and assuring a career path despite fewer first-authored papers? Also, compared with the presentation of research results, the generation of important data and methods has been undervalued as a scientific contribution.

Some journals, particularly in the biological sciences, are including statements describing the contributions of each author, and guidelines to organize these descriptions are emerging [*Brand et al.*, 2015; *McNutt et al.*, 2017]. This is a partial solution to recognizing work, and is also an attempt by journals to help users identify researchers to contact when there are questions, but it is also a clear indication that the norm of shared responsibility has waned. Allowing authors to claim responsibility for only part of a paper rather than the whole allows some authors to be lazy about evaluating the work they are joining as co-authors. For example, it is not unheard of for an editor to see two or more submissions with disparate conclusions but that include some overlapping authors. This is both a form of an unethical joint submission and an authorship ethics issue: authors have a responsibility to share knowledge of their other submitted work with all of their other authors and to declare fully their work on any submission.

The pressure for publications and credit, which include financial awards to scientists in many countries, has driven a growth in other fraudulent practices by authors, including wholesale plagiarism, preferential self-citation, and attempts to subvert the peer-review system. In an egregious case, more than 150 papers were retracted from a number of journals when it was revealed that authors and submission agencies had created fictitious reviewer accounts in editorial databases, in some cases developing profiles years in advance [*Ferguson et al.*, 2014]. Many publishers are now using plagiarism detection software and are requiring more complex validation of user accounts.

8.3.2. Shifting Metrics

Many of these authorship problems stem in part from the way research impact is measured and how those measurements affect scientific careers. In turn, these problems feed into the evaluation of journals, which becomes a factor in their integrity. A large corrupting factor has been the use of the Journal Impact Factor (JIF) or other related metrics as part of the evaluation of journals. The JIF is a metric reported originally by Thomson Reuters (now Clarivate Analytics) based on the citations to publications in a journal over 2 years, divided by the number of research publications in the following year (there are several other related indices). It represents an average of a highly skewed distribution, and the errors and lack of transparency in producing the index have been well described [*Arnold and Fowler*, 2011]. Worse, it is regularly misused in evaluations of scientists and institutions (for which the average has little predictive value), as highlighted by a coalition of publishers, scientists, and organizations (http://www.ascb.org/dora/).

A separate and larger problem is that it induces publishers to act in ways that are not in the best interests of science and integrity. The JIF is promoted widely by journals as one measure indicating their relative value to librarians, advertisers, and authors [*Anderson*, 2014a]. These stakeholders provide major sources of revenue and prestige to journals and publishers, so obtaining a higher JIF has strong rewards. Numerous cases have emerged of journals actively working to manipulate their JIF, while promoting it widely as if it were a fair measure. In essence, publishers, who have a leadership role in promoting standards of research conduct, are engaging in what they would consider to be fraud if it were to occur in a research report in their journals, essentially manipulating data to get a desired result. One approach, documented in an exposé in the *Chronicle of Higher Education* [*Monastarsky*, 2005] and later elsewhere [*Wilhite and Fong*, 2012], is direct or indirect pressure by editors on authors to pad citations to their own journal. Partly to account for this, Thomson Reuters developed a self-citation index or in extreme cases will even exclude journals from a rating. In response, a group of editors formed a secret cartel to promote citations among a small set of journals [*Davis*, 2012a] at the exclusion of their competitors.

Commentary or editorial pieces in a journal can provide citations, which is beneficial especially if they are not counted as research pieces (increasing the size of the denominator) in the JIF. Thomson Reuters uses the size of piece and number of references in their classification of whether an article is a commentary or research paper; and publications may be manipulating their articles to take advantage, allowing the JIF to drive scholarship [see also *PLOS Medicine Editors*, 2006].

Similarly, journals can manipulate their JIF by including more of the types of publications, such as reviews and methodology papers, that tend to be highly cited. Most egregiously, they may favor content from highly cited fields or subfields. For example, in 2010 a *Nature* editor indicated in a rejection letter that the journal was preferentially rejecting space physics papers because the field was not highly cited. (At about the same time, *Nature* began publishing proportionally more biology papers, which tend to be highly cited.) This led to a Facebook petition by that community in which researchers indicated that they would not submit to *Nature*, perhaps not the most effective boycott since the journal indicated it did not want many papers in this field. (The Facebook page is no longer available, but see *Socas-Navarro* [2010] and *Solanki* [2010]. A recorded presentation by a *Nature* editor in 2010 also refers to this in the question and answer session (minutes 50 to about 65) here: http://www.iac.es/info.php?op1=23&op2=119&op3=41&id=250.) This was resolved when two researchers recalculated the citation rate of space physics papers and convinced *Nature* that their original assessment was in error [*Solanki*, 2010]. Coincidentally, *Nature* released an editorial stating that it does not take actions to manipulate the JIF [*Nature*, 2010].

Another way to elevate a JIF is for a journal to hold papers that are likely to be highly cited until the beginning of a year, because papers early in the year have more time to rack up citations during that year and thus provide much greater weight for the final JIF. More broadly, journals may publish more papers in early months and fewer later in the year. That this is even being discussed as an acceptable practice (and it is being discussed in editorial meetings [*Davis*, 2015]), let alone implemented, simply illustrates the corruption and perversion by the JIF of the normal responsibility of publishers toward their primary goal of advancing science. Even at best, Clarivate, journals, and publishers promote and use it without expressing its uncertainties and with multiple significant digits, knowing that others such as libraries and advertisers may not understand the system's biases [*Clark and Hanson*, 2017; *Hanson*, 2017]. Institutions, funders, tenure committees, and more are also significantly culpable in this harm.

In sum, the JIF is a prime example of how metrics can drive decisions in scholarly publishing that are not in the interest of integrity, science, and scholarship, and worse, how their willful manipulation at a variety of levels risks eroding integrity in exactly those institutions that are needed for leadership in fostering integrity. Even if Clarivate no longer promoted the JIF, additional metrics are emerging and would continue to emerge, such as social media attention, and these can also be manipulated by authors and journals, and already are [*Bornmann*, 2014]. For example, services are available that enhance social media mentions for a fee. Journals are promoting other services such as Loop and Kudos to researchers aimed at enhancing visibility of research papers and raising "impact" measurements.

Societies should insist on transparency in their editorial processes and must provide appropriate oversight of editorial practices and decisions, cognizant of the conflicts of interest. A simple approach that would help would be to indicate metrics on publications, perhaps listing by month acceptances and times from acceptance to publication, and to provide common clarity in author instructions on scholarship in citations and their distribution of papers. Better would be the development of common standards (say by the National Information Standards Organization) on guidelines for the use and reporting of metrics, and measures to ensure or enhance integrity and transparency.

8.4. Integrity in Communicating Science

Recent changes in science journalism and the coincidental growth of social media are adding new ethical responsibilities to scientists, societies, and publishers with respect to how published research is communicated to the public.

In the United States in particular, many traditional media outlets, including most major newspapers, have greatly reduced their science reporting staff, and in too many cases have eliminated science desks entirely [e.g., *Galperin*, 2013; *Zara*, 2013]. At the same time, universities and institutions have expanded their public information staffs and efforts [*Rosenberg*, 2016].

The scientific community has long relied on science journalists as key partners in presenting research results to the public. Indeed, many societies recognize this added value with journalism and public service awards. The best stories provide context, criticisms, and discussions of what future work is needed, and provide a broad perspective on a research field. They also can and often do temper hype, either by not reporting outlandish conclusions at all or providing alternative perspectives for context.

Reflecting the shift from news staff to press officers, many news stories in media outlets today, particularly in smaller markets, are often near-verbatim replications of press releases masquerading as news. Press releases are now created with quotes from researchers, including supposedly "independent" comments, video feeds, and images to serve this purpose—a process called "churnalism" [*Russell*, 2008].

There are several consequences of this trend. First, independent news reporting, particularly around new results, is waning. Researchers, and the public, need to appreciate that the source of a story may be the researchers' institution, funders, or publisher, with minimal additional reporting or with intentional selection of sources to favor the institution or publisher. Furthermore, press information officers work for their home institutions, not researchers, so their primary goal may be drawing attention and donations to the institution, rather than making sure a particular scientist's research is conveyed correctly. This can produce conflicts in messaging that are magnified in the churnalism process. More broadly, these changes mean that responsibility of providing appropriate context, with its embedded conflicts of interest, is de facto shifted to societies, institutions, funding agencies, researchers, and their press officers, and too often they are failing. As an illustration, a recent study found that most hype in public reports of recent biomedical results was introduced in the press releases themselves, not in subsequent reporting [*Sumner et al.*, 2014]. In a related development, the misuse of hype adverbs in scientific abstracts has increased dramatically [*Vinkers et al.*, 2015].

Top journalists today actively promote their stories on a variety of social media platforms, a shift in acceptable mores from even a few years ago. This is increasingly true also of researchers and publishers, and is a way to raise "Altmetric" scores. Altmetric, a project of the Digital Science spinoff from Macmillan, is one of several services that provide tracking of a paper's mentions in social media, press accounts, and other channels, as well as selected usage data, in an attempt to provide an indication of the broader attention received by a publication. As noted above, journals and publishers are now providing services that facilitate these enhancements. Altmetric indices also link to that enrichment and the web, including many blogs or posts from researchers as well as the public. This linking promotes discovery of more commentary and analysis of many publications, but also potentially more misinformation. At the same time, for those willing to identify authoritative sources, the discussion can be much richer and more significant than before.

Societies and publishers have a difficult role to play, but their growing role in outreach provides important ways to strengthen scientific integrity. They are increasingly providing both press releases and hosting enrichment in social media. At the least, they should be cognizant of their new responsibilities, and conflicts of interest, in promoting science as well as a specific result. Going forward, despite the abundance of information available about scientific topics and results, authoritative commentary can in many ways be more valuable and is needed. Societies in particular have large untapped resources of scientists, including leaders and fellows, and should consider how to better inform the public and policy makers given these changes. There is an opportunity to highlight the integrated knowledge that is the ultimate goal of scientific publishing.

8.5. Integrity, Revenue, and New Financial Entanglements

A variety of new business models and new or expanding revenue sources, in large part associated with Web-based publishing, are posing new conflicts of interest and ethical challenges to publishers. The largest change in business models is the growth of open-access journals and publications, but other models have also emerged or expanded. The erosion or stagnation of traditional subscription-based revenue has increased the appetite of publishers to find new revenue streams, including exploring sources they might previously have avoided. High-quality scholarly publishing has real costs, and as the number and complexity of papers keep expanding, publishers are seeking new revenue both to maintain operations as well as develop new functionality and infrastructure [*Anderson*, 2014b]. Advertising, promotions, and sponsor-driven content have thus expanded and will continue to do so. As discussed above, growing financial conflicts can erode integrity. These threats are large because the revenues are significant and financial conflicts are understood, correctly, as a direct challenge to integrity.

8.5.1. Open Access

Traditionally the revenue for most journals was from personal and institutional (e.g., library) subscriptions and perhaps author fees, or in some cases provided by support from a parent organization or funder or as part of membership. The cost of print (paper and ink) was such that it placed at least broad limits on the number of publications a journal could accept, or pages that it could print, and the editorial costs also scaled. The advent of open-access publishing has changed this equation. In this model, authors pay for publication, either to an open-access journal or to allow their paper to be open in an otherwise subscription journal. If these fees exceed a journal's expenses, the more papers a journal accepts, the larger the revenue. This creates an obvious editorial conflict of interest that contrasts with longstanding ethical guidelines in news journalism, which mandates that all paid content be labeled clearly as advertising.

Open-access journals have recognized this conflict to some degree, but it is difficult to sweep away. One approach is to withhold payment information from editors who are deciding whether to accept a paper (authors can request a hardship waiver in most journals). Author information in *PLOS ONE* and *PLOS Medicine* states that "editors are unaware of whether an author can pay" and "editors are not paid on commission." However, assuming that this actually addresses the conflict is naive (or worse, it implies that their readers are). Editors, even if not told by management, are certainly savvy enough to discern which submissions, which list funding information and author affiliations among other clues, are most likely to generate revenue. Even editors understand that overall revenue affects the economic viability of a journal and their own compensation, whether they are paid on commission or not.

The financial return on the open-access business model has led to the rapid growth of what is known as "predatory open-access publications." These prey on unsuspecting authors and will accept nearly any paper, including computer-generated nonsense papers, as has been exposed by several sting operations. In 2016, there were more than 900 such publishers in a widely cited index [*Beall*, 2017; *Hanson and Lunn*, 2017], and the number has increased by several hundred each year. Here, the open-access business model is allowing unscrupulous publishers to damage the integrity of the entire scholarly publishing enterprise. A further illustration of this conflict is the recent firing of editorial staff from a major open-access publisher when the editors appropriately demanded independence from financial pressure [*Enserink*, 2015]. Given these concerns, it is not surprising that researchers state that their main concern about submitting to open-access journals is that they are perceived as being of low quality [see *Nature's* Author Insights Survey, 2014].

Many subscription journals now allow authors to pay for open access to their publication. This adds revenue to that generated from subscriptions and is known as "double dipping" because the content is being charged for twice, once from subscribers and once from the authors. From a library or institution's perspective, the value of the subscription is essentially reduced because the articles it paid to access are available to anyone for free. To address this, some journals are reducing the price of subscriptions in proportion to the amount of author-selected open-access publications. Ideally, this payment and decision should be handled postacceptance to avoid editorial conflicts.

It is clear that open-access journals and publications will continue to grow, and these steps and others that journals and societies have taken have greatly expanded access to scholarly content. The original hope by some was that this model along with online publishing could reduce the overall cost of scholarly publishing and replace the subscription-based journal model. This has not been the case, at least yet. Most new legitimate open-access journals have been started by or in cooperation with major commercial and society publishers [*Esposito*, 2014], and gross revenue associated with scholarly publishing has continued to increase [*Auclair*, 2015].

Societies and publishers, as well as scientists, have a critical role in addressing conflicts and strengthening integrity, but they have generally not yet recognized or adequately dealt with the threats. Several steps would improve integrity in all business models. One is transparency in finances and clear statements on the separation between business and editorial decisions. A second is oversight to that separation through a publications or other committee charged with ensuring integrity. A more extreme step might be to limit the number of accepted papers annually in a transparent revenue model. Another model that helps in this case is an open peer-review model, as practiced by some societies.

8.5.2. Advertising and More

An additional expanding revenue source for some journals is online advertising and sponsorship. This will likely continue to grow as advertisers value the brand association afforded by respected journals, societies, and publishers, and in turn publishers appreciate the revenue. The challenge of online advertising is that specific ads can be linked to related content, yet this also provides the largest taint to integrity. Publishers engaging in online advertising should develop transparent guidelines for associating ads with content or web pages, and also for separating larger content decisions from advertising.

Sponsored content has also expanded recently, notably in some of the leading journals, in which an advertiser will pay a publisher to produce content under or associated with its brand, or sponsor collections of content. *Nature*, for example, has long had sponsors for its Outlook and Insight collections, which are reviews, news, and other content around specific topics of close interest to the sponsor, handled and selected by *Nature's* editors. *Science's* business office has recently been producing similar sponsored collections (in which the editors are employed by the business office, not the editorial office). This practice can potentially confuse authors who contribute to these sections, who may expect a formal publication, with a DOI and other enrichment, rather than an ad or advertorial, or not realize that the branded sponsored content is not actually part of the primary journal (and readers too). The advantage for advertisers is close association with a name brand; however, this in turn risks the integrity of the brand. *Science's* recent sponsored supplements, for example, indicate that they should be cited as from "*Science*" (and content appears as from *Science* on Google Scholar searches) even though the articles are stated as being handled editorially separately from the manuscript editors. (Two recent supplements, which were included in the print edition, were on Chinese traditional medicine, a topic rather unlikely to merit a special issue in the main journal. For criticism, see http://scienceblogs.com/insolence/2015/01/05/science-and-the-aaas-sell-their-souls-to-promote-pseudo science-in-medicine/.) Furthermore, the publisher is indicating that these are pub-lications (in format, supporting a scholarly-like citation, and more) as part of their "brand," which enhances their revenue value compared to a straight adver-tisement. Publishers of these supplements meanwhile do not do many of the now-standard curation activities of responsible publishers, such as ensuring widespread indexing, XML creation, archiving, versioning, etc.

A particularly egregious example of sponsorship was exposed several years ago via disclosure in a legal case in Australia, showing that Merck had sponsored several fake journals with Elsevier [*Grant*, 2009]. These fake journals included faux editorial boards, focused on content promoting Merck's drugs, and were dis-tributed to medical practitioners. In a different case, but also illustrative of the way publishers are exploring new revenue, the journal *Scientific Reports* tried to induce surcharges on authors for expedited handling of papers. This failed when their editors, and later the scholarly community, rejected this form of bribery as unseemly, and one resigned [*Bohannon*, 2015].

These examples illustrate the way that publishers have expanded their search and appetite for new revenue as traditional sources have waned, and that they need to address further at least the appearances of conflicts of interest (in con-trast to how aggressively they have asked authors for conflict declarations). While much of the focus has been on the open-access business model, the recurring embarrassing revelations with these other sources indicate a growing threat to integrity that the scholarly community as a whole should be aware of. Newspapers and magazines have developed guidelines around advertising to address these conflicts; at a minimum, scholarly publishers should meet or exceed these stan-dards (http://www.magazine.org/asme/editorial-guidelines).

8.6. Societies and Publishers Can Foster Integrity

Publishers grew familiar with the landscape of scholarly publishing in print journals that persisted for much of the past century. They were responsible in many ways for leading the community in developing guidelines to foster integrity in publications and defining the ethics and mores of authorship, editing, review-ing, citations, and submission, in part through organizations such as the Council for Science Editors, the Society of Scholarly Publishing, and the Council on Publication Ethics. More recently journals have expanded to lead in addressing financial and other conflicts of interest of authors and reviewers (usually through detailed COI forms). However, societies and publishers for the most part have not kept up with how recent dynamics in key aspects of scholarly publishing are raising new ethical and integrity challenges, or recognizing and clarifying their own conflicts of interest in the new publishing environment. They must catch up.

A common theme in the above analysis, and an important part of the solution to the problems discussed, is the need for transparency, including in addressing new or expanding financial conflicts. Robustly connecting publications to associated content and materials involves everyone from individual scientists to societies and publishers. This requires changes in expectations, infrastructure and culture. Trusted repositories in particular need financial support, which has been difficult. New levels of transparency in editorial practices and decisions at the portfolio level can help address financial conflicts, which are moving from individual scientists and reviewers to journals. Publishers collectively are expanding efforts to address these concerns and develop guidelines, but more work is needed, including in practices (see *Nosek et al.* [2014] and the COPE Principles of Transparency and Best Practices in Scholarly Publishing [2015]).

More importantly, progress on the larger issues requires publishers and societies to lead collaboratively and engage other stakeholders, such as funders and data repositories. They did so in developing and promoting individual standards in publishing, and some have worked to develop standards around open data, and in developing and supporting CrossRef and other organizations. Collaboration is particularly challenging when journals are competing against one another for prestige and revenue. This competition poses challenges for maintaining integrity, for example, in advertising and use or misuse of metrics, but collectively the community can have a powerful voice for raising and promoting integrity. The clear common interest and motivation for collaboration is the enhanced integrity in scholarly publishing that can result. Oversight groups should in turn step up in obtaining data on publishing. (For example, comparing papers published by month by journals might reduce some JIF manipulation or allow transparent adjustments.)

A larger cultural change is also needed in how scholarly content is used in recognizing and evaluating scientists and their institutions. Some of this change requires efforts by institutions and funders. However, societies can better recognize the efforts of those members who foster integrity in publishing by sharing their data or creating new datasets. Contributions of datasets, codes, and methodology, and active sharing of these, could for example be specifically recognized criteria for awards or fellowship or other awards. By requiring formal citations to data sets in references, and including these in the main papers, not supplements, publishers can work to include these assets in the citation records.

The scholarly publishing community has faced substantial challenges before and risen to them. Recognizing these new dynamics, and the growing importance and use of peer-reviewed research for society, are key next steps toward enhancing and ensuring integrity.

Acknowledgments

I thank Monica Bradford for providing extensive comments and suggestions on the introduction in particular, Stewart Wills for extensive input throughout, input from two reviewers, and input and editing from Laura Helmuth.

References

Anderson, K. (2014a), Exhibition prohibition—Why shouldn't publishers celebrate an improved impact factor? *Scholarly Kitchen*, 11 September 2014, http://scholarlykitchen.sspnet.org/2014/09/11/exhibition-prohibition-why-shouldnt-publishers-celebrate-an-improved-impact-factor/.

Anderson, K. (2014b), UPDATED—82 things publishers do (2014 edition), *The Scholarly Kitchen*, 21 October 2014, http://scholarlykitchen.sspnet.org/2014/10/21/updated-80-things-publishers-do-2014-edition/.

Arnold, D.N., and K.K. Fowler (2011), Nefarious numbers, *Notices of the American Mathematical Society*, 58(3), 434–437.

Auclair, D. (2015), *Open access 2015: Market Size, share, forecast, and trends*, Outsell Inc.

Beall, J. (2017), What I learned from predatory publishers, *Biochemia Medica*, 27(2), 273–279, 10.11613/BM.2017.029.

Bloom, F.E. (1988), Policy change, *Science*, 281, 175.

Bohannon, J. (2015), Editor quits journal over pay-for-expedited peer-review offer, *ScienceInsider*, 27 March 2015. http://news.sciencemag.org/scientific-community/2015/03/editor-quits-journal-over-pay-expedited-peer-review-offer-updated.

Bornmann, L. (2014), Do altmetrics point to the broader impact of research? *An overview of benefits and disadvantages of altmetrics, Journal of Informatics*, 8, 895–903.

Brand, A., L. Allen, M. Altman, M. Hlava, and J. Scott, (2015), Beyond authorship: Attribution, contribution, collaboration, and credit, *Learned Publishing*, 28, 151–155, doi:10.1087/20150211.

Bubela, T., J. Guebert, and A. Mishra (2015), Use and misuse of material transfer agreements: Lessons in proportionality from research, repositories, and litigation, *PLoS Biol*, 13(2), e1002060, doi:10.1371/journal.pbio.1002060.

Castelvecchi, D. (2015), Physics paper sets record with more than 5,000 authors, *Nature News* 15 May 2015; http://www.nature.com/news/physics-paper-sets-record-with-more-than-5-000-authors-1.17567.

Clark, M.P., and R.B. Hanson (2017), The citation impact of hydrology journals, Water Resources, *Res.*, 53, 4533–4541, doi:10.1002/2017WR021125.

COPE (2015), Principles of Transparency and Best Practices in Scholarly Publishing, https://publicationethics.org/resources/guidelines-new/principles-transparency-and-best-practice-scholarly-publishing.

Council of Science Editors (2012), White paper on publication ethics, http://www.councilscienceeditors.org/resource-library/editorial-policies/white-paper-on-publication-ethics/.

Daubert v. Merrell Dow Pharmaceuticals, Inc. (1993), https://www.law.cornell.edu/supct/html/92-102.ZS.html

Davis, P. (2012a), The emergence of a citation cartel, *The Scholarly Kitchen*, 10 April 2012, http://scholarlykitchen.sspnet.org/2012/04/10/emergence-of-a-citation-cartel/.

Davis, P. (2012b), The persistence of error: A study of retracted articles on the Internet and in personal libraries, *J. Med. Lib. Assoc.*, 100(3), 10.3163/1536-5050.100.3.008.

Davis, P. (2015), Production plummets at PLOS—But for a good reason, *The Scholarly Kitchen*, http://scholarlykitchen.sspnet.org/2015/02/11/publication-plummets-at-plos-one/.

Enserink, M. (2015), Open-access publisher sacks 31 editors amid fierce row over independence, *ScienceInsider*, 20 May 2015, http://news.sciencemag.org/people-events/2015/05/open-access-publisher-sacks-31-editors-amid-fierce-row-over-independence.

Esposito, J. (2014), The size of the open access market, *The Scholarly Kitchen*, http://scholarlykitchen.sspnet.org/2014/10/29/the-size-of-the-open-access-market/.

Ferguson, C., A. Marcus, and I. Oransky (2014), Publishing: The peer-review scam, *Nature*, 515, 480–482.

Ferguson, L. (2014), How and why researchers share data (and why they don't), *Wiley Exchanges*, 3 November 2014. http://exchanges.wiley.com/blog/2014/11/03/how-and-why-researchers-share-data-and-why-they-dont/.

Fisher, E.A. (2013), Public access to data from federally funded research: Provisions in OMB Circular A-110, Congressional Research Service Report R42983, https://www.fas.org/sgp/crs/secrecy/R42983.pdf.

Galperin, J. (2013), The continued decline of environmental journalism, Yale Center for Environmental Law and Policy, *On the Environment Blog*, 14 January 2013, http://environment.yale.edu/envirocenter/the-continued-decline-of-environmental-journalism/

Grant, B. (2009), Merck published a fake journal, *The Scientist*, 30 April 2009, http://www.the-scientist.com/?articles.view/articleNo/27376/title/Merck-published-fake-journal/.

Hanson, B., Journal Impact Factors with Uncertainties, Eos, 20 June 2017 https://eos.org/editors-vox/journal-impact-factors-with-uncertainties.

Hanson, B., and J. Lunn (2017), Avoiding predators in publishing, *Eos*, 98, 10.1029/2017EO076269.

Hanson, B., K. Lehnert, and J. Cutcher-Gershenfeld (2015), Committing to publishing data in the Earth and space sciences, *Eos*, 96, doi:10.1029/2015EO022207.

Hanson, B., A. Sugden, and B. Alberts (2011), Making data maximally available, *Science*, 331, 649, doi:10.1126/science.1203354.

IPCC (2015), Principles and Procedures, https://www.ipcc.ch/organization/organization_procedures.shtml.

Interuniversity Consortium for Political and Social Research (ICPSR) (2013), Sustaining Domain Repositories for Digital Data: A Call for Change from an Interdisciplinary Working Group of Domain Repositories, http://tinyurl.com/domainrepositories.

Klein, M., H. Van de Sompel, R., Sanderson, H., Shankar, L. Balakireva, et al. (2014), Scholarly context not found: One in five articles suffers from reference rot, *PLoS ONE*, 9(12), e115253. doi:10.1371/journal.pone.0115253.

Marcia McNutt, Monica Bradford, Jeffrey Drazen, R. Brooks Hanson, Bob Howard, Kathleen Hall Jamieson, Veronique Kiermer, Michael Magoulias, Emilie Marcus, Barbara Kline Pope, Randy Schekman, Sowmya Swaminathan, Peter Stang, Inder Verma, Transparency In Authors Contributions And Responsibilities To Promote Integrity In Scientific Publication, bioRxiv 140228; doi: https://doi.org/10.1101/140228.

Marcus, A., and I. Oransky (2014), What studies of retractions tell us, *J Microbiol Biol Educ*, 15, 151–154.

Maunsell, J. (2010), Announcement regarding supplemental material, *J Neurosci*, 30(32)10599–10600 http://www.jneurosci.org/content/30/32/10599.full.

McNutt, M., K. Lehnert, B. Hanson, B.A. Nosek, A.M., Ellison, and J.L. King (2016), Liberating field science samples and data, *Science*, 351, 1024–1026, doi:10.1126/science. aad7048.

Merali, Z., and J. Giles (2005), Special report: Databases in peril, *Nature* 435, 1010–1011, http://www.nature.com/nature/journal/v435/n7045/full/4351010a.html doi:10.1038/4351010a.

Monastarsky, R. (2005), The number that's devouring science, *Chronicle of Higher Education*, 14 October 2005.

Nature (2010), Nature's choices, *Nature*, 463, 850, doi:10.1038/463850a.

Nature Publishing Group (2014), Author insights survey 2014, http://figshare.com/articles/ MSS_Author_Insights_2014/1204999.

NISO (2013), Recommended practices for online supplemental journal article materials, http://www.niso.org/apps/group_public/download.php/10055/RP-15-2013_ Supplemental_Materials.pdf.

Nosek, B. et al. (2015), Promoting an open research culture, *Science*, 348, 1422–1425.

Pearce, F. (2010), Debate heats up over IPCC melting glaciers claim, *New Scientist*, 11 January 2010, http://www.newscientist.com/article/dn18363-debate-heats-up-over-ipcc-melting-glaciers-claim.html.

PLOS Medicine Editors (2006), The impact factor game, *PLOS Medicine*, 3, e291 doi:10.1371/journal.pmed.0030291.

Rosenberg, M. (2016), America now has nearly 5 PR people for every reporter, double the rate from a decade ago, *Muck Rack Daily*, 14 April 2016, https://muckrack.com/ daily/2016/04/14/america-now-has-nearly-5-pr-people-for-every-reporter-double-the-rate-from-a-decade-ago/.

Russell, C. (2008), Science reporting by press release: An old problem grows worse in the digital age, *Columbia Journalism Review: The Observatory*, 14 November 2008 http:// www.cjr.org/the_observatory/science_reporting_by_press_rel.php.

Socas-Navarro, H. (2010), *Nature*'s policies regarding solar physics, *SolarNews*, 11 May 2010 http://dl.dropbox.com/u/359387/Nature_SolarPhysics.txt and http://solarnews. nso.edu/2010/20100515.html#section3.

Solanki, S., *Nature* will start accepting solar papers again, *SolarNews*, 2 July 2010. http:// solarnews.nso.edu/2010/20100713.html#section4.

Stodden, V., M. McNutt, D.H. Bailey, E. Deelman, Y. Gil, B. Hanson, M.A. Heroux, J.P.A. Ioannidis, M. Taufer (2016), Enhancing reproducibility for computational methods, *Science*, 354, 1240, http://science.sciencemag.org/content/354/6317/1240.

Sumner, P., S. Vivian-Griffiths, J. Boivin, A. Williams, C.A. Venetis, A. Davies, et al. (2014), The association between exaggeration in health related science news and academic press releases: Retrospective observational study, *BMJ*, 349, g7015.

Tenopir, C., S. Allard, K. Douglass, A.U. Aydinoglu, L. Wu, et al. (2011), Data sharing by scientists: Practices and perceptions, *PLoS ONE*, 6(6), e21101. doi:10.1371/journal. pone.0021101.

Vine, T.H., et al. (2014), The availability of research data declines rapidly with article age, *Current Biology*, 24, 94–97 10.1016/j.cub.2013.11.014.

Vinkers, C. H., J.K. Tijdink, and W.M. Otte (2015), Use of positive and negative words in scientific PubMed abstracts between 1974 and 2014: Retrospective analysis. *British Medical Journal*, 351, h6467, doi: 10.1136/bmj.h6467.

Ware, M., and M. Mabe (2012). *The STM report: An overview of scientific and scholarly journal publishing*, International Association of Scientific, Technical and Medical Publishers, the Netherlands.

Wilhite, A.W., and E.A. Fong (2012). Coercive citation in academic publishing, *Science*, 335(6068), 542–543, doi:10.1126/science.1212540.

Zara, C. (2013), Remember newspaper science sections? They're almost all gone. *International Business Times*, 10 January 2013, http://www.ibtimes.com/remember-newspaper-science-sections-theyre-almost-all-gone-1005680.

Zen, E. (1988), Abuse of coauthorship: Its implications for young scientists, and the role of journal editors, *Geology*, 16, 292.

9

9. SCIENTIFIC INTEGRITY AND ETHICAL CONSIDERATIONS FOR THE RESEARCH DATA LIFE CYCLE

Linda C. Gundersen

Abstract

Data can be a global engine for wealth, well-being, culture, and science. Data's creation, quality, and accessibility are affected strongly by the integrity and ethical standards under which they are collected and preserved. This chapter focuses on providing guidance for the scientific integrity and ethics of data, with an emphasis on data used in the geosciences, and within the context of the research data life cycle. The scientific process has always been driven by data: from initial observation, through hypothesis development and testing, to final theory generation and the release of new knowledge. The quality and reusability of knowledge, and its utility for informing new science or making critical decisions about Earth's future, are dependent on the ethics and integrity of each step of the underlying research data life cycle. The increasingly complex and interdisciplinary nature of the geosciences requires the integration of data and knowledge from disparate disciplines and scales. This kind of interdisciplinary science can only thrive if scientific communities work together on common data standards and vocabularies and adhere to high standards of scientific integrity, ethics, data quality, collection, curation, and sharing.

Retired, U.S. Geological Survey, Ocean View, Delaware, USA

Scientific Integrity and Ethics in the Geosciences, Special Publications 73,
First Edition. Edited by Linda C. Gundersen.
© 2018 American Geophysical Union. Published 2018 by John Wiley & Sons, Inc.

9.1. Introduction

The analysis of data is the primary engine for the scientific process; without it, all ideas, hypotheses, and theories remain simply that, untested and unknown. The careful design of surveys and experiments to test scientific ideas, coupled with the careful management of data during collection, analysis, interpretation, and preservation, will make or break a successful study. It seems obvious that sound scientific practice requires exceptional integrity in every step of the data process. Not only is it required, but it is also the ethical thing to do. Data without integrity can destroy reputations, damage economies, and impact people and the environment in ways we may not anticipate. The Office of Research Integrity (http://ori.hhs.gov/) and Retraction Watch (http://retractionwatch.com/) are just two of the entities that document the growing number of data fraud cases where data of great significance to society have been fabricated and falsified. An ethical perspective is also needed throughout the entire research data life cycle, considering such issues as privacy, safety, confidentiality, bias, long-term data stewardship, and societal impact. The term *scientific integrity* as used in this chapter can be defined as the adherence to ethical principles (honesty, excellence, objectivity, fairness, beneficence) and the professional standards and practices essential for the responsible practice of science.

When we conduct science there is an inherent social contract we make with all other scientists, with the funders of science, and with society. Most users and producers of science trust that the research process has integrity; that honest, objective, and true observations and interpretations are being made; and when published, the end result is a reliable contribution: verifiable, repeatable, reusable, and defensible. If the process is not done ethically and with integrity, then the resulting data may, for example, have inaccuracies, bias, inappropriate analysis, poor communication of error, or disclose confidential information. The cascading effects are numerous and can erode the very fabric of the scientific enterprise as well as be potentially harmful to society in any number of ways. Some examples in the geosciences include the inaccurate risk analyses of an active volcano, leading to poor choices in construction, insurance, and emergency response and resulting in loss of life; underestimating the amount of sea level rise, causing a community to build inadequate protection and leading to destructive inundation; and the incorrect assessment of a mineral resource, causing the bankruptcy of a company and its investors.

The International Council of Science [2015] (http://www.icsu.org) has established the Principle of Universality of Science, defined as follows:

The free and responsible practice of science is fundamental to scientific advancement and human and environmental well-being. Such practice, in all its aspects, requires freedom of movement, association, expression, and communication for scientists, as well as equitable access to data, information, and other resources for research. It requires responsibility at all levels to carry out and communicate scientific work with integrity, respect, fairness, trustworthiness, and transparency, recognizing its benefits and possible harms.

How then, do we take responsibility for the integrity and ethics of our data and for sharing it in an ethical manner? Many organizations acknowledge the huge deluge of data that scientists, decision makers, and the public are confronted with and the attendant problems of ensuring the integrity and management of data [*Hey and Trethefen*, 2003; *Fox and Harris*, 2013]. Most organizations conclude that in order for science and society to benefit from this bounty, we must invest in sustained stewardship of data, big and small, guided by data management principals and rigorous quality control [*Beagrie and Houghten*, 2014]. Further, scientists must learn to manage their data toward the goal of making it understandable, accessible, and publishable. This will involve investing research time and funds into training, data management, and data preservation.

A commonly used data management framework is the data life cycle model deployed by libraries, repositories, businesses, universities, and scientists for a variety of data. These models serve well in assisting with data and quality management but do not fully provide guidance on integrity and ethics. Only a few resources address integrity in data management such as the Office of Research Integrity's website that provides a basic training module in data management ethics [*Northern Illinois University*, 2005]. This chapter tries to fill the gap by presenting key ethics and integrity questions paired with the major components of a simple research data life cycle that can be used by the researcher during the scientific process to help ensure the scientific integrity and ethics of their data, and ultimately of their science. A "research" data life cycle focuses mainly on the responsibility of the researcher and the data activities of the scientific process, and may be part of a larger data lifecycle management framework for a repository or an institution.

9.2. The Research Data Life Cycle

The concept that data has a life cycle and must be managed as part of the scientific process and as part of an institution's or business's responsibility was introduced in the 1990s as the digital age emerged and the Internet opened a world of data at the touch of a key. Recent examinations of current literature on research data life cycle models by the Committee on Earth Observation Satellites [*CEOS*, 2011] and Ball [2012] reveal strong agreement on the general aspects of the data life cycle such as collection, analysis, publication, and reuse but less agreement on the complexity of data management plans, archiving of the final product, the degree of curation needed, or the many attendant kinds of metadata that accompany each step. Many published data life cycle models for science data reflect basic scientific integrity practices such as quality control and the legal/ethical issues of privacy and confidentiality. However, few provide the deeper and broader considerations of the ethics of the scientific enterprise or the impacts on society that may result from how the data are obtained, analyzed, preserved, communicated, or used. Further, what are the ethical responsibilities of the individual scientist in preserving their scientific record? The concept and ideals of "open science" and "open data" [*Gezelter*, 2009; *Molloy*, 2011] coupled with concerns about reproducibility have spurred scientific journals and institutions to re-examine their policies regarding publishing, data accessibility, peer review, and author requirements resulting in stronger rules for transparency and accessibility of methods and data [*McNutt*, 2014; *Nosek et al.*, 2015; *Hanson*, 2018, this volume]. These rules challenge scientists and research institutions with taking more responsibility for curating, preserving, and making accessible their data as a matter of public trust and scientific responsibility. There are now a number of supporting organizations for open science, such as the Center for Open Science (http://centerforopenscience.org/top/), which provides a free research work flow environment called the Open Science Framework (https://osf.io/), and Project Jupyter (http://jupyter.org/), which provides community input, review, applications, and support for sharing documents that contain live code, equations, visualizations, and text for numerous computer languages, including Python and R. The National Science Foundation is actively funding open cyberinfrastructure for the geosciences through EarthCube [*Black et al.*, 2014; http://earthcube.org/], for the environmental sciences through DataONE [*Michener et al.*, 2011; www.dataone.org], and for the life and biological sciences through the IPlant Collaborative [*Goff et al.*, 2011], which was recently expanded and renamed Cyverse (http://www.cyverse.org/).

For the purpose of this chapter we will use a combination of several published data life cycle models and data management policies to create a simple research data life cycle that will be used to frame the integrity and ethics guidance provided in this chapter. These include the following:

1. The U.S. Geological Survey Data Management Policy [*Faundeen et al.*, 2013; *USGS*, 2015] is the recently released policy for management of the extensive earth and biological data this agency provides.

2. DataONE (https://www.dataone.org/best-practices), the Data Observation Network for Earth, provides best practices for data management for a network of environmental data repositories.

3. The Digital Curation Center (http://www.dcc.ac.uk/) focuses on data curation in detail and provides expert advice to institutions on data management and curation.

Figure 9.1 shows the research data life cycle used in this chapter. The center of the figure shows the three actions, document, secure, and manage quality, that are common to all six stages. The six stages include A. Plan; B. Acquire; C. Transform; D Communicate; E. Preserve; and F. Reuse. The scientific integrity and ethical considerations for each of these stages are described in Section 9.3 below with a focus on the individual scientist's responsibilities within the scientific process and also within the science community, whether that community is a laboratory, project team, institution, or the relevant domains of science. Each stage

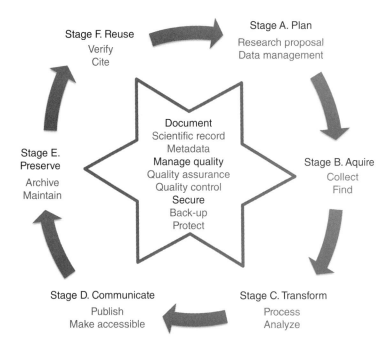

Figure 9.1 The research data life cycle as described in this chapter. Document, manage quality, and secure are actions common to all stages.

of the research data life cycle is also paired with a series of scientific integrity and ethical questions designed to encourage ethical thinking and improve scientific integrity during the scientific process.

There are several important concepts related to the scientific record that are essential to understand before diving into the individual stages of the research data life cycle. These include *curation, metadata*, and *provenance*. Digital *curation* is ongoing and dynamic, it involves adding value to data throughout the research data life cycle in order to maintain quality; record the data's history of collection, processing, and analyses; enable data discovery; and provide for accessibility and reuse of the data. (See more at http://www.dcc.ac.uk/digital-curation/what-digital-curation#sthash.7bnG5ZuR.dpuf.) One aspect of well-curated data is the metadata record. Essentially, metadata is data about the data, typically structured in a manner amenable to computer processing and understanding. Metadata can include the data author, location, instrumentation, the meaning of data fields in the database, a description of the data and its intended use, publications, analytical work flows, and standards used. If data produced in a study is well curated and has a good metadata record included, then its *provenance*, its origins and transformations through time for a specific research purpose, can be defined and the data can be put into context [*Hills et al.*, 2015]. Without curation, metadata, and provenance, it is difficult to reliably use data or reproduce other scientist's results. Reproducibility is the ability not only to replicate an experiment but to produce similar results using new or different data. Reproducibility is a key element of the scientific method and one of the principal uses of the scientific record. Providing enough of a scientific record for results to be understood and reproduced is an ethical responsibility of scientists to the scientific community, to the funders of science, and to society as a whole.

9.3. Scientific Integrity and Ethical Considerations for the Research Data Life Cycle

In this section, each stage of the data life cycle is described, key references are provided for further information, and a series of scientific integrity and ethical questions are provided to help guide scientists and scientific teams in embedding scientific integrity and ethical practices into their study.

9.3.1. Actions Common to All Stages: Document, Manage Quality, and Secure

Data documentation, quality management, and security must happen throughout the research data life cycle, and they can be embodied in three critical science integrity and ethical questions that must be asked at each stage:
1. What documentation (scientific record, metadata) will I record for each stage of the research data life cycle?
2. How will I manage the quality of the research process and the data produced?
3. How will I provide continuous security for the data to prevent loss, corruption, or inappropriate access?

These three questions and appropriate variations have been imbedded in the list of questions for Stages A through F.

9.3.1.1. Document
Creating a comprehensive, accurate, and understandable scientific record is the most critical step in conducting science. Documentation may include materials such as laboratory and field notes, methodology, software, samples, and analytical procedures and work flows. Throughout the research data life cycle, documentation must be updated to reflect all actions taken upon the data. This includes data acquisition, processing, and analysis, as well as recording metadata. As previously described, metadata include all the essential information needed for another scientist (or yourself) to understand the nature, quality, context, and history of the data, and to utilize the data responsibly. Metadata should be

considered a natural and essential part of the scientific record. The USGS has developed a simple questionnaire to help you get started on what metadata is: http://www.usgs.gov/datamanagement/documents/MetadataQuestionnaire.pdf. Most scientific fields have, or are creating, metadata standards and machine-readable formats for use within their specific scientific community. Many are also incorporating common metadata standards along with linked open data principles [*Heath and Bizer*, 2011] into their data to enhance interoperability. The Digital Curation Center (DCC) provides a list of metadata standards and tools available by scientific field (http://www.dcc.ac.uk/resources/metadata-standards). Some of the geoscience-based communities have developed discipline-based metadata standards, schema, and data templates, including the Federal Geographic Data Committee (FGDC, http://www.fgdc.gov/metadata/geospatial-metadata-standards/), the National Ocean and Atmospheric Administration (NOAA, http://www.ncddc.noaa.gov/metadata-standards/), and the Interdisciplinary Earth Data Alliance (IEDA, http://www.iedadata.org/publication/policy), as well as others that can be found on the aforementioned DCC site. Where and how the data are to be published will be an important consideration for creating the required metadata, as most data repositories have metadata standard requirements. Updated and complete metadata are critical to maintaining data quality, so consideration has to be made regarding updating metadata as a database evolves. The scientific record also consists of primary sources of data such as field and laboratory notes and physical samples, as well as derived sources of data such as published research and conference papers and published data sets and presentations. These are discussed later in the stages of the research data life cycle below.

9.3.1.2. Manage Quality

Scientists commonly refer to quality assurance and quality control (QA/QC) protocols and methods, usually in the laboratory environment. However, QA/QC is important to maintain the quality of the science and data throughout the entire scientific process, from collection to preservation. Quality assurance protocols are used to make sure data collection, analysis, and processing are done with precision and accuracy using standards, calibration, and well-documented methods that may be specific to a field of science or instrument. Quality control methods are used to test for the quality of the resulting data, looking for problems such as noise, contamination, or errors that may have happened during sampling or analysis and providing documentation for any problems found.

9.3.1.3. Secure

Steps must be taken to protect data from loss, corruption, and unauthorized access. This includes routinely backing up data and other digital resources (such as evolving software or models) and keeping multiple protected copies of raw data and any subsequent versions for recovery when problems arise. Security protections such as passwords, encrypted hard disks or files, or providing a restricted access room are best practices for keeping data secure. Sensitive or confidential data may be subject to specific policy, laws, or rules of the institution or agency providing funding, and must be adhered to. Plans will be needed on what data to keep, publish, or restrict access to, based on legal, privacy, or confidentiality issues. Security plans need to be agreed to and practiced by all involved in a project.

Particular attention needs to be paid when research involves human subjects, the geographic distribution of human-related attributes such as income or disease, the location of sensitive animal species that may be protected by law, and the location of culturally important artifacts that may be protected by law. If a project utilizes citizen science or community data through applications such as Twitter, the privacy of volunteers and contributors, especially children, must be protected by law. Examples of ethical considerations for how to deal with these kinds of data and some of the policy and law related to these data can be found in *Quinn* [2015], *Bowser et al.* [2014] and the Code of Ethics of the American Association of Geographers [2009].

The use of traditional knowledge from native or indigenous peoples requires working within the laws and ethical guidelines of the individual community or tribe. At a minimum, working with traditional knowledge requires protection of people's privacy and their intellectual property rights, requires written agreement with native communities on use of information, and protection and respect for the cultural and religious significance of the information. Examples of guidance on ethical considerations for these data can be found at www.nativescience.org and in the work of the Climate and Traditional Knowledges Workgroup [2014].

9.3.2. Stage A: *Plan*

The planning stage is the first opportunity to critically examine and incorporate scientific integrity and ethical practices into the study. If this step is done correctly, not only will it improve the odds of success, it may prevent serious losses during the scientific process while yielding science of benefit to society. The National Research Council [2002] issued a report on research integrity stating that "the responsible conduct of research is not distinct from research; on the contrary, competency in research *encompasses* the responsible conduct of that research and the capacity of ethical decision making." In other words, to conduct research well, one must conduct it ethically and with integrity. A well thought out research proposal that includes a detailed data management plan can help achieve this goal. The format of a research proposal generally depends on the funding source, but the goal or purpose of the study is critical. Most funding sources now request that the science have a benefit to society as well advancing science. For example, the National Science Foundation [2014a] requires that proposals be responsive to its strategic plan, which includes the statement: "NSF maintains a strong focus on carrying out our mission in a way that is sufficiently flexible to meet the changing requirements of the research and education enterprise as well as to address emerging and pressing societal challenges." The Singapore Statement, created by the World Integrity Conference [*Steneck et al.*, 2018, this volume], has as its final item the following statement regarding a researcher's ethical obligation to society. "Societal Considerations: Researchers and research institutions should recognize that they have an ethical obligation to weigh societal benefits against risks inherent in their work." Taking an ethical approach to the planning of scientific research therefore requires that we ask the ethical question: "Does this work benefit or harm society?" Scientists must also be careful not to bias their science by engaging in advocacy, whether at the planning stage of their work or in the communication and publishing stage. *Steneck* [2013] *Bobrowsky et al.*, [2018, this volume], and *Tuana* [2018, this volume] examine these challenges and provide guidelines to consider, some of which are provided in Section 9.3.5, "Stage D: Communicate."

Authorship and questions of intellectual property are best discussed upfront in the planning stage. Every member of the team should know their specific role in the generation of ideas, conducting the research, managing the data, writing up results, and the final creation of publications. Honorary authorships are inappropriate.

In the United States, scientists funded by the NSF and other federal agencies are also required to have a data management plan as part of the research proposal [*NSF*, 2014b]. A good data management plan describes the actions needed to collect the data; the QA/QC standards for testing, analyses, and processing; the format and outlet for storing, protecting, backup, and publishing of the data, metadata creation, and maintenance; protocols and methodologies for analysis and collection; and ultimately the plan for long-term preservation of the data. Scientists need to find out if their university, research institution, government agency, or publishing outlet has requirements and standards related to data management and preservation, or rules governing the publication or deposition of data in the respective institution's long-term repository. Finally, science project and data management plans should be updated as the science project evolves. Some good tools for data management plans include Interdisciplinary Earth Data Alliance (IEDA, http://www.iedadata.org/compliance/plan), the USGS (http://www.usgs.gov/datamanagement/plan.php), and DataONE (https://www.dataone.org/sites/all/documents/DataONE_BP_Primer_020212.pdf).

The Open Science Framework (https://osf.io/) provides a complete project and data environment.

9.3.2.1. Science Integrity and Ethical Questions for Stage A, Plan

- What problem am I trying to solve?
- What are the benefits and possible consequences of conducting this work or solving this problem? If there are consequences, how can they be mitigated or avoided?

- Have I considered the problem and method for solving it objectively? Do I have a bias or conflict of interest related to the proposed research and/or its outcome?
- What data will be needed and how can it be collected in an unbiased manner?
- Can the data be collected by not affecting living organisms or the health of the environment, or coworkers?
- If the data involve humans or other living organisms, are there laws and protocols that need to be understood and followed to ensure no harm or loss of privacy occurs?
- Who needs to be informed of the study to ensure appropriate permission, privacy law, or other legal considerations are adhered to in carrying out the work and in the use, storage, or reuse of the data?
- What instruments, laboratories, and analyses, methods, and models will be needed and what restrictions, quality management, or safety issues will I need to address?
- What format and standards will I use for the data and metadata? Does a community template exist that I should use? Are there open access standards that I should use? Is there an ISO (www.iso.org) standard?
- What data already exists and how will I gather it, ensure its provenance and quality, and provide appropriate attribution, confidentiality, and security?
- How will I curate the digital record of my work including recording, storing, protecting, and preserving the data and samples during and after collection?
- How will I publish and provide long-term preservation and accessibility of the research and supporting data, and what are the intellectual property rights I need to consider, including authorship, citation, or restrictions?
- What are the rules of behavior, sharing, and reporting, for the research environment I am creating, or working in, and how will I support these rules?

9.3.3. Stage B: *Acquire*

The careful recording of methods and data is the foundation for any study and constitutes a critical, sometimes overlooked, part of the scientific record. Through good documentation, results can be verified, mistakes traced, and techniques improved. Documentation may take place in physical logs and notebooks, on maps, on physical samples, or it may be digitally recorded. Protecting documentation and backing it up is critical to maintaining the integrity of the research. Maintaining both digital and paper versions of raw data as well as primary sources such as notebooks is critical to protect these resources from loss. Appropriate quality assurance steps, from proper calibration of instruments to clean glassware, are necessary to provide the expected level of quality data needed. Testing for data quality using quality control techniques, such as statistical analyses, checking data for noise, typos, or artifacts, and including sample blanks and duplicates in analyses, helps ensure the integrity of data and its results. A scientist once asked me, "Is keeping a sloppy lab or sloppy notebook an ethical violation?" Likely, the answer is yes if the act of being sloppy compromises results or the safety of the lab, or prevents accurate recording of what has been done. Additionally, creating a safe, positive, and harassment-free research environment is ethically essential.

During acquisition, curating data starts in earnest, and how well the acquisition, processing, and analysis of data are documented may determine if the data can be published, preserved, reproduced, and reused. Portions of the data documentation will be included in the metadata and will vary by the type of data. The time of day a rock is sampled from an outcrop might not matter, but the exact time of day, temperature, humidity, and other factors will greatly affect a gas or liquid sample. Communities of scientists and repositories have set standards for data format, metadata, protocols, and quality control during collection. It is incumbent on the scientists to know and learn these standards and formats. Consider registering the physical samples you collect, especially if they are to go in a sample archive. For geological materials, you can register the sample and have it assigned an International Geo Sample Number [*Lehnert and Klump*, 2012]. Life Science Identifiers are utilized in the life sciences for biological samples and objects [*Clark et al.*, 2004]. Research Resource Identifiers are persistent and unique identifiers for registering biomedical resources such as cell lines or antibodies (https://scicrunch.org/resources).

9.3.3.1. Science Integrity and Ethical Questions for Stage B, Acquire

- How will I sample or collect the data without bias? Will I use random, stratified, or other kinds of sampling design?
- What is the proper instrumentation for the data I want to collect? What sampling and collection protocols are required? What calibrations, standards, and maintenance are needed?
- If collecting data as part of a collaborative team or shared instrument, what are the rules and regulations regarding the instrument, data stream, standards, sharing, and processing?
- How will I number, document, and organize the data?
- What metadata will I record during collection?
- How will I calculate uncertainty and error in the data and methods?
- What QA/QC methods will be needed to ensure the quality and integrity of the data as it is collected?
- What approach will I take for safe and responsible sampling, protection from toxic materials, dangerous sampling, acknowledging rarity of samples, protection of the environment, and safety of people in the field and laboratory?
- How will I ethically sample live specimens, sensitive or personal data related to people and places, and rare or unique specimens or features?
- What are the ideal working conditions and when will ambient conditions (weather, temperature, etc.) be inappropriate or unsafe for sampling?

- How will I record and preserve observations? Have I developed a template or does a template exist?
- How will the data need to be formatted for further analysis, publication, and archiving?
- How will I ethically collaborate with stakeholders such as communities, businesses, governments, tribes, or the public?
- How will I protect the privacy and confidentiality of data and adhere to agreements or legal responsibilities for sensitive and licensed data?
- What plans do I have for maintaining laboratory organization, cleanliness, safety, and protection of people, samples, documentation, and experiments?
- Who needs to be legally informed of the work I am doing? What permissions do I need?
- What kinds of community or courtesy communication do I need to do for stakeholders or the public?

9.3.4. Stage C: *Transform*

During and after collection, raw data may be subjected to a number of transformations ranging from basic processing using simple calculations that produce usable units of measure, to sophisticated analyses resulting in complex models or utilizing sophisticated work flow processes. The raw and processed data require complete metadata to ensure results can be verified and duplicated and to assist the scientist in assuring quality and integrity. Multiple copies of the original, untransformed raw data should always be kept in secure locations on a variety of media types to ensure it can always be accessed and referred to. The protocols, methods, models, and work flows utilized for analysis should be well documented, whether they are new methods or a well-known published standard. Consider publishing any new methods, software, models, or work flows created during the research. Test and validate new methods, software, models, or work flows utilizing informal peer review and by conducting tests using other data or subsets of data. As data is analyzed, trends and anomalies emerge and QA/QC measures should be used to detect problems with data related to initial collection, transcription, or transformations. Anomalous data can reveal critical information about a system's behavior and should not be left out of a data set or publication unless it is clear that the data is erroneous because of problems in collecting or analyzing. In general, all data should be published. All potential authors and data producers should agree beforehand on the criteria for not including data in a published data set or summary of results, and these exclusions should be included in the text of the publication.

9.3.4.1. *Science Integrity and Ethical Questions for Stage C,* Transform

- What constitutes my raw data and what calibrations, standards, constants, units of measure, and other simple transformations will be made to the data?

- What analyses, analytical work flows, experimental methods, or modeling approaches will I need to document, reference, and make available as part of publication?
- Are the analyses, methods, and models appropriate for the data and what I am trying to understand?
- Am I or my team appropriately trained to conduct these analyses?
- What metadata will I record during each step of analysis?
- How will I calculate uncertainty and error for the various methods and analyses?
- How will I determine if an outcome is significant?
- What QA/QC methods will be needed to ensure the quality and integrity of the analyses and methods I deploy?
- Am I using duplicate samples, control standards, or testing subsets of data for reproducibility before publication.
- What constitutes fabrication and falsification of data, and how will I control for it in my laboratory or across my team?
- Knowing that all data results are important and should be represented, under what circumstances would I consider data not usable? How will I represent any data excluded for analyses in a publication?
- What error or uncertainty is introduced through these transformations and how will they be represented in the final results?
- What are the limitations and assumptions of the software, models, or analyses used or created, and how will they be represented in the final results?
- Have all software and models been tested, reviewed, or validated?
- How will I share, secure, and manage the results of analysis during this stage?

9.3.5. Stage D: *Communicate*

The ability to prepare and disseminate quality scientific results and data to the public, decision makers, and other scientists is a critical part of the scientific process. There are common scientific integrity codes of conduct related to publishing and authorship that all scientists should adhere to. The Committee on Publication Ethics has created a widely accepted "Guidelines for Publication Practice" [*COPE*, 1999] that has excellent scientific integrity guidance for authors, peer review, and publishing. Many professional societies, journals, agencies, and institutions have adopted new codes of conduct for research and publication in the last decade.

Communicating data and results has changed dramatically in the last 10 years. The advent of digital professional publications, social media, web publishing, digital data repositories, and other digital, often instantaneous outlets, provides numerous ways to share results and data. At the same time, competition among journals, media outlets, and scientists has created numerous problems related to everything from fraudulent authorship and peer review to sensationalism [*Hanson*, 2018, this volume; *Arrison and Nerem*, 2018, this volume]. Data peer review is still an evolving practice [*Mavernik et al.*, 2014]. The trend toward open science is also changing our perceptions of what it means to author, peer review, publish, and share science. We are in a complex transition where traditional publication methods are rapidly trying to evolve to keep up with digital innovation, the demands of big data, and increased competition. The Transparency and Openness Promotion (TOP) Guidelines [*Nosek et al.*, 2015] are beginning to be adopted by journals worldwide and should be a catalyst for changing data accessibility. Implementation and adoption of the highest level TOP guidelines will require everyone, home institutions, funding institutions, publishing outlets, and scientists, to recognize that data stewardship, and the accompanying activities needed to make data reusable and accessible, is a necessary part of the science process and will need to be funded and rewarded. The new guidelines will require scientists and journals to be open and transparent about the science they are publishing. In addition, the underlying data and methods will need to be accessible, published, and include the necessary metadata and provenance information. Many online journals include the ability to publish corrections and updates to data, metadata, and the text of publications. Any correction or update should become a part of the scientist's own scientific record and the published record.

Communication of science through digital media can fall prey to sensationalism and the omission of important supporting information. Authors should, when possible, review press releases, summaries, and postings of the results of their work before release. Scientists need to understand the fine line between conducting impactful science relevant to society and wanting one's science to have impact on a particular outcome. Exaggeration of ideas and results is unethical, and all publications, from press releases to blog posts, need to include the limitations of the data and conclusions. A recent statement by the American Association for the Advancement of Science [*Molina et al.*, 2014] with regard to climate change provides a good example of the balance that is needed:

> As scientists it is not our role to tell people what they should do or must believe about the rising threat of climate change. But we consider it to be our responsibility as professionals to ensure, to the best of our ability, that people understand what we know.

Steneck [2013] offers excellent guidance to avoid the pitfalls of advocacy and misrepresenting the science that bears repeating:
1. Limit communication to area(s) of expertise
2. Present information accurately and in clear, understandable terms
3. Disclose interests
4. Point out weaknesses and limitations
5. Mention opposing scientific views

9.3.5.1. Science Integrity and Ethical Questions
for Stage D, Communicate

- Where will the research be published, presented, or communicated? Are these the publication outlets that will reach the intended audience and are reliable, quality publications?
- What are the data and method transparency and publication requirements of the publication outlets chosen?
- Will the data and methods be published with the results, deposited in a repository, or be citable?
- Have I adequately evaluated the previous work on the subject and included appropriate references? What data has been reused from another source and needs to be cited?
- Are all conclusions effectively supported by the data?
- Who contributed tangible ideas, data, and/or research to the project and deserves authorship? Are there any "honorary" authorships that need to be removed?
- How will all authors be involved in and responsible for writing the paper, creating illustrations, responding to peer review, and approving the final product for publication?
- Do any authors have conflicts of interest that must be disclosed?
- Are there intellectual property considerations such as patents or copyrights?
- What privacy, ownership, and confidentiality issues need to be considered for data release and accessibility?
- Have all authors avoided plagiarism or self-plagiarism?
- Is there time-sensitive information of importance to people or the environment such as threats, risks, hazards, etc. that requires swift release, preapproval of an institution, or collaboration with government officials for release?
- Are public press releases, presentations, or publications devoid of sensationalism and advocacy, and are they presenting an accurate depiction of the data and limitations?
- Are opinions versus fact clearly labeled in any communications?
- Have negative results, uncertainty, assumptions, limitations, and error been communicated clearly in publications, press releases, public presentations, or other communication of the outcomes?

9.3.6. Stage E: *Preserve*

Preservation of data goes beyond publication and ensures the long-term accessibility of the underlying research data and metadata in a reliable repository, utilizing the appropriate archival standards for the kind of data published and the scientific field of inquiry. Preservation of data is a critical ethical step in the responsible stewardship of the scientific record. Data preservation standards for the

Earth sciences, and efforts to support data sharing and repositories for the sciences in general, have improved greatly in the past 20 years but are far from ideal. Efforts such as the World Data System (http://www.icsu-wds.org/), Group on Earth Observations (http://www.earthobservations.org/), Earth Science Information Partners [*ESIP*, 2012; http://www.esipfed.org/], and the aforementioned DataOne, EarthCube, and IEDA, as well as many others provide standards, access, and/or long-term repositories for Earth science data [*Downs et al.*, 2015]. The Registry of Research Data Repositories (http://www.re3data.org/) provides a list of over 1500 repositories in the sciences and works with DataCite (https://www.datacite.org/), which provides user guides for finding a repository, finding data, and citing data. The recently created COPDESS [*Hanson et al.*, 2015] provides an Earth science repository catalogue and is working across journals to promote better accessibility and preservation of Earth science data and increased use of citation. Broad coalitions of scientists, libraries, business, and others such as Force 11 have developed citation principles to enable data and materials to be not only cited but machine readable and searchable [*Data Citation Synthesis Group*, 2014]. All of this can be very daunting, and working with data scientists is strongly recommended. The COPDESS Statement of Commitment [2015] makes this recommendation to authors: "Earth and space science data should, to the greatest extent possible, be stored in appropriate domain repositories that are widely recognized and used by the community, follow leading practices, and can provide additional data services." Journals may have recommendations for the repositories they prefer, and data may need to be converted to a standard or format accepted by the relevant scientific community and the domain repository it is intended for.

9.3.6.1. Science Integrity and Ethical Questions for Stage E, Preserve

- How will the data, samples, and methods be protected and preserved in the short term for further use by the laboratory, team, or university?
- What protection from the elements, backup, and security will be needed?
- How will the data, samples, and methods be preserved in the long term and made accessible to the scientific community and/or the public?
- When will the data, samples, and methods be released for open access and deposited in a long-term repository?
- What data, samples, and methods will be archived and preserved? What decision-making process will be used to determine the data to be kept and data to be discarded?
- What required metadata, archival standards, and formats will be needed?

9.3.7. Stage F: Reuse

A fundamental part of any study is finding out what research has been done before and what data is available to inform the current study. Inadequately assessing and understanding the previous work conducted on a subject is unethical and will damage the integrity of the research. As indicated previously, significant data are becoming accessible across the globe through numerous repositories. However, the quality and usefulness of these data vary. If the originator of the data has done proper curation and metadata, then the data's provenance and usefulness will be clear. Provenance is critical to determining if the data is appropriate for use. Utilizing poorly curated or preserved data or data that have questionable origins, confidentiality, or privacy restrictions can compromise integrity and is not ethical. Acknowledgment or citation of reused data is an important part of documenting the scientific record as well as respecting the intellectual property of other scientists.

9.3.7.1. Science Integrity and Ethical Questions for Stage F, Reuse

- Is the source of the data and their provenance known and reputable?
- Are the data of known and documented quality?
- Are there errors or duplications, or have the data been transformed from their original state?
- Does the metadata provide sufficient understanding of the meaning of the data, its limitations, and appropriate uses?
- Are the data complete, representative, and comparable to the other data being used?
- Is the use of these data defensible and appropriate for their intended purpose?
- Are there any issues regarding permissions, confidentiality, or privacy?
- Are the data legally available for use and citable?

9.4. Summary

As scientists, we are experiencing technological advances and the production of data at an unprecedented rate. Scientific integrity codes of conduct are beginning to reflect more data stewardship principals, but many are still deficient in dealing with the multitude of ethical challenges that accompany rapid technological change. Although new technology is improving and in some areas revolutionizing our ability to do science, it is also creating opportunity for scientific fraud and misconduct. The concept of open science is as much an ethical response to these issues as it is a necessary response to maintaining the integrity of science. Ethically we need transparency in the scientific process to ensure that underlying data have integrity and can be reused with confidence in the future. A recent attempt to reproduce past science [*Open Science Collaboration*, 2015] had limited success; generally studies with the strongest data and correlations were reproducible. The authors concluded that the tendency to publish only new science rather than studies that help confirm previous important results or theories probably contributed to the low reproducibility rate. The study left many questions to be answered regarding why reproducibility rates were low, but concludes that reproducibility is critical to progress in science.

The new TOP [*Nosek et al.*, 2015] publication standards for documentation and publication of data and methods will help mitigate some of the issues of fraud and misconduct and hopefully inspire scientists to further test results and theories by making more data and methods available. However, we need to go deeper, through education of students and early career scientists, fundamental changes in our scientific process that promote openness and accessibility, and funding to support data management and publication. As noted by McNutt *et al.* [2016], we need to move beyond data and samples being "available by request" and empower new research applications and scientific integrity through transparency, accessibility, and reproducibility. The purpose of this chapter was to begin that deeper discussion of our everyday scientific habits and how they can be informed by integrity and ethics to be responsive to the needs of the digital age and open science. This, coupled with thoughtful, rigorous adherence to data life cycle management; the use of good curation, preservation, and publication practices; and the collaboration among scientific domains to create and use common standards for the interoperability of data and knowledge will create a much stronger scientific enterprise and pave the way for scientific advancement.

References

American Association of Geographers (2009), Statement on Professional Ethics, http://www.aag.org/cs/about_aag/governance/statement_of_professional_ethics

Arrison, T., and R.M. Nerem (2018), Fostering integrity in research: Overview of the National Academies of Sciences, Engineering, and Medicine Report, in *Scientific integrity and ethics in the geosciences*, edited by L.C. Gundersen, John Wiley & Sons, Hoboken, NJ.

Ball, A. (2012), *Review of data management lifecycle models (v. 1.0)*. REDm-MED project document redm1rep120110ab10, University of Bath, UK, http://opus.bath.ac.uk/28587/

Beagrie and Houghten (2014), The Value of Data Sharing and Curation, Charles Beagrie Ltd, Victoria University, and JISC 2013. (http://repository.jisc.ac.uk/5568/1/iDF308_-_Digital_Infrastructure_Directions_Report%2C_Jan14_v1-04.pdf).

Black, R., A. Katz, and K. Kretschmann (2014), Earthcube: A community-driven organization for geoscience infrastructure, *Limn. and Ocean. Bull.*, 23 (4), doi:10.1002/lob.201423480a.

Bobrowsky, P., V.S. Cronin, G. Di Capua, S.W. Kieffer, and S. Peppoloni (2018), The emerging field of geoethics, in *Scientific integrity and ethics in the geosciences*, edited by L.C. Gundersen, John Wiley & Sons, Hoboken, NJ.

Bowser, A., A. Wiggins, L. Shanley, J. Preece, and S. Henderson (2014), Sharing data while protecting privacy in citizen science, *ACM Interactions*, XXI, 1.

Clark T., S. Martin, and T. Liefeld (2004), Globally distributed object identification for biological knowledgebases, *Briefings in Bioinformatics*, Mar 5 (1), 59–70. PMID:15153306

Climate and Traditional Knowledges Workgroup (2014), Guidelines for Considering Traditional Knowledges in Climate Change Initiatives, http://climatetkw.wordpress.com/

Committee on Earth Observation Satellites (2011), Data life cycle models and concepts, CEOS.WGISS.DSIG.TN01, Issue 1, http://ceos.org/ourwork/workinggroups/wgiss/documents/

COPE (Committee on Publication Ethics) (1999), Guidelines on good publication practice, in *The Cope Report 1999*, pp. 43–47, BMJ Books, London, http://publicationethics.org/files/u7141/1999pdf13.pdf

COPDESS (Coalition on Publishing Data in the Earth and Space Sciences) (2015), Statement of commitment from Earth and space science publishers and data facilities, http://www.copdess.org/statement-of-commitment/.

Data Citation Synthesis Group (2014), Joint Declaration of Data Citation Principles, edited by M. Martone, FORCE11, San Diego, CA, https://www.force11.org/group/joint-declaration-data-citation-principles-final

Downs, R. R., R. E. Duerr, D. J. Hills, and H. K. Ramapriyan (2015), Data stewardship in the Earth sciences, *D-Lib*, 21(7/8), doi:10.1045/july2015-downs.

ESIP Stewardship Committee (2012), Data citation guidelines for data providers and archives, edited by M.A. Parsons et al., Fed. of Earth Sci. Inf. Partners, 10.7269/P34F1NNJ DOI:10.7269/P34F1NNJ#_blank

Faundeen, J.L., T.E. Burley, J.A. Carlino, D.L. Govoni, H.S. Henkel, S.L. Holl, V.B. Hutchison, E. Martín, E.T. Montgomery, C.C. Ladino, S. Tessler, and L.S. Zolly (2013), The United States Geological Survey Science Data Life Cycle Model, U.S. Geological Survey Open-File Report 2013–1265, 10.3133/ofr20131265

Fox, P., and R. Harris (2013), ICSU and the challenges of data and information management for international science, *Data Science Journal*, doi:10.2481/dsj.WDS-001.

Gezelter, D. (2009), What exactly is Open Science? The Open Science Project, www.openscience.org/blog/?p=269

Goff, S.A., M. Vaughn, S. McKay, E. Lyons, A. Stapleton, D. Gessler, N. Matasci, L. Wang, M. Hanlon, A. Lenards, A. Muir, N. Merchant, S. Lowry, S. Mock, M. Helmke, A. Kubach, M. Narro, N. Hopkins, D. Micklos, U. Hilgert, M. Gonzales, C. Jordan, E. Skidmore, R. Dooley, J. Cazes, R. McLay, Z. Lu, S. Pasternak, L. Koesterke, W. Piel, R. Grene, C. Noutsos, K. Gendler, X. Feng, C. Tang, M. Lent, S. Kim, K. Kvilekval, S. Manjunath, V. Tannen, A. Stamatakis, M. Sanderson, S. Welch, K. Cranston, P. Soltis, D. Soltis, B. O'Meara, C. Ane, T. Brutnell, D. Kleibenstein, J. White, J. Leebens-Mack, M. Donoghue, E. Spalding, T. Vision, C. Myers, D.B. Lowenthal, B. Enquist, B. Boyle, A. Akoglu, G. Andrews, S. Ram, D. Ware, L. Stein, and D. Stanzione (2011), The iPlant Collaborative: Cyberinfrastructure for plant biology, *Frontiers in Plant Science*, 2, doi: 10.3389/fpls.2011.00034.

Hanson, B. (2018), The new landscape of ethics and integrity in scholarly publishing, in *Scientific integrity and ethics in the geosciences*, edited by L.C. Gundersen, John Wiley & Sons, Hoboken, NJ.

Hanson, B., K. Lehnert, and J. Cutcher-Gershenfeld (2015), Committing to publishing data in the Earth and space sciences, *Eos*, 96, doi:10.1029/2015EO022207.

Heath, T., and C. Bizer (2011), *Linked data: Evolving the Web into a global data space*, 1st ed. Synthesis Lectures on the Semantic Web: Theory and Technology, 1(1), 1–136, Morgan & Claypool.

Hey, T., and A. Trefethen (2003). The data deluge: An e-science perspective, in *Grid computing: Making the Global Infrastructure a Reality*, edited by F. Berman, G.C. Fox, and T. Hey, John Wiley & Sons, New York.

Hills, D.J., R.R. Downs, R. Duerr, J.C. Goldstein, M.A. Parsons, and H.K. Ramapriyan (2015), The importance of data set provenance for science, *Eos*, 96, doi:10.1029/2015EO040557.

International Council of Science (2015), Freedom, Responsibility, and the Universality of Science [Brochure, 32 p.], http://www.icsu.org/publications/cfrs/freedom-responsibility-and-universality-of-science-booklet-2014/CFRS-brochure-2014.pdf

Lehnert, K., and J. Klump (2012), The geoscience Internet of Things, *Geophysical Research Abstracts*, 14, EGU 2012-13370, http://meetingorganizer.copernicus.org/EGU2012/EGU2012-13370.pdf

Mayernik, M.S., S. Callaghan, R. Leigh, J. Tedds, and S. Worley (2015), Peer review of datasets: When, why, and how, *Bull. Am. Meteorol. Soc.*, 96, 191–201, doi:10.1175/BAMS-D-13-00083.1.

McNutt, M., (2014), Reproducibility, *Science*, 343(6168), 229, doi:10.1126/science.1250475.

McNutt, M., K.A. Lehnert, B. Hanson, B.A. Nosek, A.M. Ellison, and J.L. King (2016), Liberating field science samples and data, *Science*, 351(6277), 1024, doi:10.1126/science.aad7048.

Michener, W., T. Vision, J. Kunze, P. Cruse, and G. Janée (2011), Dataone: Data observation network for earth-preserving data and enabling innovation in the biological and environmental sciences, *D-Lib Magazine*, 17, 3, http://webdoc.sub.gwdg.de/edoc/aw/d-lib/dlib/january11/michener/01michener.html

Molina, M., J. McCarthy, D. Wall, R. Alley, K. Cobb, J. Cole, S. Das, N. Diffenbaugh, K. Emanuel, H. Frumkin, K. Hayhoe, C. Parmesan, and M. Shepherd, (2014), What we know: The reality, risk, and response to climate change. AAAS What We Know Initiative Report, http://whatweknow.aaas.org/wpcontent/uploads/2014/07/whatweknow_website.pdf

Molloy, J.C. (2011), The Open Knowledge Foundation. Open data means better science, *PLoS Biol*, 9(12), doi:10.1371/journalpbio.1001195.

National Research Council, Institute of Medicine (2002), *Integrity in scientific research: Creating an environment that promotes responsible conduct*, National Academies Press.

National Science Foundation (2014a) Investing in Science, Engineering, and Education for the Nation's Future, NSF Report 14043, http://www.nsf.gov/pubs/2014/nsf14043/nsf14043.pdf

National Science Foundation (2014b). NSF Proposal and Award Policy and Procedures Guide, http://www.nsf.gov/pubs/policydocs/pappguide/nsf15001/nsf15_1.pdf

Northern Illinois University (2005), Responsible Conduct in Data Management [website training module], https://ori.hhs.gov/education/products/n_illinois_u/datamanagement/dmotopic.html

Nosek, B.A., G. Alter, G.C. Banks, D. Borsboom, S.D. Bowman, S.J. Breckler, S. Buck, C.D. Chambers, G. Chin, G. Christensen, M. Contestabile, A. Dafoe, E. Eich, J. Freese, R. Glennerster, D. Gorof, D.P. Green, B. Hesse, M. Humphreys, J. Ishiyama, D. Karlan, A. Kraut, A. Lupia, P. Mabry, T.A. Madon, N. Malhotra, E. Mayo-Wilson, M. McNutt, E. Miguel, E.L. Paluck, U. Simonsohn, C. Soderberg, B.A. Spellman, J. Turitto, G. VandenBos, S. Vazire, E.J. Wagenmakers, R. Wilson, and T. Yarkoni (2015), Promoting an open research culture, *Science*, 348, 1422–1425, doi:10.1126/science.aab2374.

Open Science Collaboration (2015), Estimating the reproducibility of psychological science, *Science*, 349(6251), doi:10.1126/science.aac4716.

Quinn, M.J. (2015), *Ethics for the information age*, 6th ed., Pearson Education Inc.

Steneck, N.H. (2013), Responsible Advocacy in Science: Standards, Benefits, and Risks, AAAS Report, Workshop on Advocacy in Science, http://www.aaas.org/report/report-responsible-advocacy-science-standards-benefits-and-risks

Steneck, N.H., T. Mayer, M.S. Anderson, and S. Kleinert (2018), The origin, objectives, and evolution of the World Conferences on Research Integrity, in *Scientific integrity and ethics in the geosciences*, edited by L.C. Gundersen, John Wiley & Sons, Hoboken, NJ.

Tuana, N. (2018), Understanding coupled ethical-epistemic issues relevant to climate modeling and decision support science, in *Scientific integrity and ethics in the geosciences*, edited by L.C. Gundersen, John Wiley & Sons, Hoboken, NJ.

U.S. Geological Survey (2015), USGS Data Management Policy, http://www.usgs.gov/datamanagement/index.php

Section IV: Ethical Values and Geoethics

10

10. UNDERSTANDING COUPLED ETHICAL-EPISTEMIC ISSUES RELEVANT TO CLIMATE MODELING AND DECISION SUPPORT SCIENCE

Nancy Tuana

Abstract

The aim of this chapter is to demonstrate that values are a component of all aspects of scientific practice and, as such, scientific integrity requires that scientists be able to identify and analyze the role and impact of values both within research as well as in decision-support science. Coupled ethical-epistemic values are introduced in order to clarify their role in scientific practice as well as to provide a framework for critical reflection on value choices. The first section examines the ways epistemic values are embedded in scientific practice and illustrates that these epistemic choices and the values that undergird them can have important and currently underappreciated ethical import. The second section concerns the role of values in the domain of the broader impacts of scientific research and in particular impacts on decision support. Providing decision makers with the knowledge they need to make responsible decisions often requires attention to coupled ethical-epistemic values. The conclusion of the chapter is that training in coupled ethical-epistemic analysis adds an important dimension to scientific integrity education often missing in more traditional approaches.

Department of Philosophy; and Rock Ethics Institute, The Pennsylvania State University, University Park, Pennsylvania, USA

Scientific Integrity and Ethics in the Geosciences, Special Publications 73,
First Edition. Edited by Linda C. Gundersen.
© 2018 American Geophysical Union. Published 2018 by John Wiley & Sons, Inc.

In a recent discussion of the Intergovernmental Panel on Climate Change's (IPCC) treatment of uncertainties, the authors stress that "characterizing uncertainty in the assessment of evidence is common practice when communicating science to users" [*Adler and Hadorn*, 2014, 663]. Providing decision makers and society in general with relevant information is an important component of all sciences, as well as engineering. As is often the case, and is particularly true in instances where there are complex uncertainties, it is crucial that scientists clearly communicate the degrees of confidence regarding that information. Given that the role of the IPCC is to provide decision makers with an overview and assessment of the most recent scientific understanding of climate change, as well as assessments of technological and socioeconomic factors, the IPCC developed guidance notes for the characterization and communication of uncertainties (Guidance notes for assessing uncertainties were first developed in conjunction with the Third Assessment Report in 2000 [*Moss and Schneider*]. These were updated in conjunction with the Fourth Assessment Report [*Le Treut et al.*, 2007] and most recently in 2010 for the Fifth Assessment Report [*Mastrandrea et al.*].) The IPCC acknowledges that the characterizations of uncertainties is a *deliberative process*. The guidelines are designed to provide "a traceable account" of expert judgments and suggestions for avoiding overconfidence in assessments of the level of uncertainty. However, and as Adler and Hadorn [2014, 665] underscore,

> The extent that this procedure can be considered reliable, in that it delivers an inter-subjective assessment of uncertainty, is contingent on agreements based on value judgements, for instance on spatiotemporal resolution, parameterization of models, or significance level for accuracy in empirical studies. Value judgements at every step of the assessment process gives rise to dissensus.

While the complexity of uncertainties in climate science, from the geophysical to human behavior, pose a challenge to scientific assessment as well as to communication and decision-support science, such challenges are hardly unique to climate science. Sciences dealing with ecosystem dynamics, population genetics, hydrology, and geomorphology, to name just a few, all deal with uncertainties of nonlinear systems and, in many cases, uncertainties due to coupled natural-human systems. The aim of this chapter is to clarify the complex manner in which values are a component of scientific practice. The point of doing so is to demonstrate that careful attention to values is not only relevant to scientific integrity and ethics in the domain of the responsible conduct of research, but it is also a key component of responsible scientific research and of attention to the broader impacts of science. In other words, *scientific integrity requires attention to values in all domains of scientific practice.*

This chapter builds on experiences with the National Science Foundation–funded Sustainability Research Network, Sustainable Climate Risk Management network (SCRiM, scrimhub.org). Led by geoscientist Klaus Keller and centered at Penn State, SCRiM links a transdisciplinary team of scholars at 19 universities and 5 research institutions across 6 nations to answer the question, "What are sustainable, scientifically sound, technologically feasible, economically efficient, and ethically defensible climate risk management strategies?" This effort foregrounds attention to value decisions through *coupled ethical-epistemic analyses*, an original and cross-cutting theme of the network. SCRiM is committed to providing tools and analyses that focus on examining how trade-offs between mitigation, adaptation, and geoengineering can be made through science and policy procedures that are ethically and epistemically responsible. A second focus of the SCRiM network is to develop climate risk management strategies that are effective in assisting stakeholders and decision makers to better understand the complex trade-offs and uncertainties involved in making decisions on how to manage climate risks. Since coupled ethical-epistemic issues are a cross-cutting theme of SCRiM, this project includes the identification of new areas of concern as well as the identification of new or refined scientific questions triggered by ethical assessments. Given that most approaches to these trade-off decisions do not make the value-based dimensions of problem framing and strategies transparent, attention to coupled ethical-epistemic analysis adds an important and unique dimension to the research of the SCRiM network.

Coupled ethical-epistemic analysis adds an important dimension to scientific integrity often missing in more traditional research integrity approaches. While not discounting the importance of attention to issues in the responsible conduct of research (*Steneck* [2007] provides an excellent guide), the more robust approach described in this chapter provides a venue for identifying the types of value decisions that are embedded in research models and methods. This approach also serves as a resource for understanding the ethical implications of research trajectories in science. As I previously argued, coupled ethical-epistemic analysis is a key component of scientific integrity in that it will lead to "better science both in the traditional sense of advancing knowledge by building on and adding to our current knowledge as well as in the broader sense of science for the good of, namely, scientific research that better benefits society" [*Tuana*, 2013b, 1955].

The first section of the chapter will focus on a discussion of the role of coupled ethical-epistemic analysis for identifying and examining the roles of values that are embedded in scientific research. The second part will focus on coupled ethical-epistemic analyses of the broader impacts of scientific research, and, in particular, impacts on decision support.

10.1. Coupled Ethical-Epistemic Issues: Embedded Ethics

For some scientists it is anathema to claim that there are values embedded in scientific research practice. But values guide all human activity, including scientific activity. In the domain of science, for example, values help us determine what is important to know.

Too often it is mistakenly assumed that the inclusion of values will negatively bias scientific deliberation. The "objectivity ideal" in science is often framed in terms of practicing science such that, at least in the ideal case, scientific research and the information thereby produced is not influenced by values at either the community or the individual level. It is true that scientific practice can be distorted by values. For example, many scholars have documented this type of distortion in the context of biases resulting from corporate sponsorship of research [e.g., *Brandt*, 2013; *Grifo et al.*, 2013; *Markowitz and Rosner*, 2013; *Proctor*, 2012] or gender or race biases in science [*Fausto-Sterling*, 1985; *Gould*, 1981; *Haraway*, 1989]. However, as Paul Thagard illustrated in *The Cognitive Science of Science: Explanation, Discovery, and Conceptual Change* [2012], values legitimately affect scientific deliberations at all stages of research, including the discovery phase, assessments of findings, and practical applications. As Thagard explains:

> The decisions that scientists and others need to make about what projects to pursue, what theories to accept, and what applications to enact will unavoidably have an emotional, value-laden aspect. Normatively, therefore, the best course is not to eliminate values and emotions, but to try to ensure that the best values are used in the most effective ways (p. 300).

A central goal of enhancing scientific integrity through coupled ethical-epistemic analyses is to render transparent the values at play, both at the individual and at the community levels, in order to be able to analyze their impact and assess whether, to underscore Thagard, *the best values are used in the most effective ways.*

To better understand the roles of values in scientific practice it is helpful to appreciate the venues in which they can be relevant. Values can enter into scientific practices through at least five avenues:

1. The choice of a scientific research problem;
2. The determination and selection of data that are relevant to the phenomena under investigation and provide evidence in support of relevant inferences;
3. The likelihood of a hypothesis or theoretical explanation providing an account of the phenomena;
4. The reliability of the evidential connections between the hypothesis or explanation and the data/evidence as determined in #2 above;
5. The application of the results of the research either to other research problems or to addressing social problems (decision-support).

The first and fourth venues are more typically relevant to questions of the broader impacts of science, which will be discussed in the next section. What motivates a scientist to select a particular research problem over others can be due to various types of values, from ethical values (this is knowledge needed for the good of society) to aesthetic values (this is a complex and exciting problem) to epistemic values (this is a problem that must be solved if we are to fully understand dynamics in the broader field). As I will discuss in Section 10.2, how the results of research are applied to other problems brings in both epistemic and ethical values in determining responsible ways to apply the results. In this section, I will focus on the second and third venues: that is, what counts as relevant data, evidential connections, and confidence levels for hypotheses or theoretical explanations.

Scientific integrity, in the broader sense that I am developing and advocating, requires being aware of the values that are embedded in each of the above five components of a particular scientific research practice in order to ensure that (a) the individual researcher, the research group, or the scientific community is/are doing research grounded in the right values in the sense of values that are (i) the most relevant to the research and (ii) the most likely to lead to responsible knowledge production, and (b) that these values are used in the most effective ways, both in terms of (i) their import for data and hypothesis selection and (ii) how the values are communicated to users of scientific information. What this means for integrity training is that scientists and engineers need to develop the skills needed to identify values and analyze their impact on research and decision-support. Although my focus in this chapter is directed at scientists, the broader conception of scientific integrity detailed in the chapter is relevant for engineers as well. (Scientific integrity training developed in conjunction with the NSF grant Graduate Pedagogy for Ethical Dimensions of Coupled Natural and Human Systems Research (EESE- 1135327) and sponsored by the Pennsylvania State University's Rock Ethics Institute (rockethics.psu.edu) has been developed for engineers as well as for scientists based on this more robust model.)

Different types of values can be involved in the above four domains of scientific research. Values here are understood to be beliefs about what is good or bad, desirable or undesirable, which serve as guidelines or principles for behavior. To claim that something is a value is not the same as indicating that it is something that is an individual or group preference or disposition. It is a claim that the value is itself something that is good or desirable and thus is something that should be seen as a value by others. While the determination of which values are most relevant for a particular issue or practice can be very controversial, there is general consensus about what is to be included as values. Social psychologists [e.g., *Haidt and Joseph*, 2004] have argued that basic ethical values, or what they call moral intuitions, are common across cultures. However, how they are prioritized and when and how they are applied lead to significantly different principles and practices. Among the ethical values seen as common to cultures are respect, loyalty, reciprocity, liberty, care, fairness, and purity.

This chapter will focus on two types of values, namely, epistemic values and ethical values. *Epistemic values* have long been acknowledged to be a component of scientific research. The philosopher of science Ernan McMullin defined an epistemic value as a value that will "promote the truth-like character of science, its character as the most secure knowledge available to us of the world we seek to understand. An epistemic value is one we have reason to believe will, if pursued, help toward the attainment of such knowledge" [1983, 18]. Epistemic values are those that support or encourage responsible knowledge practices. Thomas Kuhn [1977], for example, identified epistemic values such as simplicity, scope, theoretical elegance, and fruitfulness as guidelines for choosing between competing models or theories. *Ethical values* are nonepistemic values that serve as guidelines for behavior. Ethical values serve to define what is right and wrong and thereby guide the behavior of individuals as well as groups and institutions, serve as aspirational goals and ideals, and are that against which we evaluate our own behavior, the behavior of others, and the ethical integrity of communities, institutions, and societies. Ethical values include justice, sustainability, trustworthiness, well-being, utility, fairness, and caring.

It is important to note that utility as interpreted in the utilitarian ethical framework is that which is the ultimate ethical value and should thus be maximized. It has been defined in various ways, including pleasure, happiness, and well-being. However, it is important to distinguish this conception of utility from that common to much of economic analyses that define utility as "market value" or willingness to pay. Market value is not a judgment of the ethical value or worth of a thing. It is simply a measure of the price at which the supply of an item equals demand for it. It has been used as a way to elicit the value that grounds willingness to pay, e.g., "How much is an individual or community willing to pay for clean air or water?" However, willingness to pay is a factor of many other aspects of an individual's or community's situation and hence is not a direct measure of value in this sense [cf. *Sen*, 1991].

It is a relatively uncontroversial claim to say that epistemic values function in the practice of science. What is more controversial is to claim that nonepistemic values, that is, ethical values have an important role in the responsible practice of science. A closer examination of the role of epistemic values can help to clarify the role of ethical values in the practice of responsible science and illuminate the reasons why it is essential that training in scientific integrity include training in coupled ethical-epistemic analysis.

Thagard [2012] provides the following cognitive map of epistemic values relevant to science (see Figure 10.1).

Predictive power, objectivity, robustness of evidence, simplicity, coherence, convergence of evidence, scope, and explanatory power are typical epistemic values embraced by the scientific community. While such values are commonly shared by members of the scientific community, there can be trade-offs and disagreements within the scientific community regarding how they ought to be ranked in a particular context. To illustrate this point, consider climate modeling.

Predictive power and scope are often in tension in climate modeling. Confidence levels decrease when shifting from outcomes projected by global models to models that have been downscaled in order to develop regional climate models [*Flato et al.*, 2013]. However, a global model is often not sufficient for policy makers in a particular country or region to make decisions about adaptation investments. In the latter case, predictive accuracy, while considered important, would be balanced with the value of offering information relevant to the decision makers. This illustrates that a shift in scope at the expense of reliability can be seen as an ethically sound choice. The question, however, regarding how much one can lower predictive reliability and still provide helpful information is a

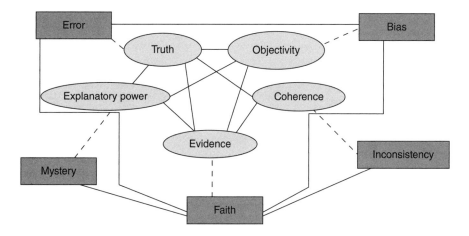

Figure 10.1 Ovals represent positive values; rectangles represent negative values. Solid lines indicate mutual support and dotted lines indicate incompatibility. Source: Adapted from *Thagard* [2012]. Used with permission of MIT Press.

decision that often brings together epistemic values with ethical values. Concerns, for example, about the severity of impacts can lead to decisions about what degree of confidence is needed to responsibly engage in the research or to offer it as a component of decision support science. The philosopher of science Richard Rudner [1953], for example, argues that given that there is always a degree of uncertainty governing the relationship between theories and evidence,

> the scientist must make the decision that the evidence is *sufficiently* strong or that the probability is *sufficiently* high to warrant the acceptance of the hypothesis. Obviously, our decision with regard to the evidence and how strong is "strong enough" is going to be a function of the *importance*, in the typically ethical sense, of making a mistake in accepting or rejecting the hypotheses....*How sure we need to be before we accept a hypothesis will depend on how serious a mistake would be.* [emphasis in original]

Rudner's example of the value judgments related to the sufficiency of evidence reflects a coupling between ethical and epistemic values (see Figure 10.2). Such attention to the role of nonepistemic values in science has led to important studies of the ways ethical values can in fact be a relevant and appropriate factor in the evaluation and justification of scientific claims [*Cranor*, 1990; *Douglas*, 2009; *Elliott*, 2011; *Kitcher*, 1993; *Longino*, 1990; *Machamer and Wolters*, 2004]. Scientific integrity thus includes attention to how the epistemic values are prioritized within a particular research domain. Doing so can include ethical values, for example, considerations of the ethical salience of an error in terms of the impact either on other research or on its application.

Coupled ethical-epistemic values can be a component of decisions at different scales. They can be a component of decisions that an individual scientist or

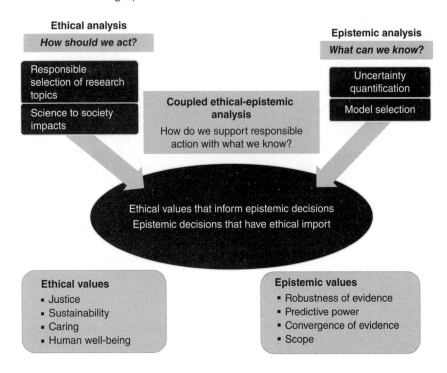

Figure 10.2 A visualization of coupled ethical-epistemic analyses.

research group must make or they can be relevant to decisions that involve the scientific research community at large. And there can be disagreements within the scientific community regarding the prioritization of values. To be a responsible scientist thus requires the skills needed to understand not only the nature of the values at play but also the reasons for the disagreements in how to prioritize them or even which are relevant. The responsible practice of science also requires that scientists also possess the ability to make values-informed decisions about which are the *best* values for the research problem.

A recent salient example of a scientific community-wide epistemic decision that involved coupled ethical-epistemic values concerns the inclusion of data about ice sheet melt in sea level rise projections. The question of how best to predict and communicate knowledge regarding future sea level rise has been, for example, a controversial component of the IPCC reports. As just one example, scientists involved in the Fourth Assessment Report [*Solomon et al.*, 2007] had to decide whether data on ice sheet melt was sufficiently robust to include in the sea level rise projection models. Here scientists had to balance robustness with predictive power, given that a decision not to include ice sheet data would result in the models being overconfident in very likely underreporting future levels of sea level rise [cf. *Rahmstorf*, 2010]. There are a number of unknowns and uncertainties regarding ice sheet dynamics as well as concerning the interaction between ice shelves and the ocean [cf., *Solomon et al.*, 2007]. One component of this is the complexity of the problem. The mathematical models for ice sheet dynamics are, for example, very difficult. Their accuracy also requires more extensive observations than were available at that time, such as the depth of the bedrock under ice sheets or the distribution of water under the ice sheets. In addition, the physics in the models remains incomplete, with limited understanding of the interactions between the ocean and the ice shelves, or how changes at the bed of the ice sheets may affect ice flow. While the sea level change findings of the Fifth Assessment Report [*Church et al.*, 2013] include information on significant improvements in modeling and observations since the Fourth Assessment Report in 2007, the question of what level of robustness of data is sufficient to include that data in the model has remained a point of controversy among the scientific community [*Purkey et al.*, 2014; *Chang et al.*, 2014; *Schaeffer et al.*, 2012]. This is a controversy involving epistemic values.

As we can see even with this brief sketch, decisions about how to responsibly balance epistemic values, in this case epistemic robustness and predictive power, are not always clear-cut and scientists can legitimately disagree about how well a criterion is met or how to rank the importance of different epistemic values. Yet these decisions may be coupled with ethical values as the decisions can have an impact on the usability of the information for decision support. Since the IPCC reports are designed to provide information for policy makers to base decisions about adaptation as well as mitigation strategies, clarity about value disagreements is thus crucial information for policy makers for them to understand potential trade-offs, such as the trade-off between reliability and predictive power discussed above.

10.2. Coupled Ethical-Epistemic Values: Broader Impacts

The previous discussion illustrates that the epistemic value choices made by scientists can have ethical import when this information is used by the wider public, particularly when scientific information is being used for policy making. The case of sea level rise is an example of this tight connection between the two domains and provides an example of a coupled ethical-epistemic issue. It also serves as an example of the ethical dimensions of the broader impacts of scientific research.

The broader impacts of scientific research is an important domain in which coupled ethical-epistemic issues arise. The label for this domain is based on the NSF Second Merit Criterion. This additional review criterion was formally introduced in 1997 by the NSF and added to the review criterion of intellectual merit in an effort to enhance the 1950 goal of using this federal funding "to promote the progress of science; to advance the national health, prosperity, and welfare; to secure the national defense" [NSF Act of 1950]. The broader impacts of scientific research had been a long-standing focus of the NSF; the addition of the second criterion was not a change from this goal but rather was designed to emphasize its importance [cf. *Schienke et al.*, 2009].

Appreciation of the coupling of ethical and epistemic values and the related issues that can result in the domain of the broader impacts of scientific research is essential to the NSF's view of the role and purpose of ethics education, namely:

> Ethics education is particularly critical to the science and engineering community as it faces an increasingly competitive funding environment; rising collaboration with international colleagues who may follow different guidelines; and growing recognition of the relevance of science and engineering to social, economic, and ethical issues of wide public and political interest [*NSF*, 2008].

The NSF Proposal and Award Policies and Procedures Guide [2013] includes the explanation that attention to broader impacts is an essential component of a funding application in that the "NSF values the advancement of scientific knowledge and activities that contribute to societally relevant outcomes" (13-1, II-9). The NSF Broader Impacts criterion is a useful frame for thinking about the impacts of science on society; however, it is important to consider not only the benefits but also the risks of proposed research. In other words, the broader impacts criterion as the NSF envisioned it is designed to provide an opportunity for scientists and engineers to examine how their research impacts society, and provides an occasion to consider who benefits from such research (and who does not) and who or what might be harmed by such research (and who or what will not). That is, it provides the occasion for scientists to consider ethical values linked to the justice impacts of their research.

Coupled ethical-epistemic analyses relevant to broader impacts raise value questions in at least two ways. First, ethical values are involved in consideration of which benefits are most desirable as well as which harms are to be avoided. Second, researchers should analyze how the benefits (or harms) of the research will and should be distributed across sectors of society; that is, they should consider questions of distributive justice as well as, at least in some instances, issues of environmental justice. While it is certainly not the case that scientists have control over how the products of their research will be deployed once they are available to the public, this type of broader impact concern is often relevant to avenues 1 and 4 listed in section 10.1.

To briefly illustrate this linkage regarding the first category, that is, that these issues can be relevant to choosing and framing a research problem, consider two choices of emphasis relevant to climate science research: (a) Is the focus of the research on slow-onset impacts from climate change or on more abrupt changes? (b) Does the model focus primarily on nonlinearities in the physical system and ignore or minimize nonlinearities in the social system? [cf. *Moser*, 2005]. In both areas, choice of emphasis or even how to balance these disparate foci will have broader impacts on what is and is not known relevant to adaptation and mitigation choices.

Working in the domain of decision-support science provides a clear example of the fourth category, namely, how coupled ethical-epistemic issues arise in the application and communication of scientific information in the service of decision support. I will briefly examine the issue of communication of results and then turn to decision-support science.

How scientists communicate the results and limitations of their research to the public is an obvious instance of a coupled ethical-epistemic issue. To demonstrate how epistemic values can be coupled to ethical values, consider again the previous discussion of the decision, based on epistemic values, to exclude ice dynamic data from the West Antarctic Ice Sheet in sea level rise models in the IPCC Fourth Assessment Report [*Solomon et al.*, 2007] (see Figure 10.3.) Once this epistemic decision is made, it results in a coupled ethical issue, namely, how to represent and communicate this decision in information that might be used by the broader public. That is, it raises ethical issues regarding how to effectively signal what has been omitted and the significance of doing so for those who intend to use the information as a component of decision making.

The ethical implications of the AR4 decision to exclude uncertainty relating to ice-dynamical changes is spelled out by Stefan Rahmstorf, an expert on the role of oceans in climate change. "For the past six years since publication of the AR4, the UN global climate negotiations were conducted on the basis that even without serious mitigation policies global sea-level would rise only between 18 and 59 cm, with perhaps 10 or 20 cm more due to ice dynamics. Now [in AR5] they are being told that the best estimate for unmitigated emissions is 74 cm, and even with the most stringent mitigation efforts, sea level rise could exceed 60 cm by the end of century. It is basically too late to implement measures that would very likely prevent half a meter rise in sea level." Rahmstorf argues that the broader impacts of the "conservative estimates" of the IPCC AR4 is that they "lulled policy makers into a false sense of security, with the price having to be paid later by those living in vulnerable coastal areas" [2013].

Coupled ethical-epistemic issues are profoundly implicated in decision-support science. Prediction based quantitative analysis is most traditionally formulated as best-estimate predictions of the future, often as a probability density function (PDF) over likely states of the world. These, combined with stakeholder preferences, are often used in decision-support contexts to rank decision options. In many decision situations that involve complex uncertainties and multiple decision makers and stakeholders, there will often be disagreements regarding both levels of risk tolerance as well as decision objectives. To address such a decision-support context with a best-estimate prediction based on expected utility

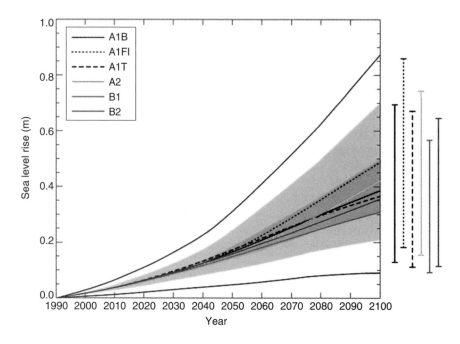

Figure 10.3 Global average sea level rise 1990 to 2100 for the SRES [Special Report on Emissions scenarios]. Thermal expansion and land ice changes were calculated using a simple climate model calibrated separately for each of seven AOGCMs [atmosphere-ocean general circulation models], and contributions from changes in permafrost, the effect of sediment deposition, and the long-term adjustment of the ice sheets to past climate change were added. Each of the six lines appearing in the key is the average of AOGCMs for one of the six illustrative scenarios. The region in dark shading shows the range of the average of AOGCMs for all 35 SRES scenarios. The region in light shading shows the range of all AOGCMs for all 35 scenarios. The region delimited by the outermost lines shows the range of all AOGCMs and scenarios, including uncertainty in land-ice changes, permafrost changes, and sediment deposition. *Note that this range does not allow for uncertainty relating to ice-dynamical changes in the West Antarctic ice sheet....* The bars show the range in 2100 of all AOGCMs for the six illustrative scenarios. (Source: *Church et al.*, 2001, Figure 11.12. Original caption, my emphasis. Reproduced with permission of IPCC.)

maximization, while a common approach, raises a series of coupled ethical-epistemic issues as articulated by Lempert *et al.* [2013]:

> Under conditions of deep uncertainty, such best-estimate predictions can promote overconfidence, so that decision makers fail to give sufficient attention to important, but perceived unlikely, threats and opportunities. Best-estimate predictions can also inhibit deliberations among individuals with differing expectations and values, because by definition best-estimate predictions privilege some expectations and values over others (p. 1).

Decision-support thus requires attention to the ways in which the selection of methods and tools, choices involving epistemic values, has broader impacts on the decision making context, and thus involves coupled ethical-epistemic issues. Epistemic decisions concerning, for example, how uncertainties are represented [*Lempert et al.*, 2006] or considerations as to whether, when, and how to aggregate impacts [*Smith et al.*, 2001, *Tuana*, 2013a] can have significant ethical import in the decision making context.

10.3. Concluding Remarks

The goal of this chapter is to elucidate why scientific integrity requires values analysis in all domains of scientific research. Coupled ethical-epistemic analyses serve as a tool for ensuring scientific integrity in these complex value dimensions. To act with integrity as a scientist in order to advance *good science* is a threefold charge. It requires (1) "getting the science right" in the sense of providing well-grounded information based on the best models and frameworks. But it also requires (2) "getting the right science" in the sense of focusing research on the information that is most epistemically *and* ethically salient for the problem at hand. Furthermore, it requires (3) "getting it right" in the sense of providing information in ways that users of that information understand and that promote the most effective use of that information.

This may sound like a daunting charge. But it is one that can and should be addressed through enhanced scientific integrity training. To help ensure, as Thagard [2012] advocates, that *the best values* are used in the most effective ways, integrity training should include education in values identification and analysis. An important first step toward achieving the type of scientific integrity required to address coupled ethical-epistemic issues is *values transparency*. The goal here is to make clear the values, both epistemic and ethical, that are part of scientific discovery, explanation, and application. Only in this way can these values be subject to the same careful and critical examination found in other dimensions of scientific practice. Fully responsible research in science and engineering thus requires skills in values clarification as well as training in critical reasoning skills designed to exam the role and impact of potentially relevant epistemic and ethical values.

As a guide for and an incentive to developing this broader training in scientific integrity, I close with a list of the types of questions to which scientists trained in coupled ethical-epistemic issues will be able to effectively respond. I urge readers to add to this list. Training to address questions such as these will help to ensure that all components of *good science* noted above will be sustained.

1. What epistemic and/or ethical values are embedded in this instance of scientific practice?
2. What do we need to know to determine which epistemic and/or ethical values are relevant?

3.   Are there any epistemic and/or ethical values embedded in the model or account that are not relevant and/or are potentially harmful or distorting?
4.   What is the best way to rank the salience of relevant epistemic and/or ethical values?
5.   Are there any epistemic and/or ethical value disagreements within the scientific community and/or between the scientific community and the public relevant to this instance of scientific research and practice?
6.   How are standards of proof, errors, or uncertainties handled in a given analysis and what epistemic values are relevant to these choices? Are there ethical implications of these decisions?
7.   What epistemic values underlie considerations of empirical adequacy and how consistent are results, over how many runs? What are the ethical implications of these considerations?
8.   What is the scope of the study or the model? Are some dimensions of the analysis oversimplified? What is the ethical import of scope decisions?
9.   What epistemic and/or ethical values are embedded in the classification typologies and variable definitions?
10.   What epistemic and/or ethical values influenced method selection?
11.   What epistemic and/or ethical values are relevant to the choice of research questions?
12.   What is the likely impact of information from this research and how will that impact be distributed, i.e., who will benefit, who will be harmed?
13.   What epistemic and/or ethical values are relevant to how information from this research is communicated?
14.   What epistemic and/or ethical values are relevant to decision support science in this domain?

While I am a strong advocate of including such skill development within the scientific integrity training of scientists and engineers, I have also made the case that the complexity of this work argues for the value of embedding philosophers specifically trained in coupled ethical-epistemic analyses in scientific teams as we have done in the SCRiM network [*Tuana*, 2013b]. I view this not as an either/or choice but as an opportunity to better enhance the ability of scientists to "get it right."

Acknowledgments

This work was partially supported by the National Science Foundation through the Network for Sustainable Climate Risk Management (SCRiM) under NSF cooperative agreement GEO-1240507 and the NSF EESE grant, Graduate Pedagogy for Ethical Dimensions of Coupled Natural and Human Systems Research (EESE- 1135327). Any opinions, findings, and conclusions or recommendations expressed in this material are those of the author and do not necessarily reflect the views of the National Science Foundation. I thank Klaus Keller,

Robert J. Lempert, Lauren Mayer, Tom Richard, Stephanie Vasko, Bryan Cwik, and Erich Schienke for their inputs in our many discussions about these topics. The work was enhanced thanks to the feedback of the anonymous reviewer. Any errors or omissions are my responsibility.

References

Adler, C.E., and G.H. Hadorn (2014), The IPCC and treatment of uncertainties: Topics and sources of dissensus, *WIREs Climate Change*, 5, 663–676, doi:10.1002/wcc.297.

Brandt, A.M. (2012), Inventing conflicts of interest: A history of tobacco industry tactics, *American Journal of Public Health*, 102(1), 63–71. doi:10.2105/AJPH.2011.300292 DOI: 10.2105%2FAJPH.2011.300292#pmc_ext.

Chang, W., P.J. Applegate, M. Haran, and K. Keller (2014), Probabilistic calibration of a Greenland Ice Sheet model using spatially-resolved synthetic observations: Toward projections of ice mass loss with uncertainties, *Geoscientific Model Development*, 7, 1933–1943, doi:10.5194/gmd-7-1933-2014.

Church, J.A., J.M. Gregory, P. Huybrechts, M. Kuhn, K. Lambeck, M.T. Nhuan, D. Qin, and P.L. Woodworth (2001), Changes in sea level, in *Climate change 2001: The scientific basis. Contribution of Working Group I to the Third Assessment Report of the Intergovernmental Panel on Climate Change*, J.T. Houghton, Y. Ding, D.J. Griggs, M. Noguer, P.J. van der Linden, X. Dai, K. Maskell, and C.A. Johnson (eds.), Cambridge University Press, Cambridge and New York, http://www.grida.no/climate/ipcc_tar/wg1/pdf/TAR-11.pdf

Church, J.A., P.U. Clark, A. Cazenave, J.M. Gregory, S. Jevrejeva, A. Levermann, M.A. Merrifield, G.A. Milne, R.S. Nerem, P.D. Nunn, A.J. Payne, W.T. Pfeffer, and D. Stammer and A.S. Unnikrishnan (2013), Sea level change, in *Climate change 2013: The physical science basis. Contribution of Working Group I to the Fifth Assessment Report of the Intergovernmental Panel on Climate Change*, edited by T.F. Stocker, D. Qin, G.-K. Plattner, M. Tignor, S.K. Allen, J. Boschung, A. Nauels, Y. Xia, V. Bex and P.M. Midgley, Cambridge University Press, Cambridge and New York.

Cranor, C. (1990), Some moral issues in risk assessment, *Ethics*, 101, 123–143.

Douglas, H. (2009), *Science, policy, and the value-free ideal*, University of Pittsburgh Press, Pittsburgh, PA.

Elliott, K. (2011), *Is a little pollution good for you? Incorporating societal values in environmental research*, Oxford University Press, New York.

Fausto-Sterling, A. (1985), *Myths of gender: Biological theories about women and men*, Basic Books.

Flato, G., J. Marotzke, B. Abiodun, P. Braconnot, S.C. Chou, W. Collins, P. Cox, F. Driouech, S. Emori, V. Eyring, C. Forest, P. Gleckler, E. Guilyardi, C. Jakob, V. Kattsov, C. Reason, and M. Rummukainen (2013), Evaluation of climate models, in *Climate change 2013: The physical science basis. Contribution of Working Group I to the Fifth Assessment Report of the Intergovernmental Panel on Climate Change*, T.F. Stocker, D. Qin, G.-K. Plattner, M. Tignor, S.K. Allen, J. Boschung, A. Nauels, Y. Xia, V. Bex, and P.M. Midgley (eds.). Cambridge University Press, Cambridge and New York.

Gould, S. J. (1981), *The Mismeasure of Man*, W. W. Norton & Company, New York.

Grifo, F., M. Halpern, and P. Hansel (2012), Heads they win, tails we lose: How corporations corrupt science at the public's expense, The Scientific Integrity Program of the

Union of Concerned Scientists, UCS Publications, Cambridge, MA. Available at http://www.ucsusa.org/scientific_integrity.

Haidt, J., and C. Joseph (2004), Intuitive ethics: How innately prepared intuitions generate culturally variable virtues, *Daedalus*, 133(4), 55–66, doi:10.1162/0011526042365555 DOI:10.1162%2F0011526042365555.

Haraway, D. (1989), *Primate visions: Gender, race, and nature in the world of modern science*, Routledge, New York.

Kitcher, P. (1993), *The advancement of science: Science without legend, objectivity without illusions*, Oxford University Press.

Kuhn, T.S. (1977), Objectivity, value judgment and theory choice, in *The essential tension: Selected studies in the scientific tradition and change*, University of Chicago Press, Chicago, pp. 356–367.

Lempert, R.J., D.G. Groves, and J.R. Fischbach (2013), Is it ethical to use a single probability density function? RAND working paper, online at http://www.rand.org/content/dam/rand/pubs/working_papers/WR900/WR992/RAND_WR992.pdf

Lempert, R.J., D.G. Groves, S.W. Popper, and S.C. Bankes (2006), A general, analytic method for generating robust strategies and narrative scenarios, *Management Science*, 52, 514–528, 10.1287/mnsc.1050.0472.

Le Treut, H., R. Somerville, U. Cubasch, Y. Ding, C. Mauritzen, A. Mokssit, T. Peterson, and M. Prather (2007), The IPCC assessments of climate change and uncertainties in *Historical Overview of Climate Change, in Climate change 2007: The physical science basis*. Contribution of Working Group I to the Fourth Assessment Report of the Intergovernmental Panel on Climate Change, S. Solomon, D. Qin, M. Manning, Z. Chen, M. Marquis, K.B. Averyt, M. Tignor, and H.L. Miller (eds.). Cambridge University Press, Cambridge and New York.

Longino, H. (1990), *Science as social knowledge*, Princeton University Press.

Machamer, P., and G. Wolters (eds.) (2004), *Science, values, and objectivity*, University of Pittsburgh Press, Pittsburgh, PA.

Markowitz, G., and D. Rosner (2013), *Deceit and denial: The deadly politics of industrial pollution*, University of California Press, Oakland, CA.

Mastrandrea M.C., C.B. Field, T.F. Stocker, O. Edenhofer, K.L. Ebi, D.J. Frane, H. Held, E. Kriegler, K.J. Mach, P.R. Matschoss (2010), Guidance note for lead authors of the IPCC Fifth Assessment Report on Consistent Treatment of Uncertainties, Intergovernmental Panel on Climate Change, Geneva, Switzerland, available at http://www.ipcc.ch.

McMullin E. (1983), Values in science, *PSA: Proceedings of the Biennial Meeting of the Philosophy of Science Association*, 4, 3–28.

Moser, S. C. (2005), Impacts assessments and policy responses to sea-level rise in three U.S. states: An exploration of human dimension uncertainties, *Global Environmental Change* 15, 353–369, doi:10.1016/j.gloenvcha.2005.08.002.

Moss R.H., and S.H. Schneider (2000), Uncertainties in the IPCC. TAR: Recommendations to lead authors for more consistent assessment and reporting, in *Guidance Papers on the Cross Cutting Issues of the Third Assessment Report of the IPCC*, R. Pachauri, T. Taniguchi, and K. Tanaka (eds.), World Meteorological Organization, Geneva.

National Science Foundation Act of 1950 (Public Law 810507).

National Science Foundation (2008), Status Update on NSF Implementation of Section 7009 of the America COMPETES Act (ACA): Responsible Conduct of

Research Advisory Committee for Business and Operations Spring Meeting, 29–30 May 2008, https://www.nsf.gov/oirm/bocomm/bo/bfa_updates_handout2final_27may08.pdf.

National Science Foundation (2013), Proposal and Award Policies and Procedures Guide, NSF 13-1OMB Control Number: 3145-0058, http://www.nsf.gov/pubs/policydocs/pappguide/nsf13001/gpg_index.jsp

Proctor, R. N. (2012), *Golden holocaust: Origins of the cigarette catastrophe and the case for abolition*, University of California Press, Oakland, CA.

Purkey, S.G., G.C. Johnson, and D.P. Chambers (2014), Relative contributions of ocean mass and deep steric changes to sea level rise between 1993 and 2013, *Journal of Geophysical Research: Oceans*, 119, 7509–7522, 10.1002/2014JC010180

Rahmstorf, S. (2010), A new view on sea level rise, *Nature Reports Climate Change*, 4, 44–45, doi:10.1038/climate.2010.29.

Rahmstorf, S. (2013), Sea level in the Fifth IPCC Report, online at *Real Climate*, 15 October 2013, http://www.realclimate.org/index.php/archives/2013/10/sea-level-in-the-5th-ipcc-report/#ITEM-15875-4.

Rudner, R. (1953), The scientist *qua* scientist makes value judgments, *Philosophy of Science*, 10, 1, 1–6.

Schaeffer, M., W. Hare, S. Rahmstorf, and M. Vermeer (2012), Long-term sea-level rise implied by 1.5°C and 2°C warming levels, *Nature Climate Change*, 2, 867–870.

Schienke, E.W., N. Tuana, D.A. Brown, K.J. Davis, K. Keller, J.S. Shortle, M. Stickler, and S.D. Baum (2009), The role of the NSF broader impacts criterion in enhancing research ethics pedagogy, *Social Epistemology*, 23(3–4), 317–336.

Sen, A. (1991), Utility: Ideas and terminology, *Economics and Philosophy*, 7, 277–283 doi:10.1017/S0266267100001425. doi:10.1038/nclimate1584.

Smith, J.B., H. Schellnhuber, M.M. Qader Mirza, S. Fankhauser, R. Leemans, L. Erda, L. Ogallo, B. Pittock, R. Richels, C. Rosenzweig, U. Safriel, R.S.J. Tol, J. Weyant, and G. Yohe (2001), Vulnerability to climate change and reasons for concern: A synthesis, in *Climate change 2001: Impacts, adaptation and vulnerability, Contribution of Working Group II to the Third Assessment Report of the Intergovernmental Panel on Climate Change*, edited by J.J. McCarthy, O.F. Canziani, N. A. Leary, D.J. Dokken, and K.S. White, Cambridge University Press, Cambridge and New York.

Solomon, S., D. Qin, M. Manning, Z. Chen, M. Marquis, K.B. Averyt, M. Tignor, and H.L. Miller (eds.) (2007), *Contribution of Working Group I to the Fourth Assessment Report of the Intergovernmental Panel on Climate Change*, Cambridge University Press, Cambridge and New York.

Steneck, N.H. (2007), *ORI: Introduction to the responsible conduct of research*, U.S. Department of Health and Human Services Dept., Office of Research Integrity. Available at http://ori.hhs.gov/ori-introduction-responsible-conduct-research.

Thagard, P. (2012), *The cognitive science of science: Explanation, discovery, and conceptual change*, MIT Press, Cambridge, MA.

Tuana, N. (2013a). Gendering climate knowledge for justice: Catalyzing a new research agenda, in *Research, action and policy: Addressing the gendered impacts of climate change*, edited by M. Alston and K. Whittenbury, Springer Press.

Tuana, N. (2013b), Embedding philosophers in the practices of science: Bringing humanities to the sciences, *Synthese: An International Journal for Epistemology, Methodology and Philosophy of Science*, 190(11), 1955–1973, doi:10.1007/s11229-012-0171-2.

11. THE EMERGING FIELD OF GEOETHICS

Peter Bobrowsky[1,5], Vincent S. Cronin[2,5], Giuseppe Di Capua[3,5],
Susan W. Kieffer[4,5], and Silvia Peppoloni[3,5]

Abstract

The geosciences need practitioners who possess an ethical conscience and the desire to act responsibly. Ethically responsible geoscientists will achieve success and satisfaction by carrying out excellent research and professional activities, and by maintaining honest and open collaborations with colleagues. Such individuals will be able to contribute to building a resilient society, be better prepared to face global economic and environmental challenges and be willing to take concrete actions for the conservation of the geo-environment. Geoethics provides ethical, social, and cultural values for the scientific community and for society as a whole. Geoethics represents a new vision of a world in which it is possible to maintain a more balanced relationship between humans and nature, considering modern economic and social development expectations. This chapter illustrates some aspects of geoethics, provides an overview of its basic values and themes, and highlights prominent global issues that involve geoethics, including climate change, geo-risks, land management, exploitation of geo-resources, and sustainability. The International Association for Promoting Geoethics (IAPG) provides a multidisciplinary platform for discussion, a place where multidisciplinary collaboration can strengthen the development of geoethics from a scientific and philosophical perspective, in order to better introduce geoethical values into society.

[1] Geological Survey of Canada, Sidney, British Columbia, Canada

[2] Department of Geosciences, Baylor University, Waco, Texas, USA

[3] Istituto Nazionale di Geofisica e Vulcanologia (Italian Institute of Geophysics and Volcanology), Rome, Lazio, Italy

[4] Department of Geology, University of Illinois, Champaign, Illinois, USA

[5] International Association for Promoting Geoethics, (www.geoethics.org).

Scientific Integrity and Ethics in the Geosciences, Special Publications 73,
First Edition. Edited by Linda C. Gundersen.

11.1. Introduction

Many scientists and prominent individuals claim that the impact of people on the natural evolution of the planet has reached a level of alarm [*Sachs*, 2014, *Pope Francis*, 2015]. Although scientific and engineering advances often give society the capability of finding solutions to problems, humans and institutions frequently pay little attention to the negative effects of their actions, particularly those that may arise in the medium and long term. For several years, the scientific community has debated which geo-environmental indicator/parameter should be used to formalize the Anthropocene [*Crutzen*, 2002] as a new geologic epoch, an interval of time during which the global and local impacts of man-made actions on the dynamics of the Earth are visible in the geologic record [*Lewis and Maslin*, 2015; *Zalasiewicz et al.*, 2015].

In addition to the stratigraphic aspect, there is also a cultural aspect to be considered. Humans recognize themselves as an additional force of nature capable of producing or inducing changes on the geosphere and the biosphere. Thus, they are aware that their actions can have an impact on the Earth's ecosystems. This awareness suggests that humans also recognize the clear responsibility and necessity to act to protect those ecosystems and to ensure the long-term safety for our species. Only a respectful approach to the geological and biological processes of the planet that takes into account different temporal scales of occurrence can avert the risk of entering into a state of nonsustainability for our growing needs of development, with inevitable but nonpredictable consequences for the social, political, and economical balances.

From this perspective, the geosciences play a fundamental role in defining the limits of sustainability of the planet and in leading to behavior required to respect these limits. The Earth is made up of a set of complex, unique, and mutually necessary systems: the biosphere (the realm of all living organisms), and the geosphere (the land surface, the solid Earth, the hydrosphere, the cryosphere, and the atmosphere). These systems strongly interact with each other on a wide range of scales in the space-time continuum. The need to integrate many different skills to effectively address studies of Earth systems thus becomes evident. The awareness of how our technical actions impact the previously mentioned systems brings us to the ethical issue of determining what will best ensure the present and future preservation of life for all species, by seeking a balance between scientific-technological progress and the conservation of geodiversity and biodiversity.

Geoscientists have the particular and unique skillsets necessary to solve problems created by the exploitation of the Earth's mineral, energy, and water resources, the defense against natural hazards, the protection and enhancement of geodiversity, and the compilation of geo-environmental knowledge for society. However, the technical and professional skills used by geoscientists in solving complex problems can be most effective when accompanied by training that carefully considers the social and ethical implications of geological practice [*Mogk and Geissman*, 2014; *Mogk et al.*, 2018 this volume]. If the behavior of geoscientists is not guided by geoethical principles, even scientifically competent individuals may not find and choose solutions that are equally respectful of the natural dynamics of our planet. Professional research and practice require geoscientists to consider the quality of the data, respect the scientific method, maintain a cooperative attitude, and adopt critically constructive but collaborative approaches towards their scientific and professional communities.

Geoethics aims to provide a framework of values within which geoscientists can act in a conscious and responsible way within the profession while serving society. But geoethics can be much more. It can become a common way of thinking about the planet in a sustainable manner [*Selle*, 2014], one that is widespread and shared in different cultural, social, and economic contexts, a functional way through which to build (or rebuild) a healthier balance between humans and nature, thus contributing to a "mature" Anthropocene [*Bohle*, 2015]. Geoethics can provide a new understanding of geosciences as tools for appropriate management of the Earth. Geosciences are not only a collection of useful scientific and technical data and knowledge, they also represent a significant cultural resource capable of influencing our future while still considering the current and future complex of global problems affecting societies. A sustainable world can also be economically beneficial to society as a whole.

11.2. Overview

11.2.1. The Origins of Geoethics

In the 1940s, Aldo Leopold (1887–1948) proclaimed a need to develop a new relationship between humans and the natural environment and identified the concept of "conservation" as the ethical criterion upon which to base this new relationship, since "Conservation is a state of harmony between men and land" [*Leopold*, 1949]. In the 1960s, the development of environmental awareness resulted in the birth of several environmental movements that acted and continue to act with a perspective geared primarily to safeguarding the biosphere, although with important differences in their approaches to saving the Earth.

In the years that have followed, environmentalism has gradually become an articulated phenomenon, one that proposes different strategies towards protecting nature and biodiversity primarily as a result of different visions [*Marshall*, 1993]. "Conservation ethics" considers that nature must be preserved in a relationship of subordination to the needs of mankind. Because natural resources are functional to human life, society must exploit them in a sustainable way, through actions that allow their rational and prudent use, all the while being aware of their limits and potential for exhaustibility.

The "ecologic extension," a more eco-centric approach, points out that the value of nature to mankind must be balanced by the intrinsic value of nature itself. Therefore, this approach proposes taking strong and stringent actions to preserve the natural environment and requires a limited exploitation of natural resources. Finally, the "Libertarian Extension," an extreme eco-centric approach, pushes for the adoption of policies strongly oriented towards the preservation of nature, severely if not completely limiting the ability of humans to act upon the Earth.

Environmental ethics oscillates between extreme positions, from anthropocentrism [*Passmore*, 1974] to eco-centrism [*Næss*, 1973; *Devall and Sessions*, 1985]. The former promotes the vision of humans as separate from nature, whereas the latter considers humans an integral part of nature.

Geoethics rediscovers, expands, and enhances the cultural dimension of geoscience knowledge [*Frodeman* 1995; *Frodeman and Baker*, 2000; *Peppoloni and Di Capua*, 2012] as a basic element of a holistic vision that goes beyond the dualism between humans and nature. Geoethics emphasizes the ethical criterion of responsible human action towards the Earth [*Peppoloni and Di Capua*, 2012], not only in its biotic components (biodiversity) but also in the abiotic (geodiversity).

The geologist Antonio Stoppani (1824–1891) introduced the concept of the Anthropozoic Era [*Stoppani*, 1873]. He identified humans as a new "geological force" and thus as an integral and essential part of nature. As an early popularizer of geological knowledge, Stoppani thus became one of the pioneers of geoethics.

Geoethics recognizes the contingency of human evolution on the planet [*Pievani*, 2009, 2012], identifies *Homo sapiens* as a geological force acting on the geological and biological environments, and assigns to humans an ethical responsibility that arises from the consciousness of being a modifier of Earth systems.

Near the end of the last century the more modern concept of the Anthropocene finally crystallized [*Crutzen*, 2002]. An intense scientific debate about this concept ensued. Because geology adheres to temporal and stratigraphic limits [*Lewis and Maslin*, 2015; *Zalasiewicz et al.*, 2015], the discipline of geoethics is required in order to understand the inevitable consequences (positive and negative) of human progress, especially when we consider the enormous amount of energy and natural resources used for economic development and the wide environmental spaces occupied and modified by human beings. The incredible increase in population and the legitimate aspirations of every individual to improve his or her own material conditions can find in geoethics a frame of reference for values rooted in geology. Geological thinking has the peculiar characteristic of being able to analyze space and time in a perspective that goes beyond the human experience but that accepts the presence of people on the planet as a natural consequence of the unpredictability of natural evolution [*Pievani*, 2009, 2012].

Geoethics provides a framework for any human action on the geosphere. The responsibility that derives from this framework is the basis of the semantic meaning of the word *ethics* [*Peppoloni and Di Capua*, 2015]. Therefore, the word *geoethics*, as used starting in the early 1990s [*Savolainen*, 1992; *Cronin*, 1992], signifies the duty of mankind to behave responsibly and become the natural consciousness of the planet. To place humans at the center of the discussion on geoethics does not represent a new form of anthropocentrism but rather stresses that only by accepting their responsibility initially toward themselves can humans become fully aware of their role as an active geological force. Then changes in the dynamics of Earth systems can be made, when necessary, through responsible behaviors and proper practices toward geodiversity and biodiversity [*Ellis and Haff*, 2009].

As guardians and developers of geoscience knowledge and given their particular sensitivity towards natural systems, geoscientists must assume the responsibility of promoting a new way of thinking about human lives in relation to Earth systems. The definition of geoethics proposed by Peppoloni and Di Capua [2015] summarizes this vision: "Geoethics consists of research and reflection on the values which underpin appropriate behaviors and practices, wherever human activities interact with the geosphere." It deals with the ethical, social, economic, and cultural implications of using Earth sciences for societal benefits, represents an opportunity for geoscientists to consider their activities under an ethical perspective, and provides an avenue for increasing societal awareness of problems related to geo-resources exploitation and energy supplies, geo-environmental changes, and geo-hazards [*Lucchesi and Giardino*, 2012; *Peppoloni and Di Capua*, 2012].

11.2.2. Values

Geoethics addresses how geoscientists should manage their individual consciences, their behavior towards their colleagues, their stewardship of the Earth, the responsible use of geo-resources, and the mitigation of geo-hazards as well as the involvement of the public in education on geosciences and Earth's processes and their involvement in the development of important public policies. The responsible behavior of geoscientists is a key point of geoethics [*Peppoloni and Di Capua*, 2012 and 2016]. But responsible behavior requires a conscious decision. A choice is possible only when the freedom to choose exists. In other words, freedom of action is a founding value of geoethics. Geoethics and freedom are intimately connected. One cannot act responsibly and ethically if one is not free to choose between possible alternatives. Therefore, in conducting their activities geoscientists should consider if they are free from compromises and (or) conflicts of interest, including political, social, and psychological pressures. Geoscientists must conduct occasional self-assessments to determine if conditions allow them to make ethical choices in their activities. The individual dimension of each geoscientist is the basis of discussion within geoethics, to which the dimension related to the collaborative work is included.

Because geoscientists know the complexity of Earth systems, they are aware of the need to investigate and understand those systems through multidisciplinary, interdisciplinary, and transdisciplinary approaches [*Choi and Pak*, 2006]. Multiple approaches to geoscience problem solving are necessary to ensure completeness in analysis, an important requirement in examining problems that include many variables. The value of multiple approaches is that they involve the integration of different knowledge, skills, and experience as well as the sharing of scientific methods through professional collaboration, where openness to the comparison between colleagues with different ideas and the sharing of data, procedures, and results are routine. The same is true of geoethics as a "discipline".

This particular sensitivity to and way of approaching nature make the geoscientist an indispensable figure within society. The commitment to serve the public's best interests is a fundamental duty of the responsible geoscientist. This commitment also implies an ethical duty to educate people about the importance of geoscience and transfer the information needed to build a knowledgeable society. Geoscientists possess knowledge that can benefit all. They can identify fundamental cultural and technical turning points in the sustainable and prudent management of geo-resources and energy and can suggest ways to mount effective defenses against geo-risks [*Di Capua and Peppoloni*, 2014; *GSL*, 2014; *Lambert et al.*, 2013].

Sustainability is now one of the reference values of our modern culture [*WCED*, 1987]. Prolonged use of a resource well represents the need to find a balance between environmental protection and the development of the society. When it comes to geo-resources, true sustainability is in many cases either impractical or impossible. Nevertheless, in a wider vision, the concept of sustainability should not be related to a single resource (i.e., oil, gas, water) but rather to Earth systems as a whole. Sustainability should be the working paradigm for society as it focuses on ways to reduce the use of nonrenewable resources and to increase the use of renewable resources. In this perspective, shifts in the exploitation of resources ideally should not have an impact on the entire Earth systems, especially if they may jeopardize lives of humans and other living organisms. Geoscientists therefore have an ethical obligation to propose multidisciplinary solutions that take into account the natural environment in a balanced global manner.

11.2.3. Themes

The field of study of geosciences is very broad and concerns all Earth systems. Geoethics first of all proposes itself as a discipline aimed at reflecting upon and discussing in scientific terms the cultural, ethical, and societal implications of the various geo-disciplines in relation to their different sectors of interest and application, with active contribution and input from other disciplines such as philosophy, sociology, and economics. In particular, geoethics performs the following:

- Highlights the social role played by geoscientists and their responsibilities, focusing on the ethical, cultural, and economic repercussions that their choices may have on society and the environment [*Peppoloni and Di Capua*, 2012; *Wyss and Peppoloni*, 2015]
- Promotes the cooperation of geoscientists with philosophers, sociologists, economists, biologists, chemists, psychologists, news media, and others in order to favor mutual cultural exchange, including their participation in the intellectual life of society and the development of their critical thinking [*Cervato and Frodeman*, 2012; *Peppoloni*, 2012a and 2012b]
- Fosters the proper and correct dissemination of scientific studies, data, and results and promotes responsibility, integrity, expertise, and professionalism in all geoscientific research and practice, by providing guidance in promoting honesty and integrity in collaborative activities [*AGI*, 2015; *Mayer*, 2015; *Peppoloni*, 2015; *Peppoloni et al.*, 2015] and for defining data policies to permit wider circulation of information while at the same time respecting intellectual and public property
- Organizes effective teaching tools to develop awareness, share values, and increase responsibility among geoscientists, especially those in the early stages of their careers [*Mogk and Geissman*, 2014]
- Promotes geo-education as a way to emphasize the importance of geosciences and as a means of attracting young people to geoscience studies, with the ultimate goal of encouraging ethical and critical thinking and offering a new vision of the world to society
- Encourages critical analysis of the use and management of geo-resources by trying to find socioeconomic solutions within the framework of sustainability for future generations [*Lambert et al.*, 2013]
- Deals with problems related to geo-risks management, communication, and education in order to improve community resilience to natural disasters [*Di Capua and Peppoloni*, 2014; *Dolce and Di Bucci*, 2015; *Guzzetti*, 2015]
- Aims to analyze and improve the relationships between the scientific community, decision makers, the mass media, and the public [*Di Capua and Peppoloni*, 2014; *Dolce and Di Bucci*, 2015; *Guzzetti*, 2015; *Lambert and McFadden*, 2013; *Liverman et al.*, 2008]
- Promotes the educational, aesthetic, and scientific value of geological heritage and geodiversity in order to improve social awareness of the importance of protecting geological items and outcrops [*GA*, 1975, 2011] and encourages the development of geo-parks [*UNESCO*, 2006] and geo-tourism [*Dowling*, 2010; *Allan*, 2015] as tools to achieve these objectives
- Highlights the value and usefulness of geoscience knowledge in daily life by using disciplines such as medical geology [*Selinus et al.*, 2013] and forensic geosciences [*Bergslien*, 2012]

- Encourages inclusive policies in the geoscience community, particularly regarding women [*Holmes and O'Connel*, 2003; *Bell et al.*, 2003], visible minorities, and geoscientists with disabilities [*Asher*, 2001].

Geoscientists often claim social roles and spaces of communication in society as their right. Nevertheless, they also have a duty to consider ethical issues as the basis of their professional training and overcome anachronistic and damaging disciplinary divisions [*Guzzetti*, 2015]. In this manner, they will gain authority as well as trust and respect of other components of society.

11.3. Geoethics and Geoscientists

11.3.1. Role and Responsibility of Geoscientists

The study of the many interrelationships that exist within the Earth systems requires the integration of multidisciplinary approaches and different skills and methods of investigation. In fact, modern science requires that professional scientists be able to work independently and in multidisciplinary teams.

The role of the geoscientist is essential in the analysis of phenomena and processes and in their integrated and dynamic understanding, as well as in their management and supervision. Their skills allow them to frame natural phenomena and choose the more appropriate interventions within a given natural context and time dimension. Qualities such as precision, accuracy, reliability, punctuality, attention, self-sacrifice, self-denial, patience, enthusiasm, and intuitiveness are some prerequisites for a qualified geoscientist. Nevertheless, geoscientists can be responsible and aware of the ethical, social, and cultural implications of their activities only if they also possess additional virtues such as integrity, honesty, transparency, collaborative attitude, humility, and respect for the values, ideas, and scientific hypotheses of others.

Responsible geoscientists understand the importance of correctly communicating the results of their research and appreciate the importance of transferring their scientific knowledge to society. Sharing data and ideas with colleagues, decision makers, and citizens can open the way to valuable and functional relationships that will benefit the planet and humankind. Geoethics offers this possibility.

The geoscientist is the core of an ethical reference system in which individual, professional, social, and environmental values coexist. In fact, geoethics refers to the self-behavior of a geoscientist as well as to his or her behavior toward colleagues, with regard to research integrity and professionalism; but it includes also aspects of responsibility toward society and the environment.

Matteucci *et al.* [2014] are confident that the introduction of an ethical oath for geoscientists, the "Geoethical Promise," which is akin to the oath for physicians, could help develop a binding awareness of a geoscientist's professional and social responsibilities. These responsibilities arise from the possession of specific knowledge that has clear social implications and involves moral obligations to serve society to the best of one's own scientific and technological potential. The ethical obligation of modern geoscientists is based on fundamental values that should guide their activity, such as the responsibility to ensure excellence in science and in the search for truth, the promotion of sustainability, and the transfer of knowledge to colleagues and society, as a lifelong commitment.

11.3.2. Geoethics and Research

Honest and open collaboration among colleagues is a prerequisite for building modern scientific, technological, and professional networks able to address the complexity of an interdependent world.

Geoscientists often work in teams, where they manage large amounts of data. Their careers are often certified and evaluated on the basis of the number and quality of publications they produce. Personal ambitions may collide with increasing competition between professional groups, which can lead to unethical behavior [*Mayer*, 2015] or conflicts of interest [*Oreskes*, 2015; *Tollefson*, 2015]. Inappropriate attitudes toward colleagues, falsification and concealment of data and results, and conflicts of interest are just some of the unethical actions that may arise during research and professional activities and that will undermine both the general trust in science and technology and the potential to attain real progress for society.

For these reasons, the scientific community recognizes the importance of identifying tools to promote integrity among its members, to define principles, and to develop responsibilities that will govern the practice of science. Two significant statements have arisen: the Singapore Statement on Research Integrity [2010] and the Montreal Statement on Research Integrity in Cross-Boundary Research Collaborations [2013]. As expressed by the authors of the Singapore Statement, "The value and benefits of research are vitally dependent on the integrity of research. While there can be and are national and disciplinary differences in the way research is organized and conducted, there are also principles and professional responsibilities that are fundamental to the integrity of research wherever it is undertaken." (http://www.singaporestatement.org/statement.html). The Singapore Statement is a global guide for the responsible conduct of research and lists the following ethical principles and responsibilities: honesty and accountability in all aspects of research, professional courtesy and fairness in collaborative work, and good stewardship of research on behalf of others. These are the cornerstones upon which the research of geoscientists should be based. Even though the Singapore Statement relates specifically to research activity, it contains universal principles that can be followed in all activities conducted by geoscientists, whether they work in the public or the private sector. The Montreal Statement addresses the problems inherent in professional collaborations in more detail, identifying the principles behind a framework that will assure fruitful and transparent collaboration among colleagues.

These two documents do not solve all the problems of research misconduct or conflict of interest, but they are useful tools for making scientists aware of their personal behavior and their responsibilities toward their colleagues. Although some in the geoscience community seem to be unaware of the difficulties that can arise from geoethical misconduct, some sensational cases have been well documented in the scientific literature [*Mayer*, 2015] and make even more persuasive the case for considering geoethics carefully in the coming years.

11.3.3. Geoethics and Professional Societies

Training geoscientists to be aware of their professional behavior begins with honor codes in academia during the undergraduate and graduate years and continues into professional practice through membership in professional societies such as the American Geophysical Union (AGU), the European Geophysical Union (EGU), the Geological Society of America (GSA), and numerous other geoscience societies as well as the American Association for the Advancement of Science (AAAS). In order to find codes relating to personal conduct, organizational conduct, and conduct toward society and the environment, one must often look at both the codes of conduct and the mission/values/objectives of the professional societies themselves.

Organizations go beyond setting guidelines for individual behavior by encouraging members to examine their relation to society and to the environment. Action in these broader areas requires careful balancing of statements to which we lend our professional credibility and statements that might be misconstrued as political advocacy. In the past, organizations have tended to shy away from strong statements about the relationship of humans with the environment for fear of appearing to take advocacy positions. However, owing to the pressing need for action to preserve the environment, more and more societies have adopted broader statements and have become more adept at avoiding clear advocacy positions.

One example of how a professional society can be guided by ethical considerations is the statement regarding climate change on March 18, 2014, by AAAS:

As scientists it is not our role to tell people what they should do or must believe about the rising threat of climate change. But we consider it to be our responsibility as professionals to ensure, to the best of our ability, that people understand what we know: human-caused climate change is happening, we face risks of abrupt, unpredictable and potentially irreversible changes, and responding now will lower the risk and cost of taking action.

In issuing statements like this, professional societies are guided by both their codes of ethics/conduct and their organizational values/objectives.

As a second example, in its Strategic Objectives, AGU states its desires to do the following:

- increase awareness of the importance of Earth and space science issues for nonscience audiences;
- increase effectiveness and recognition of AGU among decision makers as an authoritative source of integrated, interdisciplinary Earth and space science information;
- increase awareness of the reality and consequences of global climate change among scientists and the public;
- increase the role of Earth sciences in informing policy and mitigating impacts of natural disasters;
- raise awareness of natural resource limitations and increase the application of (AGU) Earth sciences in developing solutions for the sustainability of the planet.

As a final example, the American Geosciences Institute (AGI) is the umbrella organization for 50 American professional societies such as AGU and GSA. In an effort to summarize codes of conduct, AGI recently revised its ethical guidelines [*Boland and Mogk*, 2018, this volume] to include the following:

"*In day-to-day activities geoscientists should:*

- Be honest.
- Act responsibly and with integrity, acknowledge limitations to knowledge and understanding, and be accountable for their errors.
- Present professional work and reports without falsification or fabrication of data, misleading statements, or omission of relevant facts.
- Distinguish facts and observations from interpretations.
- Accurately cite authorship, acknowledge the contributions of others, and not plagiarize.
- Disclose and act appropriately on real or perceived conflicts of interest.
- Continue professional development and growth.
- Encourage and assist in the development of a safe, diverse, and inclusive workforce.
- Treat colleagues, students, employees, and the public with respect.
- Keep privileged information confidential, except when doing so constitutes a threat to public health, safety, or welfare.

"*As members of a professional and scientific community, geoscientists should:*

- Promote greater understanding of the geosciences by other technical groups, students, the general public, news media, and policy makers through effective communication and education.
- Conduct their work recognizing the complexities and uncertainties of the Earth system.
- Sample responsibly so that materials and sites are preserved for future study.
- Document and archive data and data products using best practices in data management, and share data promptly for use by the geoscience community.

- Use their technical knowledge and skills to protect public health, safety, and welfare, and enhance the sustainability of society.
- Responsibly inform the public about natural resources, hazards, and other geoscience phenomena with clarity and accuracy.
- Support responsible stewardship through an improved understanding and interpretation of the Earth, and by communicating known and potential impacts of human activities and natural processes."

Such guidelines for personal and organizational ethical behavior are meant to help geoscientists retain professional integrity as they interact with society and the environment.

11.3.4. The Meaning of Serving Society

In a landmark presidential address to the Geological Society of America in 1992, outgoing GSA president E-an Zen contemplated the role of geologists in society [*Zen*, 1993]. First, he addressed the need to "relate scientific knowledge to society's sense of value—what is right, what is wrong, what is important..." He suggested that one of the first actions required was to show people that scientists are not the TV caricatures that they often were, and still are, portrayed as, but rather are normal human beings, sharing the concerns, impulses, and other human traits common to our species. Stressing that science is a public enterprise with public funding, he urged that scientists dialogue with the public, cultivating shared values as well as shared interests by focusing on the common stake that we have in the future.

One of the major ethical issues faced by geoscientists is finding their role in the context of ever-increasing population pressures on the planet. To illustrate how these challenges have grown, consider that in 1992, when Zen gave his speech, the population of the planet was 5.448 billion. At the time of the writing of this manuscript (November 2016), the population is 7.355 billion. In 23 years, the population of the planet has increased by nearly 1,677,000,000 people, a 31% increase. Alone among all scientists, geoscientists and geoengineers are the ones who explore, find, and extract needed resources. Extraction of these resources is an activity that often damages the environment, more so in the past and less so today with the adoption of more ethical practices. Yet, along with ecologists, geoscientists are also uniquely qualified to help protect the environment. Geoscientists and society face many ethical challenges as we strive to provide resources for the present, conserve resources for future generations, and preserve the environment for other species if not for their own intrinsic value, then because a thriving ecosystem is essential for the survival of *Homo sapiens*. As individual geoscientists acting within our narrow mandate to explore the Earth, we can contribute much to the body of knowledge about how the planet works. However, in order to serve society, we must interface with those who make decisions on how our body of geo-knowledge is used. Treading along this interface is not easy, and the slightest slip, or appearance of slip, can cost the geoscientist his or her credibility.

An interesting and successful experiment in California may set a precedent for how this balancing act can be accomplished. A partnership was developed between the U.S. Geological Survey (USGS) and the office of the Mayor of the City of Los Angeles to explore preparation for and response to the inevitable large earthquake that will impact the city. A prominent USGS seismologist, Dr. Lucy Jones, was assigned by her agency to work in the office of Mayor Eric Garcetti. Jones perceived her job as explaining the science and the consequences of taking action or of not taking action. But she always stopped short of getting involved in the decision-making process regarding actions. In this way, she retained her credibility as a scientist while gaining the ears and respect of the decision makers in Los Angeles. Mayor Garcetti mandated that Jones should put together recommendations on building resilience to earthquake damage.

The collaboration between the USGS and the City of Los Angeles resulted in a survey to determine which buildings are at risk. Among those at risk, two major categories were defined and studied. The first, the so-called "soft-first-story" buildings that were typically built before 1980 when earthquake reinforcement became mandatory, are most at risk. These structures characterize a particular style of Los Angeles apartment building, typically made of wood, in which the base level is smaller than upper levels in order to allow protected parking spaces at street level. It was recommended that mandatory retrofits be done for these buildings within five years. Longer time scales are proposed for the second category: concrete structures for which the required work is more complicated and expensive and for which tenants will have to be evacuated during retrofitting. Recommendations were also made for reinforcing water systems and their backups, parts of which cross active faults, so that water pressure would be maintained for fighting the fires that will inevitably break out during earthquakes.

On October 9, 2015, the mayor's office announced that the building retrofit ordinance passed the city council with unanimous support, and the mayor signed it into law (http://www.lamayor.org/mayor_garcetti_signs_historic_earthquake_retrofit_measure_into_law_ensures_safety_for_thousands_of_angelenos).

Interaction with decision makers is but one way in which geoscientists can serve society. Another way is in the broad area of education. Geosciences are seriously underrepresented in our science curricula. We can, and should, lobby for our fair share of the curriculum and then provide a supply of talented teachers so that future citizens and decision makers are aware not only of the resources of the planet, but also of the constraints and limitations of the planet which we inhabit.

11.4. The Geoethical Approach to Global Issues

11.4.1. Climate Change

Many geoscientists have documented evidence for short and long periods of shifting global climate. In the past, much of this research focused on covering the billions of years of Earth history preceding the appearance of humans. For years, the discipline of geosciences dominated climate change research, given the defining attributes of climate change and the boundary conditions needed to document time-based fluctuations. What is studied, where data are preserved, how they are collected, and the underlying focus required to look at issues through time made geosciences ideally predisposed for such study. As the study of climate change shifted from the distant past to those changes that have occurred during the last few centuries, however, geosciences were displaced by other disciplines: astronomy, biology, physics, chemistry, and mathematics/computer sciences. Geoscientists appeared to become lost in the transition once the link between anthropogenic input and changing climate was proposed. One can only speculate on what caused geoscientists' initial reluctance to enter the field of anthropogenic influences on changing climates.

Only in the last few years have geoscientists once again sought to reaffirm the importance and relevance of their data, knowledge, and skills with regard to this often contentious, but always relevant issue (see *IPCC*, 2014, Contents). Parameters associated with climate are generally viewed through unique components within the atmosphere, biosphere, cryosphere, hydrosphere, and lithosphere. Each of these components shares attributes, characteristics, outputs, and products that may be well preserved directly and indirectly in the sediment/rock record. Consequently, geoscientists should play a leading role in influencing the scope, direction, pace, and reliability of information gleaned, studied, analyzed, and interpreted in climate change research. Plate tectonics, solar outputs, ocean variability, carbon/oxygen shifts, temperature modeling, biological extinctions/evolution, sea-level fluctuations, glacier advances/retreats, and so on are but a few of the topics of relevance to modern-day climate change studies to which geosciences can and do contribute significant skills and knowledge.

"Climate change" is now firmly entrenched in the culture of contentious topics, heated debates, political awkwardness, public confusion, and, regrettably, scrutiny of scientific ethics. Geoscientists, like many others from related disciplines involved in climate studies, are held to higher standards of accountability, especially when one considers the great chasm that exists between climate change "skeptics" and "proponents." And when real or presumed inappropriate actions attract media attention, the end result takes on a life of its own. For instance, a few years ago, several thousand personal e-mail messages exchanged between International Panel on Climate Change (IPCC) scientists were leaked to the public social network, sparking years of arguments and accusations that focused on conspiracy and data manipulation, political agendas, unprofessional practices, and poor judgment by some climate scholars. This so-called Climategate scandal proved to be a difficult hurdle for the IPCC to overcome and detracted from the underlying message of this influential body of researchers. Few believed that information was actually being fabricated by the accused scientists in question, but the perception that respected scientists were not bound to follow the highest ethical standards lingers. The media scandal provided a sobering lesson to geoscientists. One must tacitly adhere to the formalities of the scientific method, welcome discourse and disagreement, adopt transparency of practice, and at all costs remain disconnected from judgments in conclusions.

The role of the geoscientist is clearly to provide accurate information, unbiased interpretation, and fact-based replies to queries. Confusion arises when personal opinions and beliefs merge with our scientific practice. Scientific skepticism is fundamental to our success in performing good science, and indeed, questioning all our beliefs underlies the success of the science. However, skepticism should not be confused with obstinacy and denial. Distinguishing healthy skepticism from unethically motivated denial is still a barrier to successfully communicating the valuable information managed by the climate change community. The challenges facing geoscientists in climate change research should diminish appreciably when they are working firmly within the constructs of a geoethical paradigm.

Finally, as noted elsewhere in this chapter, the role of professional geoscience organizations, learned societies, and other bodies is paramount in adhering to the principles of disciplinary geoethics. Easily accessible and widely distributed commentary in the form of "position papers" is a fundamental obligation of the geosciences. In the case of climate change discussions, groups such as the Geological Society of America (https://www.geosociety.org/gsa/positions/position10.aspx), Geological Society of London (https://www.geolsoc.org.uk/climaterecord), and the Association of Professional Engineers and Geoscientists of British Columbia (https://www.egbc.ca/getmedia/a39ff60e-80a1-4750-b6a5-9ddc1d75248a/APEGBC-Climate-Change-Position-Paper.pdf) have been instrumental in leading the way for such socially beneficial interaction through their own position papers.

11.4.2. Geo-risks

Ensuring the safety of individuals and society is a fundamental obligation of those geoscientists involved in the study of geo-hazards and geo-risks. The efficacy with which this fundamental obligation is being met can best be assessed by looking at the annual global synopsis by the Munich Re Group on the state of affairs for hazards, risks, and disasters in the preceding year. In their 2015 assessment, they confirm that the loss of lives due to natural disasters around the world in 2014 was the second lowest since 1980 [*Munich Re*, 2015]. This conforms to an overall trend in lowering the death rates from natural catastrophes in both developed and developing nations. Better informed individuals and communities, more resilient land-use practices, higher standard building codes, enhanced emergency response measures, and so on have collectively contributed to reducing the risks and impacts of many natural disasters. In this regard, geoscientists have performed well through their contributions to studying natural hazards and risk reduction through collaborative interdisciplinary work, effective community engagements, and inclusive reliance on local, regional, and national government forces [*Bobrowsky*, 2013].

The flip side of natural catastrophes and disasters is that any human losses are still an unacceptable consequence. Moreover, the clear trend in economic losses during the past few decades has been inversely proportional to the human losses. Costs from natural catastrophes continue to rise at alarming rates. What used to be measured in millions of dollars is now routinely measured in billions of dollars. One pervasive observation remains: human losses are more often borne by developing nations, whereas the economic losses are more often borne by developed nations.

In the past, the question of geoethics as related to geo-risks was rarely considered, in contrast to the more common queries surrounding geoethics as related to the resource sector for instance. Trust and complacency in the geoscience community have been tested many times, for instance, the 1985 eruption of Nevado del Ruiz volcano in Colombia, which pitted volcanologists against government officials and resulted in the death of some 23,000 individuals after warnings by the professionals were dismissed [*Bruce*, 2001]. In this case, although the opinions of the geoscientific community were ultimately proven to be correct, there were no long-term changes in the level of enhanced trust and respect of opinion accorded to geoscientists. However, after the cataclysmic 1991 eruption of Mount Pinatubo in the Philippines, the IAVCEI Committee for Crisis Protocols published a professional code of conduct for scientists during volcanic crises [*Newhall et al.*, 1999].

The significance of ethics in geosciences took a dramatic shift in the eyes of the world following an earthquake in central Italy near L'Aquila that contributed to the death of some 300 people. Ironically, the mortality associated with this event was only 1/100th that of the 1985 Colombian disaster and yet the long-term impact for the profession was several orders of magnitude greater for political and judicial reasons. In 1985, no individual or group of individuals was held accountable for the death of thousands in Colombia. But in 2009, the Italian political and judicial systems held that the group of scientists responsible for overseeing earthquake risk in the area failed in its duty [*Hall*, 2011]. The subsequent charge of criminal negligence rocked the geoscientific community. Geoscientists around the world were more concerned with the misguided belief that practitioners are expected to predict hazardous events than they would have been with issues of inappropriate guidance or poor communications [*Cocco et al.*, 2015].

Geoscientific work in hazards and risk is less vulnerable to conflicts of interest, personal bias, contrasting values, and lack of interdisciplinarity. Problems arise instead over matters of judgment, decisions, and communication. Risk reduction is normally addressed through well-established protocols based on quantitative assessments [*Jordan et al.*, 2014; *Newhall et al.*, 1999]. Because the communication of risk is necessarily transformed into qualitative descriptors upon which life-and-death decisions are often made, it is here that practitioners need the greatest reflection within a functioning paradigm of geoethical philosophy.

11.4.3. Natural Resources

As far back as the Paleolithic, early hominids relied on a rudimentary geological knowledge to use unique formations and landscapes for protection and habitation (rock shelters), specialized mineral extraction (for ochres/stains/paints essential for cave murals and art), and the unique petrological characteristics of certain rocks (such as basalt, chalcedony, obsidian, and flint) proven to be best suited for production of tools required to hunt game. Since that time, the reliance of modern societies on geosciences as a resource-based discipline has grown exponentially. Today's geoscientific practitioners are the backbone of multibillion-dollar economies centered on an unprecedented global demand for oil, gas, minerals, and water. Geoscientists by definition play the pivotal role of ensuring that society has sustained and reliable access to essential resources. The ability to properly search for, successfully locate, economically extract, and reliably provide an incredibly diverse suite of basic natural resources for human consumption is often overlooked by most outside the geoscientific community itself.

Questions concerning ethics in the resource industry are considerable because it provides a perfect storm of competing obligations, objectives, and goals. In many cases, an individual geoscientist need not be concerned about geoethical issues because he or she works in situations far removed from such interactions; but in other cases, opportunities for unethical actions appear almost unavoidable. Responsible behavior quickly enters a gray area when geoscientists are expected to work in an environment that limits freedom of choice. Company practices that focus on profits dictate operational policy and drive workforce performance, placing a burden on geoscientists. When compromises become additive, individuals are soon complicit in actions that contravene what is best for the environment or society in general. Those dealing in geologic commodities are particularly prone to the pressures of maintaining ethical practices. The inherent nature of supply and demand markets for natural resources such as coal, gas, oil, copper, iron ore, and rare-earth elements centers on prices and profits. In the frenzy to reach profits along the food chain of the resource industry, the altruistic goal of providing necessary resources for societal benefit can quickly become lost.

A turning point regarding geoethics in the resource industry occurred in 1997 with the collapse of the Canadian company Bre-X Minerals Ltd. A few years earlier, geologist Michael de Guzman announced the discovery of gold in the jungles of Borneo (Indonesia). Although the site contained no gold, de Guzman systematically manipulated assays by salting core samples. Geologists, stock analysts, and investors around the world were deceived and helped drive the penny stock shares to an extraordinary peak of $286 CDN. Once the fraudulent actions of de Guzman were exposed, the value of company shares completely collapsed, generating one of the most prominent mining scandals of all time. As a consequence of this event, the push to regulate the professional practice of geology was initiated [*Andrews*, 2014].

Self-regulation by technical and professional bodies in the resource sector works effectively to improve the standards of performance within the geoscience industry. For example, organizations such as the Prospectors and Developers Association of Canada (PDAC) help develop and promote high standards of practice, values, and ethics to their community of members. The mission of PDAC is clear: "The PDAC exists to promote a responsible, vibrant and sustainable Canadian mineral exploration and development sector. The PDAC encourages leading practices in technical, environmental, safety and social performance in Canada and internationally." In March 2007, the PDAC developed a special publication titled "Sustainable Development and Corporate Social Responsibility [CSR]: Tools, Codes and Standards for the Mineral Exploration Industry" (http:// www.eldis.org/document/A39525). This document summarizes 36 national and international CSR codes, standards, and tools that provide best practice examples for industry practitioners to adopt. Moreover, the PDAC addresses other key geoethical issues, including land access and land-use planning, Aboriginal rights and engagement, and revenue transparency, and promotes federal legislations such as the government of Canada's Extractive Sector Transparency Measures Act. The PDAC's actions provide a good example of how geoethics can be intimately woven into the fabric of an entire resource industry such as mining.

Geological resources are extremely difficult to extract. The impact on the environment can be substantial but not always permanently damaging. Minimizing impacts, reducing risks, and adopting strong mitigation and remediation practices have become standard actions for most practitioners in the resource industry. Thousands of documents now exist to regulate or guide resource industries in various phases of development from exploration to construction to closure. For instance, in Alberta, Canada, the Alberta Energy Regulator has jurisdictional responsibility for managing water and the environment with respect to energy resource activities in the province. Documents such as the 141-page "Best Management Practices for Pipeline Construction in Native Prairie Environments" help regulate the energy sector in that province to function according to the highest standards (https://www.aer.ca/documents/applications/ BestManagementPracticesPipeline.pdf). Another Canadian example is the "Guide for Surface Coal Mine Reclamation Plans" developed and overseen by the Nova Scotia government in eastern Canada (https://novascotia.ca/nse/ ea/docs/EA.Guide-SurfaceCoalMineReclamation.pdf). The guide covers everything from site monitoring and maintenance to acid rock drainage control and public safety.

Geosciences play a significant role in the management of water resources. Conflicts continue to increase over access to water, and predictions for the future suggest that such conflicts will increase in the face of more frequent droughts, political border disputes, drying reservoirs, and restrictions to subsurface aquifers. The significance of this resource has finally been recognized at the highest levels of bureaucracy. In 2010, the United Nations General Assembly recognized that access to clean drinking water is a basic human right through Resolution 64/292 (http://www.un.org/es/comun/docs/?symbol=A/RES/64/292&lang=E). Geoscientists involved in water resources will be increasingly challenged to deal with conflicting economic, social, cultural, and political pressures, and as a consequence will be burdened with dilemmas regarding acceptable standards of practice, adhering to high values, and adopting ethical decisions.

11.4.4. Engineering Geology

"Engineering Geology is the science devoted to the investigation, study and solution of the engineering and environmental problems which may arise as the result of the interaction between geology and the works and activities of man as well as to the prediction and of the development of measures for prevention or remediation of geological hazards" [*IAEG*, 1992]. This definition encompasses both the microethical domain of individual projects and the macroethical domain of engineering geology's relationship with society, providing the expertise needed to address global issues of our changing planetary ecosystem [*Herkert*, 2005].

How does an engineering geologist maintain an ethical practice? A typical engineering geologist might answer, "Just do a good job." "Be honest." "Always meet the legal requirements of the job." In the United States they might refer to the ethical codes of the Association of Environmental and Engineering Geologists (AEG), the American Institute of Professional Geologists [*AIPG*, 2003], and the Professional Practice Handbook of the AEG [*Hoose*, 1993]. Canadian engineering geologists would refer to Canadian Professional Engineering and Geoscience, Practice and Ethics [*Andrews*, 2014], which is studied prior to their Professional Practice Exam.

Focus on project-scale issues is appropriate and necessary for most of the work of a typical engineering geologist, who often works as part of a team led by engineers. Their goal is to complete the project on time and within the budget. The limited role of the geologist on such a team is to supply quantitative geological data and useful geomorphic information to the engineers for their subsequent design work. In this tightly constrained scenario, "Do a good, honest job that meets all code requirements" might be an adequate statement portraying the nominal microethics of an average engineering geologist. However, in some cases "just meeting the codes" might be insufficient to protect public safety and welfare.

We assert that engineering geologists have a range of responsibilities in service to the public that is broader than the legal responsibility to meet code requirements. Any project will have its intended product (for example, a flood control dam built in an area that is prone to flooding during storms), but that project might have broader effects that might adversely impact other people or infrastructure [*McPhee*, 1989]. The project may be entirely legal yet could place innocent people in danger. The result of a project may persist beyond the lifetimes of those who work on it and might have adverse effects that will take decades to become evident. A professional engineering geologist needs to be an advocate for the public as well as an informed steward of the Earth who considers the broader impacts and long-term effects of a project.

We assert that the ultimate client of any engineering geologist is society and that engineering geologists share in an engineer's canonical responsibility to "hold paramount the safety, health, and welfare of the public and [to] strive to comply with the principles of sustainable development" [*ASCE*, 2017; *Cronin*, 1991, 1993]. If engineering geologists fail to act ethically as responsible scientists in the public interest, their contributions "will not be sought or valued by society" [*Slosson et al.*, 1991].

Engineering geology occupies a vital niche in society, providing essential expertise as society confronts geologic hazards and decides how to mitigate or avoid such hazards. In addition to long-recognized hazards (for example, volcanism, earthquakes, floods, expansive soils, groundwater issues, and landslides), society will have to become more adaptive and resilient in the face of regional and global-scale environmental changes in the decades to come. Examples of likely challenges include persistent droughts, severe storms, more frequent and severe flooding; coastal erosion and retreat due to rising sea levels, and inundation of areas where towns, critical infrastructure, and agricultural resources now exist.

Engineering geologists must understand, accept and act upon their responsibility to help society manage geologic hazards, regardless of whether they are natural or human-induced. Engineering geologists need to maintain their ability to manage microethical issues of day-to-day project-level work. They also need to expand their vision to include the macroethical imperative to help society understand and address sustainably the regional and global challenges it faces. No other group has the training, perspective, and knowledge of Earth's history, processes, materials, and hazards to provide this essential service. Engineering geologists must learn to speak with a voice that society listens to, considers seriously, and values.

11.4.5. Geoscience Communication

We live in an age of communication. At the dawn of the 19th century, the fastest long-distance communication involved someone riding a horse. We communicated face to face or in writing. Geologic ideas and discoveries were presented in books or in a journal produced by one of the very few scientific/ philosophical societies of that era. Today, grammar-school kids in Alabama can engage geoscientists doing research in the Arctic in real time via satellite uplinks and the internet. The world-wide-web and social media makes information, good or bad, unbiased or biased, useful or not, available to anyone who has the means (and the political freedom) to connect.

The proliferation of information sources and the wide range in quality are challenges we all face. A geoscientist working in almost any active field of research is deluged by information. Ease and immediacy of communication can overwhelm us with information while sometimes depriving us of sufficient time to turn information into enduring knowledge. As with many things, this wealth of information is not evenly distributed.

The production and distribution of research papers and other forms of formal scientific communication cost money. The traditional approach has been to assess page charges to authors for publishing their papers and to charge fees to people who would like to read the papers. Subscription-access, peer-reviewed scientific communication is a filter that excludes the vast number of people who cannot afford to pay for access. The trend toward open-access and delayed-open-access publication of peer-reviewed scientific work via the web is an enormously positive development that will empower people worldwide. As diversity among scientists increases, it is likely that the range of useful scientific ideas will also increase.

Academic geoscientists are trained to communicate with other scientists at professional meetings and through peer-reviewed scientific literature. Commercial geoscientists are trained to provide their expertise to their internal and external clients, often in confidential reports. The reward systems for geoscientists are based in part on how well we communicate with our specialized technical language. Geoscientists also have a fundamental responsibility to communicate with the rest of the public about geoscience issues, using language that is clear, direct, and understandable by the average young adult. Society paid ahead its investment in geoscientists long ago by building universities and concentrating the resources needed for our specialized technical education. We have a responsibility to report back to society, sharing what we have learned about Earth's history, processes, materials, and hazards.

Geoscientists who possess reliable, unbiased, science-based knowledge are notably absent from the public square, where important matters affecting society are brought forward and discussed. It is of great concern that the public is largely ignorant about some of the greatest challenges that it faces: problems whose solutions will require geoscience expertise, such as water and energy supply, soil conservation, geologic hazards, and supply of industrial minerals. A well-educated geoscientist has a general store of knowledge about Earth that significantly exceeds that of a typical politician or member of the public. That knowledge must be shared with every new generation. It is reasonable to expect any well-educated geoscientist to help the public understand issues such as the basic implications of the depletion (extraction) of potable groundwater, or the general risks associated with living near a fault that can produce magnitude 6+ earthquakes, or why it is not a good idea to live in a 100-year flood plain, or the hazards of living above a subduction zone or downwind from an active volcano, or the consequences of sea-level rise for coastal communities and low-elevation agricultural fields.

Well-educated geoscientists have the intellectual resources to be of service to the public as it tries to manage a range of important problems. We have an ethical responsibility to provide reliable information to society about Earth. Providing expertise is part of the social compact between geoscientists and the public. The information void that will be created if we shirk our responsibilities to educate society will likely be filled by the uninformed or the inexperienced.

11.4.6. Geoeducation

When members of a community act ethically toward one another, public benefits such as peace, justice, and general welfare are more likely to predominate. This common-sense assumption has been around for as long as there has been a written record of moral philosophy more than two thousand years in the western tradition. We observe, however, that people do not always act ethically toward one another, so how does a community promote the idea that its members should behave ethically? Three steps seem to be necessary. First, ethical behavior should be affirmed by the community as the expected norm. Second, ethical behavior should be taught as well as modeled in both formal and informal educational settings. Third, unethical behavior should be identified as unacceptable, and there should be undesirable consequences for such behavior.

Ethical behavior is essential to science, whose purpose is to develop reliable knowledge about the physical world based on reproducible observation and development of testable explanations. There is no science without honesty, and truth telling is a fundamental ethical virtue. There is an ethical element present in even the most basic scientific observations: I am measuring this carefully using an appropriate standard; I repeat the measurement several times; I assess the uncertainty in my measurements; I report my method, the data, and the uncertainty for others to scrutinize. Even the simplest scientific methodology has as its purpose the discernment of some reproducible fact about the physical world within an inevitable haze of uncertainty. That idea should be conveyed and understood from the first brief lessons about science in primary school onward. Science does not ignore uncertainty; it accepts uncertainty as an essential part of the process. Whether a primary school student grows up to become a scientist or not, it is crucial that he or she understand that the practice of science involves uncertainty. Students will eventually encounter people who are absolutely certain that science is wrong, and they may believe that the Earth is not old, or organisms did not evolve, or that human activity cannot change the global environment. Science, on the other hand, is not in the business of achieving absolute certainty. Science is in the business of framing ideas in a way that they can be tested and falsified. Science is rooted in doubt, uncertainty, and skepticism, and those roots have generated our most reliable knowledge about the physical world. Scientists seek knowledge that withstands the tests of experience and time.

In the geosciences, the imperative to act ethically should be emphasized continuously from the first Earth science module in primary school until a person takes off their field hat, lays down his or her maps and compass, and turns off the laptop for the last time as a practicing geoscientist. It is particularly important that geoethics be infused throughout a novice geoscientist's undergraduate (and graduate) education. (See chapter 13 of this volume for more on teaching geoethics within a university.)

Each of us is ultimately responsible for our own behavior; however, the geoscience community is responsible for promoting ethical behavior among its members and discouraging or penalizing unethical behavior. Ethics should be a constant and primary concern for the entire alphabet soup of scientific and professional groups that serve the geosciences. These communities bear partial responsibility for the continued ethical development of their members. Knowledge of applied geoethics is a requirement of professional certification and licensure for geoscientists in many jurisdictions [for example, *AIPG*, 2003, 2015; *Williams*, 2018, this volume; *Andrews*, 2014]. Continuing education in ethics is a common requirement for renewal of professional-practice licenses for geoscientists.

K-12 education (kindergarten/preschool to high school/precollege) in science should introduce all students to the central importance of truth telling, uncertainty, and ethics in science. University geoscience courses should emphasize the ethical practice of our science throughout their curriculum. This effort should be supported by online materials developed by the broader geoscience community to enhance student understanding of geoethics. Postgraduate geoscientists should constantly work to improve their practical understanding of applied ethics and, more importantly, should model ethical professional conduct every day of their careers. They have an ethical responsibility that extends beyond their clients to all of society as we grapple with worldwide challenges related to water, energy, minerals, geologic hazards, and the effects of global change.

11.4.7. Protection of Geoheritage and Geodiversity

The concepts of geodiversity and geoheritage have been informally addressed for centuries by the geosciences, but only recently have the terms themselves taken on the need for formality and definition. Notwithstanding a plethora of definitions,

> geodiversity is the variety of rocks, minerals, fossils, landforms, sediments, water and soils, together with the natural processes which form and alter them; geoheritage comprises those elements of the Earth's geodiversity that are considered to have significant scientific, educational, cultural or aesthetic value. Geoconservation is actions and measures taken to preserve geodiversity and geoheritage for the future" (https://www.iucn.org/theme/protected-areas/wcpa/what-we-do/geoheritage).

It is within this realm of research where few issues relating to the negative aspects of geoethics arise and where most characteristically the positive outcomes of geoethics are practiced. The study of geoheritage provides ethical, social, and cultural values of reference for the scientific community and society as a whole. Our efforts in geoconservation provide an example of how the discipline aims to ensure that the "face" of the planet (rocks, landscapes, water, and so on) is adequately protected for future generations. Promoting geoeducation and geosciences to the global society is an important aim that has become increasingly well addressed by those geoscientists who have devoted their practice to geoheritage work.

In the past decade, there has been an explosion in studies, movements, organizations, and societies whose focus covers the full range of geoheritage and geodiversity. Notably at the highest levels of bureaucracy, there is the International Union for Conservation of Nature (IUCN) World Commission on Protected Areas (WCPA) Geoheritage Specialist Group (GSG) that facilitates the conservation and effective management of protected area geoheritage. There is also the International Commission on Geoheritage of the International Union of Geological Sciences (http://www.iugs.org/index.php?page=directory#GEM), which facilitates national and international awareness and understanding of underlying geoheritage concepts. Special interest groups have formed for specific goals such as conservation and preservation, the most notable of which is ProGEO (http://www.progeo.ngo/), the European Association for the Conservation of the Geological Heritage, whose purpose is "to give Earth-science conservation in Europe a stronger voice, and to act as a forum for the discussion of issues, advising and influencing policy makers."

The Geopark movement is another global phenomenon that has taken on the multidisciplinary roles of geoheritage recognition, conservation, and protection; public awareness and education; cross-border international collaboration; improving the quality of life for local populations through economic stimulus; and a general move toward greater awareness of the importance and diversity of nature. Again at the highest levels of bureaucracy, UNESCO supports the Global Geoparks Network (GGN) by providing formal accreditation to those "parks" that successfully "demonstrate geological heritage of international significance[;] the purpose of a geopark is to explore, develop and celebrate the links between that geological heritage and all other aspects of the area's natural, cultural and intangible heritages" (http://www.globalgeopark.org/index.htm). Regional efforts that complement the GGN include the European Geoparks Network and the Asia Pacific Geoparks Network.

Geoheritage, geodiversity, and geoconservation are concrete expressions of a geoethical vision of the planet: recognizing their importance as a means to restoring an inner connection between humans and the Earth is a fundamental starting point to develop best practices in land management.

11.4.8. Sustainability, Resilience, and Geoethics

Sustainability and resilience are related. Sustainability is defined in many ways, but the most generally cited definition is that of "sustainable development" from the Bruntland Commission [*WCED*, 1987]: "Sustainable development is the concept that development must meet the needs of the present without compromising the ability of future generations to meet their own needs." The resilience of a community is its capacity to prepare and plan for, recover from, or successfully adapt to adverse events (hazards), both natural and human-made, in both the short term and the long term.

In many ways, resilience and sustainability have more in common than is evident from these definitions. For example, although not obvious in the definition above, resilience has economic, infrastructure, social, and personal components that include equity in the present and the future, components that are explicit in the definition of sustainable development. Likewise, although not obvious in its definition, sustainability has a component that includes resilience, for a society that is not resilient certainly cannot be sustainable.

The great astronomer Carl Sagan (1934–1996) had much to say about science and ethics and typically had eloquent ways to express himself. Although not using the words *sustainability* and *resilience*, for example, in his 1996 book *The Demon-Haunted World*, he states (p. 291):

> It is the particular task of scientists, I believe, to alert the public to possible dangers, especially those emanating from science or foreseeable through the use of science. Such a mission is, you might say, prophetic. Clearly the warnings need to be judicious and not more flamboyant than the dangers required; but if we must make errors, given the stakes, they should be on the side of safety.

Such warnings are especially difficult for scientists to frame and defend when the dangers lie in the future, not the far future but a near future perhaps still out of reach of the experience of imagination of most people, as the issues of resilience and sustainability seem to be. It is here, at the interface of science and society on these long-term issues, that many geoethical considerations arise. What needs to be done to build resilience, and how do we prioritize the sequence in which it is done? Who loses and who wins as actions are taken to correct living situations that are not resilient or sustainable, such as settlements on flood plains or in the path of volcanic outbursts or landslides? How do we distribute finite resources among the ever-expanding population? What is our obligation to preserve resources for future generations? Who do we compensate for damages from a hazard, and how much? How do compensations for damages from a hazard help or impede our ability to reduce their impact in the future?

The leaders in building resilience typically are not and cannot be geologists, unless they abandon their scientific identity and enter the public decision-making arena. Those leaders are decision makers in a wide variety of fields ranging from engineering to management, public policy, social services, human geography, economics, sociology, and anthropology. But, we geoscientists argue, they can do their jobs better if they have been exposed to the geosciences in their education and if we can gain their attention and provide relevant information. Generally, the information that we provide should not be limited to "the facts and only the facts." Our role must also include formulating realistic scenarios [*Albarello*, 2015] about the consequences of various actions and the consequences of not acting, while always avoiding entering the decision-making process itself in order to retain our credibility.

Much of the work in building resilience and sustainability lies in the domains of the social sciences, economics, and government/NGO efforts at various levels. As an example of the talents needed, in 2013 the U.S. National Academy of Sciences established a large group of experts, the Resilient America Roundtable (http://resilientamerica.nas.edu), to examine how American cities and regions can increase their resilience to disasters. The goals of the Roundtable are these:

- improve risk communication
- identify ways to measure community resilience
- share data and information related to hazards, disasters, risk, and resilience
- develop or strengthen partnerships and coalitions within and among communities to build resilience

A partial list of the agencies to which members of the Roundtable belong illustrates the breadth of input that will bring values and ethical issues to the table:

- The Federal Emergency Management Agency (FEMA)
- Department of Homeland Security (DHS)
- National Oceanic Atmospheric Administration (NOAA)
- Red Cross
- Federal Insurance and Mitigation Administration (FIMA)
- Koshland Science Museum
- U.S. Geological Survey (USGS)
- National Association of Counties
- National Voluntary Organizations Active in Disaster
- U.S. Chamber of Commerce

11.5. Promoting Geoethics: The International Experience of the IAPG (International Association for Promoting Geoethics)

Choosing to study geoscience is the result of a passion for nature, rocks, and landscape and the curiosity about the functioning of Earth systems. Often the life of the geoscientist is considered fascinating. Many young people see in this profession a way to live in accord with their ideals: respect for the environment, protection of the planet, and service to society, particularly to people who have limited access to primary resources such as water and arable land.

Geoscience is a group of disciplines that often appeals to those with a strong framework of ideals rather than to those who want to do a quick search for a job. The teaching of ethics in the geosciences has never been a standard part of university courses, but is considered an indispensable element on which to base technical-professional training. Moreover, the need to make frustrating compromises and strong competition in the workplace can push young geoscientists away from the ideals that inspired them at the start of their professional careers. The final result is that the current geoscience community is not sufficiently aware of its cultural potential, of the cultural value of geoscience knowledge, of its role in society, and of its ethical responsibility [*Peppoloni and Di Capua*, 2012].

Promoting geoethics first needs to involve the geoscience community in the discussion about ethical aspects of the geosciences, in order to rediscover the ideals and values that attracted geoscientists at the beginning of their university studies. Such involvement would permit strengthening the ethical conscience of individuals. Geoscientists more aware of the ethical and social implications of their activities, solutions, and decisions will be better able to transfer not only best practices but also best values to society and to contribute to creating a knowledgeable society and a sustainable development for future generations.

11.5.1. The Birth of the IAPG

During the 34th International Geological Congress in Brisbane (Australia, 5–10 August 2012), the foundation of the IAPG (http://www.geoethics.org/) appeared as a logical and necessary step toward giving geoethics a scientific status and toward enlarging the debate on ethical issues to include a wider number of colleagues with different skills and from different countries. The IAPG was born to stimulate reflection on ethical problems in geoscience research and practice and to persuade the geoscience community to consider seriously the importance of conducting their activities responsibly so as to better serve society. The IAPG is an international, scientific, multidisciplinary community whose primary aim is to create awareness about the application of ethical principles to theoretical and practical aspects of geosciences. In the last few years, it has become an important space in which many geoscientists can share opinions, exchange information, and propose new ideas and visions for the future development of geosciences in an ethical and social perspective. Being primarily a scientific association, the IAPG takes responsibility for developing geoethics as a scientific discipline, guided by scientific method and reasoning and trying to avoid self-referential positions and to promote discussion. To this end, the following results and products of its activity are fundamental:
- Scientific publications under a peer-review process
- Sessions on geoethics at national and international congresses

These powerful tools enlarge the number of geoscientists involved, encourage the sharing of information, thoughts, and proposals, and develop the mature and strong reflection on geoethics necessary to give it wide visibility, full dignity, and official recognition within the geoscience community.

The year 2012 was a turning point for geoethics. The publication of the first volume on geoethics by the journal *Annals of Geophysics* [*Peppoloni and Di Capua*, 2012] was the start of an intense editorial activity that has led to significant results [*Lollino et al.*, 2014; *Peppoloni and Di Capua*, 2015; *Wyss and Peppoloni*, 2015]. The books and articles published since 2012 can offer many cues for current and future discussion. Although presently these publications do not fully treat issues or exhaust the subject, steps made so far have changed the perception of geoethics and turned it from a movement of opinion into a legitimate scientific emerging field.

11.5.2. Mission

IAPG [2012] statutes state, "Geoethics should become part of the social knowledge and an essential point of reference for every action on the land, waters and atmosphere usage that all stake-holders and decision-makers have to take into proper account." The IAPG aims to promote geoethical values through international cooperation, encouraging the involvement and discussion of geoscientists from all over the world. Defining an ethical framework on which to base geological activities is the main IAPG goal, which includes the following:

- Adhering to the scientific method
- Being open to debate and to the comparison with ideas of colleagues, even when very far from our view
- Accepting multidisciplinarity as a fundamental working attitude and organization, as an approach that allows the joining together of different professional skills for seeking and sharing solutions to problems in a complex reality
- Assuring to society a qualified long-term learning
- Respecting the principles of intellectual honesty and integrity in the research and practice of geosciences and being an example for young geoscientists and society as a whole

11.5.3. Strategies and Initiatives

In just a few years, the IAPG has become an international reality, known and appreciated for the significant amount of work it has carried out and the contents that it has developed. The results stem from a clear strategy of action and communication, which has its pillars in the following points.

11.5.3.1. Development of Geoethics as a Scientific Discipline

Talking about geoethics does not mean building opinions. Rather, it means articulating a rational reasoning based on experiences that have been analyzed using the scientific method and are related to issues of interest for the geosciences.

11.5.3.2. Building a Strong Network, Capable of Developing and Supporting Itself in Time

The spread of new ideas in society is possible if groups of individuals who share a set of values and a vision of the future are formed. The network of the

IAPG is an infrastructure capable of developing initiatives, organizing events, creating products, and sharing experience and expertise.

11.5.3.3. Realization of Partnerships with Other Organizations

Partnerships with other national and international organizations promote initiatives on geoethics widely and promptly by involving different sectors of the scientific community. The IAPG is affiliated with the International Union of Geological Sciences (IUGS), with the International Council of Philosophy and Human Sciences (ICPHS), and is recognized as an International Associate Organization of the American Geosciences Institute (AGI), as an Associate International Society of the Geological Society of America (GSA), and as an Associated Society of the Geological Society of London. The IAPG recognizes the International Council for Science (ICSU) as the coordinating and representative body for the international organization of science; this link ensures that a political frame of reference is maintained. Finally, the IAPG has agreements of collaboration, among others, with the EuroGeoSurveys, the association that joins the Geological Surveys of Europe, the European Federation of Geologists, the network of European Professional Organizations of Geologists, the American Geophysical Union, the International Geoscience Education Organisation, the Association of Environmental and Engineering Geologists, and the African Association of Women in Geosciences; these links ensure efficacious actions with regard to society.

11.5.3.4. Promotion of Scientific Publications

Publishing is an essential requirement to strengthen the scientific and educational status of geoethics within the scientific community and to give it dignity in front of all other sciences and humanities. IAPG members have promoted and edited several volumes dedicated to geoethics [*Peppoloni and Di Capua*, 2012; *Lollino et al.*, 2014; *Peppoloni and Di Capua*, 2015; *Wyss and Peppoloni*, 2015], participated in the drafting of the formula of the "Geoethical Promise" [*Matteucci et al.*, 2014], and translated into different languages the Montreal Statement on Research Integrity in Cross Boundary Research Collaborations [*Peppoloni*, 2015]. Recently, the IAPG has released the "Cape Town Statement on Geoethics" (http://www.geoethics.org/ctsg). The "Geoethical Promise" (see paragraph 11.3) can contribute to strengthening the identity of young geoscientists and improving their awareness of the importance of geoethics as a fundamental base of their work life in geosciences. The aim of translating the Montreal Statement has been to promote research integrity principles by considering linguistic and cultural differences. This means "to share and recognize common principles and values that belong to mankind as a whole regardless of cultural differences. Furthermore, it means to increase the feeling of concerted nature and unity within the entire scientific community and to promote the enhancement of cultural diversity. Cultural diversity should be an element of union and not of division, through which all the scientists in the world can strengthen their identity and their sense of belonging to the scientific community" [*Peppoloni*, 2015]. The Cape Town Statement on Geoethics sums up all the values, concepts, contents developed in the first 4-year activity of IAPG, giving a perspective for the future development of geoethics. It aims to capture the attention of geoscientists and organisations, and to stimulate them to improve their shared policies, guidelines, strategies and tools to ensure they consciously embrace (geo)ethical professional conduct in their work.

11.5.3.5. Organization of Scientific Sessions in National and International Conferences

Scientific sessions in conferences involve colleagues in the discussion on geoethical topics, stimulating their proposals, reflections, and initiatives, and to increase the visibility of geoethics among the issues addressed in scientific assemblies. Therefore, the IAPG has organized sessions at the European Geosciences Union (EGU), General Assembly from 2013 to 2017, registering over the years an increasing interest and attendance; at the International Association of Engineering Geology and the Environment (IAEG) XII Congress 2014; at the American Geophysical Union Fall Meeting in 2014, 2015, and 2016; and at the 2015 Annual Meeting of the Geological Society of America. IAPG members give invited lectures and speeches at many other conventions and conferences.

11.5.3.6. Disseminating Geoethics within the Society

This means promoting the themes and reflections on geoethical topics outside the scientific community in order to engage citizens, media, and stakeholders in a new vision of the relationship between mankind and the geosphere. To this end, for example, the Italian section of the IAPG organized two conferences/conversations between a geologist and a philosopher of science at the Science Festival of Genoa in 2013 and 2014, one of the most important events of its kind organized in Europe. The success of such events has shown that the public is attracted by cultural "contaminations" of different knowledge and is not intimidated by the complexity of the topics. Rather, topics such as the defense against natural hazards, geo-resources exploitation, environmental defense, and climate changes are central for a public that is asking for scientifically correct information from scientists. From this perspective, interviews given to the Italian media and a broadcast aired on Radio Austria on May 2015 featuring interviews with IAPG members, allowed individuals to speak about geoethics to a large part of the population for the first time. The long-term effects of these initiatives cannot be easily assessed, as they are only the first significant steps by geoscientists to culturally affect society.

11.5.4. The Future

Geoethics is an emerging field in full growth, and in recent years many ideas have been planted in hopes that they would sprout. The results obtained up to now are encouraging. What, then, are the great challenges for the future of geosciences from a geoethical point of view? And in particular, what is the future of geoethics?

11.5.4.1. Young Geoscientists and the Teaching of Geoethics

Early career geoscientists are the greatest resource for the future of geosciences. They will be the custodians of the technical and scientific knowledge of Earth systems, those who will have to transfer this knowledge to future generations and transmit the social and cultural value of geosciences for the welfare of mankind. They can do this only if the scientific community, beyond the adoption of codes of ethics/conduct, will be able to introduce the teaching of geoethics into the university systems as a fundamental component in the training of a geoscientist. Examples and techniques for teaching geoethics across the geoscience curriculum are discussed in *Mogk et al.* [2018, this volume]. Geoscience is not ethical in itself; it is simply a tool in our hands. It is essential that young geoscientists become aware that doing science also means taking responsibility for communicating the impacts of the geosciences on society and ways that society can benefit from this knowledge. What, why, and how we may teach geoethics to young geoscientists should be clarified. It is our duty to give them strong motivations on why they should act in an ethical way.

11.5.4.2. Multiple Disciplinarity and Multiknowledge

Geosciences operate in cross-boundary areas and are called on to contribute by giving solutions through a multiple-disciplinary working method in strong collaboration with other disciplines. The multidisciplinary approach brings together different skills and represents a modern way to manage complex systems like the Earth. It also means that the topics of interest for geoethics are in many cases at the intersection between different disciplines. Discussions on geoethics should be able to integrate reflections from a multi-knowledge perspective, in which geoscientists can dialogue with philosophers, sociologists, economists, biologists, chemists, archaeologists, and other scientists and professionals, being always aware of the specificity and uniqueness of their point of views and skills.

11.5.4.3. Geodiversity and Cultural Diversity

A globalized and interconnected world is an irreversible reality. The geoscientist has a cultural education which by its nature is open to global dynamics. The geoscientist knows the impact that local phenomena can have at a global scale and studies the variety of natural environments and the richness of landforms in different places on the Earth. There are no good technical solutions to environmental problems if we do not take into account the geodiversity that characterizes each area of the planet as well as the biodiversity and human cultures and traditions that very often (especially in low-income countries) have been able to preserve and transmit human behaviors in equilibrium with natural dynamics. Practicing geoscience in an ethical way means also to consider geodiversity as an essential substrate for biodiversity and as a fundamental component of the cultural richness of human communities.

11.5.4.4. Cultural Lobbying

Are geoscientists able to have an effect on the discussions that animate the public? On one hand, geoscientists are called upon to communicate science in cultural and entertainment programs; on the other hand, they often have to deal with geo-risks communication and can sometimes be legally prosecutable [*Albarello*, 2015, *Cocco et al.*, 2015]. Geoscience is not perceived as a glamorous science like physics or astronomy, and yet geoscientists have a fundamental role in building a knowledgeable society. The geoethics arena may be the space in which we can discuss and share those values that will help to develop a healthier relationship between humankind and the planet.

11.6. Summary

Geoethics is an emerging field within the geosciences and more generally within science itself. Freedom, responsibility, multidisciplinarity, service to society, sustainability, and conservation of Earth systems are the fundamental values of geoethics on which geoscientists can base their ethical duty to act responsibly toward themselves, their colleagues, society, and the planet.

The young history of geoethics is the story of a world vision in which geological knowledge, in the broad sense, is the basis of a new way of thinking and practicing within the geosciences in order to contribute to building a knowledge society and to give a rational, positive, and sustainable perspective to future generations. The IAPG (http://www.geoethics.org) is promoting this vision and the values that underpin it.

Geosciences have always been interwoven with society, so geoscientists need to be fully aware and understand the broader implications of their work if they want to serve society. The lack of an ethical benchmark can be confusing for the geoscientist and may lead to uncertainty and the inability to make decisions. Ethical reflection and awareness should be encouraged and developed in the early years of university and not left solely to later personal initiative.

A globalized world such as the one we now live in offers great opportunities for development, but at the same time our achievements may endanger the very life of our communities. The implications of an unweighted exploitation of georesources, the effects of climate change, the repercussions of nonsustainable energy policies, and a land management practice that is not respectful of natural dynamics could have negative consequences that are unpredictable in the long term.

An ethical conscience can guide geoscientists to respond in a more effective and sustainable way to the needs of society and environment. It is therefore necessary that the development of the technical and scientific possibilities of geosciences be accompanied by an equal development of an ethical conscience. Ethical reflection can prevent errors in evaluating future consequences when sciences and technologies are applied irresponsibly and not enlightened by wisdom and foresight. Geoethics is an orientation tool for geoscientists, able to provide them with the ethical dimension of their actions. A deep bond unites environmental protection and the development of human communities. In coming years, geoscientists must be able to face the enormous challenge of reconciling geoethical values with the practice of geosciences. With this aim in mind, geoscientists must be able to function without making compromises in their work, undertake the pursuit of the common good, and ensure the right balance between sustainable living conditions while respecting Earth processes.

Acknowledgments

Susan W. Kieffer acknowledges support from the Charles R. Walgreen, Jr., Foundation.

References

AGI (2015), AGI Guidelines for Ethical Professional Conduct, American Geosciences Institute, https://www.americangeosciences.org/community/agi-guidelines-ethical-professional-conduct (accessed 25 June 2017).

AIPG (2003), Code of ethics, American Institute of Professional Geologists, http://www.aipg.org/about/ethics.htm (accessed 25 June 2017).

AIPG (2015), Professional certification, American Institute of Professional Geologists, http://www.aipg.org/Licensure/stateregboards.htm (accessed 25 June 2017).

Albarello, D. (2015), Communicating uncertainty: Managing the inherent probabilistic character of hazard estimates, in *Geoethics: The role and responsibility of geoscientists*, edited by S. Peppoloni and S. Di Capua, Geological Society, London, Special Publications, 419.

Allan, M. (2015), Geotourism: An opportunity to enhance geoethics and boost geoheritage appreciation, in *Geoethics: The role and responsibility of geoscientists*, edited by S. Peppoloni and S. Di Capua, Geological Society, London, Special Publications, 419.

Andrews, G.C. (2014), Canadian Professional Engineering and Geoscience Practice and Ethics [5th edition], Nelson Education, Toronto, Canada.

ASCE (2017), ASCE Code of Ethics, American Society of Civil Engineers, http://www.asce.org/code-of-ethics/ (accessed 25 August 2017).

Asher, P. (2001), Teaching an introductory physical geology course to a student with visual impairment. *Journal of Geoscience Education*, v.49, n.2, p. 166–169.

Bell, R.E., K.A. Kastens, M. Cane, R.B. Muller, J.C. Mutter, and S. Pfirman (2003), Righting the balance: Gender diversity in the geosciences, *Eos: Transactions American Geophysical Union*, 84(31), 292, doi:10.1029/2003EO310005.

Bergslien, E. (2012), *An introduction to forensic geoscience*, Wiley-Blackwell.

Bobrowsky, P.T. (2013), *Encyclopedia of natural hazards*, Springer.

Bohle, M. (2015), Geoethics: From "blind spot" to "common good." Blog of the International Association for Promoting Geoethics, http://iapgeoethics.blogspot.it/2015/06/geoethics-from-blind-spot-to-common-good.html (accessed 23 June 2015)

Boland, M.A., and D. Mogk (2018), The American Geosciences Institute guidelines for ethical professional conduct, in *Scientific integrity and ethics in the geosciences*, edited by L.C. Gundersen, Wiley.

Bruce, V. (2001), *No apparent danger: The true story of volcanic disaster at Galeras and Nevado del Ruiz*, Harper.

Cervato, C., and R. Frodeman (2012), The significance of geologic time: Cultural, educational, and economic frameworks, *Geological Society of America Special Papers*, 486, 19–27, doi:10.1130/2012.2486(03).

Choi, B.C., and A.W. Pak (2006), Multidisciplinarity, interdisciplinarity and transdisciplinarity in health research, services, education and policy: 1. Definitions, objectives, and evidence of effectiveness, *Clin. Invest. Med.*, 29(6), 351–364.

Cocco, M., G. Cultrera, A. Amato, T. Braun, A. Cerase, L. Margheriti, A. Bonaccorso, M. Demartin, P.M. De Martini, F. Galadini, C. Meletti, C. Nostro, F. Pacor, D. Pantosti, S. Pondrelli, F. Quareni and M. Todesco (2015), The L'Aquila trial, in *Geoethics: The role and responsibility of geoscientists*, edited by S. Peppoloni and S. Di Capua, Geological Society, London, Special Publications, 419.

Cronin, V.S. (1991), Engineering geology must be dominated by a public-safety-based ethic [abs], *Geol. Soc. Am., Abs. with Prog.*, 23(5), A41.

Cronin, V.S. (1992), On the seismic activity of the Malibu Coast Fault Zone, and other ethical problems in engineering geoscience, *Geological Society of America, Abstracts with Programs*, 24(7), A284.

Cronin, V.S. (1993), A perspective on professional ethics in engineering geosciences, in *Proceedings of Symposium on Ethical Considerations in the Environmental Practice of Engineering Geology and Hydrogeology*, Association of Engineering Geologists National Meeting, San Antonio, Texas, edited by S.N. Hoose.

Crutzen, P.J. (2002), Geology of mankind, *Nature*, 415, 23.

Devall, B., and G. Sessions (1985), *Deep ecology: Living as if nature mattered*, Gibbs M. Smith, Salt Lake City.

Di Capua, G., and S. Peppoloni (2014), Geoethical aspects in the natural hazards management, in *Engineering geology for society and territory: Volume 7. Education, professional ethics and public recognition of engineering geology*, edited by G. Lollino, M. Arattano, M. Giardino, R. Oliveira, and S. Peppoloni, Springer.

Dolce, M., and D. Di Bucci (2015), Risk management: Roles and responsibilities in the decision-making process, in *Geoethics: Ethical challenges and case studies in Earth science*, edited by M. Wyss and S. Peppoloni, Elsevier, Waltham, MA, ISBN 978-0-12-799935-7.

Dowling, R.K. (2010), Geotourism's global growth, *Geoheritage*, 3(1), 1–13, 10.1007/s12371-010-0024-7.

Ellis, E.C., and P.K. Haff (2009), Earth science in the Anthropocene: New epoch, new paradigm, new responsibilities, *EOS Trans.*, 90(49), 473.

Frodeman, R. (1995), Geological reasoning: Geology as an interpretive and historical science. *Geological Society of America Bulletin*, 107, 960–968, doi:10.1130/0016-7606.

Frodeman, R., and Baker, V. (2000), *Earth matters: The earth sciences, philosophy, and the claims of community*, Prentice Hall, ISBN 978-0-13-011996-4.

GA (1975), Geological fieldwork code, Geologists' Association, London, http://www.geologistsassociation.org.uk/downloads/Code%20of%20conduct/Code%20for%20fieldwork%20combined.pdf (accessed 25 June 2017).

GA (2011), A code of conduct for rock coring, Scottish National Heritage and the Geologists' Association London, http://www.geologistsassociation.org.uk/downloads/Code%20of%20conduct/GARockCoringGuide.pdf (accessed 25 June 2017).

GSL (2014), Geology for society, Geological Society of London, https://www.geolsoc.org.uk/geology-for-society (accessed 25 June 2017).

Guzzetti, F. (2015), Forecasting natural hazards, performance of scientists, ethics, and the need for transparency, *Toxicological & Environmental Chemistry*, doi:10.1080/02772248.2015.1030664.

Hall, S. (2011), Scientists on trial: At fault? *Nature*, 477, 264–269.

Herkert, J.R. (2005), Ways of thinking about and teaching ethical problem solving: Microethics and macroethics in engineering, *Science and Engineering Ethics*, 11(3), 373–385.

Holmes, M.A., and S. O'Connel (2003), Where are the women geoscience professors? *Papers in the Earth and Atmospheric Sciences*, Paper 86, http://digitalcommons.unl.edu/geosciencefacpub/86 (accessed 25 June 2017).

Hoose, S.N. (1993), *Professional practice handbook* (3rd ed.), Association of Engineering Geologists, Special Publication No. 5.

IAEG (1992), Statutes Article 2, The definition of engineering geology, International Association for Engineering Geology and the Environment, http://iaeg.info/media/1008/iaeg-statutes.pdf (accessed 25 June 2017).

IAPG (2012), Constitution of the International Association for Promoting Geoethics, http://www.geoethics.org/constitution.html (accessed 25 June 2017).

IPCC (2014), Climate change 2014: Synthesis report. Contribution of Working Groups I, II, and III to the Fifth Assessment Report of the Intergovernmental Panel on Climate Change (Core Writing Team, R.K. Pachauri and L.A. Meyer (eds.)), IPCC, Geneva, Switzerland.

Jordan, T.H., W. Marzocchi, A.J. Michael, and M.C. Gerstenberger (2014), operational earthquake forecasting can enhance earthquake preparedness, *Seismological Research Letters*, 85(5), 955–959, doi:10.1785/0220140143.

Lambert I., R. Durrheim, M. Godoy, M. Kota, P. Leahy, J. Ludden, E. Nickless, R. Oberhaensli, W. Anjian, and N. Williams (2013), Resourcing future generations: A proposed new IUGS initiative, *Episodes*, 36(2), 82–86.

Lambert, I., and P. McFadden (2013), Scientific advice underpinning decisions on major challenges, *Episodes*, 36(1).

Leopold, A. (1949), *A Sand County almanac*, Oxford University Press.

Lewis, S.L,. and M.A. Maslin (2015), Defining the Anthropocene, *Nature* 519, 171–180, doi:10.1038/nature14258.

Liverman, D.G.E, C.P.G. Pereira, and B. Marker (2008), *Communicating environmental sciences*, Geological Society of London, Special Publication, 305.

Lollino, G., M. Arattano, M. Giardino, R. Oliveira, and S. Peppoloni (Eds.) (2014), *Engineering geology for society and territory: Volume 7. Education, professional ethics and public recognition of engineering geology*, Springer.

Lucchesi, S., and M. Giardino (2012), The role of geoscientists in human progress, in *Geoethics and geological culture. Reflections from the Geoitalia Conference 2011. Annals of Geophysics (Special Issue)*, 55(3), 355–359, edited by S. Peppoloni and G. Di Capua, doi:10.4401/ag-5535.

Marshall, A. (1993), Ethics and the extraterrestrial environment, *Journal of Applied Philosophy*, 10(2), 227–236.

Matteucci, R., G. Gosso, S. Peppoloni, S. Piacente, and J. Wasowski (2014), The "geoethical promise": A proposal, *Episodes*, 37(3), 190–191.

Mayer, T. (2015), Research integrity: The bedrock of the geosciences, in *Geoethics: Ethical challenges and case studies in Earth science*, edited by M. Wyss and S. Peppoloni, Elsevier, Waltham, Massachusetts, ISBN 978-0-12-799935-7.

McPhee, J. (1989), Los Angeles against the mountains, in *The control of nature*, edited by J. McPhee, The Noonday Press, New York.

Mogk, D., and J. Geissman (2014), Developing resources for teaching ethics in geoscience, *Eos*, 95(44).

Mogk, D., J. Geissman, and M.Z. Bruckner (2018), Teaching geoethics across the geoscience curriculum: Why, when, what, how, and where? in *Scientific integrity and ethics in the geosciences*, edited by L.C. Gundersen, John Wiley & Sons, Hoboken, NJ.

Montreal Statement on Research Integrity (2013), 3rd World Conference on Research Integrity, Montreal, 5–8 May 2013, http://www.cehd.umn.edu/olpd/MontrealStatement. pdf (accessed 18 June 2017).

Munich Re (2015), *Topics Geo. Natural catastrophes 2014. Analyses, assessments, positions*. Munich. Retrieved from: https://www.munichre.com/site/mram-mobile/get/documents_ E-1601714186/mram/assetpool.mr_america/PDFs/3_Publications/Topics_Geo_2014.pdf (accessed 25 June 2017).

Næss, A. (1973), The shallow and the deep, long-range ecology movement: A summary, *Inquiry* 16, 95–100.

Newhall C., S. Aramaki, F. Barberi, R. Blong, M. Calvache, J.L. Cheminee, R. Punongbayan, C. Siebe, T. Simkin, S. Sparks, and W. Tjetjep (IAVCEI Subcommittee for Crises Protocols) (1999), Professional conduct of scientists during volcanic crises, *Bulletin of Volcanology*, 60, 323–334.

Oreskes N., D. Carlat, M.E. Mann, P.D. Thacker, and F.S. vom Saal (2015), Viewpoint: Why disclosure matters, *Environ. Sci. Technol.*, 49(13), 7527–7528. doi:10.1021/acs. est.5b02726.

Passmore, J.A. (1974), *Man's responsibility for nature: Ecological problems and Western traditions*, Gerald Duckworth & Co Ltd, London, ISBN 978-0-71-560756-5.

Peppoloni, S. (2012a), Ethical and cultural value of the Earth sciences, Interview with Prof. Giulio Giorello, in *Geoethics and geological culture. Reflections from the Geoitalia*

Conference 2011. Annals of Geophysics (Special Issue), 55(3), 343–346, edited by S. Peppoloni and G. Di Capua.

Peppoloni, S. (2012b), Social aspects of the Earth sciences, Interview with Prof. Franco Ferrarotti, *Geoethics and geological culture. Reflections from the Geoitalia Conference 2011. Annals of Geophysics (Special Issue)*, 55(3), 347–348, edited by S. Peppoloni and G. Di Capua.

Peppoloni, S. (2015), Sharing ethical principles through cultural diversity: Translations of the Montreal Statement on Research Integrity, IAPG, http://www.geoethics.org/translationsMS.html (accessed 10 July 2017).

Peppoloni S., P. Bobrowsky, and G. Di Capua (2015), Geoethics: A challenge for research integrity in geosciences in *Integrity in the global research arena*, edited by N.H. Steneck, T. Mayer, M. Anderson, and S. Kleinert, World Scientific, ISBN: 978-9-81-463238-6.

Peppoloni, S., and G. Di Capua (2012), Geoethics and geological culture: Awareness, responsibility and challenges, in *Geoethics and geological culture. Reflections from the Geoitalia Conference 2011. Annals of Geophysics (Special Issue)*, 55(3), 335–341, doi:10.4401/ag-6099, edited by S. Peppoloni and G. Di Capua.

Peppoloni, S. and G. Di Capua (2015), The meaning of geoethics, in *Geoethics: Ethical challenges and case studies in Earth science*, edited by M, Wyss and S. Peppoloni (pp. 3–14), Elsevier, Waltham, Massachusetts, ISBN 978-0-12-799935-7.

Peppoloni, S. and G. Di Capua (2016), *Geoethics: Ethical, social, and cultural values in geosciences research, practice, and education*. In: Wessel G. & Greenberg, J. (Eds.). *Geoscience for the Public Good and Global Development: Toward a Sustainable Future*. Geological Society of America, Special Paper 520, pp. 17–21, doi: 10.1130/2016.2520(03).

Pievani, T. (2009), The world after Charles R. Darwin: Continuity, unity in diversity, contingency, *Rend. Fis. Acc. Lincei*, 20, 355–361.

Pievani, T. (2012), Geoethics and philosophy of Earth sciences: The role of geophysical factors in human evolution, in *Geoethics and geological culture. Reflections from the Geoitalia Conference 2011. Annals of Geophysics (Special Issue)*, 55(3), 349–353, doi:10.4401/ag-5579, edited by S. Peppoloni and G. Di Capua.

Pope Francis (2015), Laudato si' (Praise be to you: On care for our common home), Encyclical, http://w2.vatican.va/content/francesco/en/encyclicals/documents/papa-francesco_20150524_enciclica-laudato-si.html (accessed 25 June 2017).

Sachs, J.D. (2014), Sustainable development goals for a new era, *Horizons*, 1, 106–119.

Sagan, C. (1996), *The demon-haunted world*, Random House, New York.

Savolainen, K. (1992), Education and human rights: New priorities, in Adult Education for International Understanding, Human Rights and Peace, Report of the Workshop held at UIE, Hamburg, 18–19 April 1991, *UIE Reports*, 11, 43–48.

Selinus, O. (Ed.) (2013), *Essentials of medical geology* (revised edition), Springer, XIX, ISBN 978-94-007-4375-5.

Selle, M.J. (2014), Interview to Silvia Peppoloni, in Geoethics Provides Vision for Best Practices, *Hartenergy, The Official 2014 Offshore Technology Conference Newspaper*, May 5, Day 1, p. 20.

Singapore Statement on Research Integrity (2010), 2nd World Conference on Research Integrity, Singapore, 21–24 July 2010, http://www.singaporestatement.org/statement.html (accessed 25 June 2017).

Slosson, J.E., J.W. Williams, and V.S. Cronin (1991), Current and future difficulties in the practice of engineering geology, *Eng. Geol.*, 30, 3–12.

Stoppani, A. (1873), *Corso di geologia Vol. II*, G. Bernardoni e G. Brigola, Editori, Milano.

Tollefson, J. (2015), Earth science wrestles with conflict-of-interest policies, *Nature*, 522, 403–404. doi: 10.1038/522403a.

UNESCO (2006), Global Geoparks Network, United Nations Educational, Scientific and Cultural Organization, http://unesdoc.unesco.org/images/0015/001500/150007e.pdf (accessed 25 June 2015).

WCED (1987), *World commission on environment and development: Our common future*, Oxford University Press, ISBN 978-0-19-282080-8.

Williams, J. (2018), The National Association of State Boards of Geology (ASBOG): Involvement in geoscience professional ethics, in *Scientific integrity and ethics in the geosciences*, edited by L.C. Gundersen, John Wiley & Sons, Hoboken, NJ.

Wyss, M., and Peppoloni, S. (2015), *Geoethics: Ethical challenges and case studies in Earth science*, Elsevier, Waltham, MA, ISBN 978-0-12-799935-7.

Zalasiewicz, J., C.N. Waters, M. Williams, A.D. Barnosky, A. Cearreta, P. Crutzen, E. Ellis, M.A. Ellis, I.J. Fairchild, J. Grinevald, P.K. Haff, I. Hajdas, R. Leinfelder, J. McNeill, E.O. Odada, C. Poirier, D. Richter, W. Steffen, C. Summerhayes, J.P.M. Syvitski, D. Vidas, M. Wagreich, S.L. Wing, A.P. Wolfe, A. Zhishengw, and N. Oreskes (2015), When did the Anthropocene begin? A mid-twentieth century boundary level is stratigraphically optimal, *Quaternary International* (Available online 12 January 2015), 10.1016/j.quaint.2014.11.045 (accessed 23 June 2015).

Zen, E-an (1993), The citizen-geologist. GSA Presidential Address 1992, *GSA Today*, 3(1), 2–3.

Section V: Scientific Integrity, Ethics, and Geoethics in Education

12

12. EXPERIENTIAL ETHICS EDUCATION

Vance S. Martin and Donna C. Tonini

Abstract

Throughout human history most people have learned in an experiential way, as observers of and participants in their local environments. While many learners and practitioners across the globe consider "modern" education as a teacher-led, textbook-based method of learning, delivered en masse, that approach has only been in practice for about the last 150 years. However, this form of education has been criticized for not infusing the learning environment with experience. Dewey (1938) considered experience as central to the educational process, and Freire (1970) encouraged active participation in learners as opposed to being passive receptors of knowledge. Both educators espoused using the community as a schoolhouse from which to gain and share experience. Drawing on the idea of a "community of practice", Lave and Wenger, 1991 offer a modern update to apprenticeship, framing "Legitimate Peripheral Participation" as a means by which new learners gain experience. A learner begins as an observer and minor actor (apprentice), and then progress to a more involved actor-participant level (journeyman), before they become a full-fledged participant (master) in the craft. Parents view a similar process as children go through developmental stages towards adulthood. As academics, we experience similar stages as graduate students progress from their first class to their dissertation to their first post-doctoral job. However, what are graduate students learning as they progress through a doctoral program? Certainly they are learning content as they progress towards mastery. Yet, it is one

National Center for Professional & Research Ethics, University of Illinois at Urbana-Champaign, Illinois, USA

Scientific Integrity and Ethics in the Geosciences, Special Publications 73,
First Edition. Edited by Linda C. Gundersen.

thing to learn content, and another thing to learn how to apply knowledge and skills in a responsible and ethical manner. In the past, the tendency of those in both the hard and soft sciences to reproduce studies and publish in refereed journals offered a check and balance to research and scholarship, lending a self-correcting nature to the field. However, the growth of PhDs in these fields coupled with the intensifying pressure to publish to stay competitive in both academia and industry has resulted in an increase in undesirable research related behaviors, which risk a loss of legitimacy and credibility in the sciences. High levels of undesirable research behavior and high-profile cases of research misconduct threaten research integrity and "damage to institutions' reputations, and loss of public trust in the research process" (Martinson, Thrush & Crain, 2013, p. 2). Academia, industry, and the public they serve all stand to benefit from a rigorous and systematic incorporation of ethics into the programs, products, and mindsets of all stakeholders involved in the research process. The question is, how do these stakeholders, especially graduate students master the inculcation of ethics into their approach towards their studies? Mumford (2007, 2008) discusses a 2 day ethical training program which yielded long-term retention of the ethical material. This experiment was conducted on first year graduate students who showed retention over 6 months. In this article we will propose Mumford's work as a basis for the ethical content which graduate students should learn. We will then discuss applicable learning theories to describe the potential best way to convey the material, initially and over time. We will finally examine the importance and potential of creating an ethical culture within the university.

12.1. Introduction

As researchers in an ethics center, my colleague and I were recently required to update our institutional review board (IRB) training online, the renewal of which is mandated every three years. As we were making our way through bland text and staid examples, we realized that we were not really internalizing any of the lessons. It was not that the screens were conveying information that was useless; rather, it was being delivered in a way that did not resonate with our situations and experiences. Instead of cultivating a passion for ethical practice and culture, it was driving us to click through the boxes as quickly as possible to "get through it and renew it." It occurred to us that if academic institutions wanted to inculcate the practice of ethics into researchers' and scholars' everyday approach toward their fields and their work, a connection between ethical theory and practice needed to be made through the use of experiential teaching methods that link theoretical grounding to the lived experiences of the learners. Niemi

[2015], in his paper "Six Challenges for Ethical Conduct in Science," echoes that thinking, stating:

> As all experienced teachers know, just presenting important learning contents to students is far from guaranteeing that they will be remembered a month later. We need to make sure that the importance of these challenges "sinks in" and this can best be achieved by creating situations in which students encounter them in personal and concrete ways (p. 14).

It is our intent to explore ways of teaching and learning ethics that do encourage the subject matter to "sink in," ensuring that lessons are better understood, retained, and carried out in practice as students progress in their careers.

The field of research ethics is complex and spans the responsible conduct of research, including human subjects protection, the ethical treatment of animals, and ethical use of research to benefit the greater good and avoid harm to others. The need for the creation of an ethical culture spans all fields and departments, regardless of discipline or institutional setting. Thus, this piece has applicability to all fields that have ethical components, be they social or natural science disciplines that date back thousands of years, or newer sciences that are emerging and being established as new fields of study.

For this chapter we were asked to write about how to think about educating researchers on ethics in the new field. We envisioned this as the opportunity to start from scratch, thinking about some of the current work we have been involved in and coupling this with educational literature to ground it. In creating a new area of study, such as the geosciences, it is desirable to develop an internal culture of ethical behavior, as well as an ethical outlook on global problems. Knowledge and self-understanding of participants is also needed in addition to content and pedagogy. In cultivating this internal culture of ethical behavior, we will explore how the use of experiential teaching methods achieves those aims. In this essay we will first discuss learning theory and how the experiences of graduate students and academics influence how they will act and behave. This chapter will then examine various methods for teaching ethics by reviewing both current practice and literature. We will conclude with insight from our work at the National Center for Professional and Research Ethics (NCPRE) and provide recommendations on the teaching of ethics education.

The modern educational system in the United States, and much of the world, can be characterized as a tri-level system of elementary, secondary, and higher education. Elementary and secondary schooling is mandatory for all children in the United States, while higher education is reserved for a subset of those who graduate from secondary school. Prior to this system, which has been in effect for about 150 years, there were schools and tutors, but they were often limited to a very small minority of either promising students or children of the elite.

Today when we discuss how best to teach something, we often think of this tri-level system and a setting with an expert who controls the time, content, and delivery of information. Educational debate often centers on discussions of how best to use the time, what content should be conveyed, or how to deliver the information.

An area that is discussed far less frequently is how people learn most effectively, not just within a prescribed school setting but overall. There are other ways of thinking about education beyond the modes of learning that occur inside of compulsory education. One of these is what Lave and Wenger [1991] refer to as legitimate peripheral participation, which focuses on the learning that occurs within a social setting, whether hierarchical or peer. An individual who is a novice is placed in a situation with one or more experts, and through observation and practice over time begins to become an expert. This approach has been a time-honored tradition as an apprenticeship model for blacksmiths, woodworkers, printers, and other tradespeople, as well as professionals such as physicians. While we may see this method of learning as a remnant of the past, parents raise their children in this manner by modeling behavior and expecting their progeny to follow their example. Teachers are also trained using an apprentice-type model, whereas they observe master teachers during their first few years of training, and nearing graduation, take over a classroom for a semester or more under the guidance of a master teacher. Professional academics also go from novice graduate student to master professor over their careers, through observation of advisors, participation in research, and peer interaction.

This layering, building, or scaffolding of knowledge has been an area of study for educational psychologists for many years. Attempting to understand how knowledge is "built" has been a great undertaking. Anderson et al. [1978] explain the scaffolding of information in how students learn a concept then construct new understanding through added layers of difficulty as the instructor creates levels and scaffolds to connect new knowledge to prior knowledge to encourage learning. Vygotsky [1978] theorized that a student possesses one level of knowledge, and with aid from an expert, the student is able to move to the next level. The key to this progression is that there is a "zone of proximal development," whereby the student is close to the next level of understanding and aided by the expert [Vygotsky, 1978]. This idea is the basis for constructivism as a learning theory that focuses on the knowledge and meaning constructed through experience [Martin, 2011]. As a subset of constructivism, social constructivist theory examines how knowledge is conveyed through social interactions with friends, teachers, or parents [Berger and Luckmann, 1967]. Through these interactions we learn social cues and how to navigate within our environment, whether it be to chew with your mouth closed, how to ask a question, or if it is ethical to write down incorrect data in a lab book to support a preheld thesis.

Educational scholars such as John Dewey, one of the most prolific and long-lived educators, continually wrote about the importance of experience as a central part of education [Dewey, 1938]. Other international educators such as Paulo Freire also felt that being a participant in the educational process was key to learning. These theorists helped form schools of thought that take the focus away from the passive learning that might occur in a "traditional" classroom and refocus on the importance of being an active participant, active in the educational experience, and learning a role over time as progress toward mastery occurs.

For each of these scholars a key component is the social interaction that occurs. Thus, in addition to the traditional learning environment, lived experiences are another way in which we all learn. In a graduate program, desired academic practices are modeled, observed, and recreated as students learn how to write scholarly pieces, engage in critical discussion, and practice ethical research. Knowledge is built over time; new graduate students are not expected to write a dissertation on day one, nor should we expect them to understand all the ethical implications of their actions in academia on day one. Students learn from their experiences and actions as well as from the books they read and lectures they attend.

However, what are graduate students learning as they progress through a doctoral program? Certainly they are learning content as they progress toward mastery. Yet it is one thing to learn content and another thing to learn how to apply knowledge and skills in a responsible and ethical manner. In the past, the tendency of those in both the hard and soft sciences to reproduce studies and publish in refereed journals offered a check and balance to research and scholarship, lending a self-correcting nature to the field. However, the growth of PhDs in these fields coupled with the intensifying pressure to publish in order to stay competitive in both academia and industry has resulted in an increase in undesirable research related behaviors, which risk a loss of legitimacy and credibility in the sciences. High levels of undesirable research behavior and high-profile cases of research misconduct threaten research integrity, "damage to institutions' reputations, and loss of public trust in the research process" [*Martinson et al.*, 2013, 2]. Academia, industry, and the public they serve all stand to benefit from a rigorous and systematic incorporation of ethics into the programs, products, and mindsets of all stakeholders involved in the research process. The question is, how do these stakeholders, especially graduate students, master the inculcation of ethics into their approach toward their studies?

12.2. Ethics Education

The need for the inclusion of ethics in the classroom, laboratories, and intellectual spaces of our postsecondary institutions is well documented [*Mumford et al.*, 2007; *Mumford et al.*, 2008; *Roland*, 2007]. Feedback from a recent symposium focused on the improvement of the scientific endeavor also indicates that graduate students and postdoctoral researchers have expressed the necessity of "rigorous and honest scientific communication" [*McDowell et al.*, 2015, 5]. This symposium sought to synthesize current issues impeding the progress of scientific research, and identified the need for better training in research integrity and support in those endeavors from sponsoring institutions as critical to encouraging the pursuit of scientific truth with honesty and integrity [*McDowell et al.*, 2015]. The Association for Practical and Professional Ethics (APPE), an organization advancing the scholarship, education, and practice of practical and professional ethics, also supports the training of new faculty and professionals to foster ethical conduct in the workplace (http://appe. indiana.edu/).

While there is consensus in the demand for ethics education, and support for its inclusion in postsecondary academic and training programs, there is divergence in the approaches and means by which ethics are incorporated into the educational sphere. First of all, there are multiple venues in which ethics-based offerings are found, from college courses grounded in philosophy of moral theory, such as the University of Illinois at Urbana-Champaign's "Ethics and Engineering" (https://publish.illinois.edu/ecephil316/), to short-term executive development courses that feature practical skills and tools, such as Bentley University's "Managing Ethics in Organizations" (http://www.bentley.edu/ centers/center-for-business-ethics/managing-ethics-organizations). Ethics education is also featured in forums such as the Ethics@Illinois Seminar Series hosted by the National Center for Professional and Research Ethics, which focuses on fostering dialogue around the responsible conduct of research (http://ethicscenter. csl.illinois.edu/programs/ethics-awareness-week/). Ethics instruction is even making an appearance in massive online open courses (MOOCs), where offerings such as "Practical Ethics" and "Technology and Ethics" are available for enrollment on the Coursera platform (https://www.coursera.org/courses?query= ethics). Second, there are various models and instructional methods being employed to teach ethics, which run the gamut from standard lectures and presentations to the use of online technology and the arts to convey key concepts and ideas. The remainder of this chapter will be dedicated to a discussion of the important elements required for effective ethics education and training in general, and an exploration of how these components are employed by the various instructional methods in use to teach ethical behavior.

12.3. Experiential Ethics Education Model

There are important elements to consider when developing an educational model focused on the teaching of ethics. Research indicates that to maximize the impact and increase the longevity of effects from training, the overall instructional method should be based on the active participation of the instructor and the participants. According to Mumford *et al.* [2007, 2008] and Gunsalus [2015b], actively involving students in participatory learning, through the use of cooperative and iterative discussion and action and employing engaging instructional activities, while tying these approaches to learning goals and outcomes, improves participants' understanding of the material. As discussed in the introduction, the use of scaffolding, building, and layering materials, ideas, and concepts improves both comprehension and retention of the information and increases the likelihood that the experience will carry through into practice [*Mumford et al.*, 2008]. Each of these ideas, participatory learning, engaging activities and learning goals, will be discussed in turn.

12.4. Active Participation

Active participation and active learning are large ideas within educational circles, and the definitions and usage vary, depending upon the user. For this chapter, we focus on three of the four areas noted by Prince [2004]: collaborative, cooperative, and problem-based learning. In full semester courses, collaborative learning is shown to enhance academic achievement, learning outcomes, and student retention [*Prince*, 2004]. According to Prince, collaborative learning focuses on having less individual work and more work in collaboration with fellow learners. This approach relates to the idea of legitimate peripheral participation from *Lave and Wenger* [1991], where students learn from their social setting. Similar to collaboration, cooperative learning also improves academic performance, attitudinal outcomes, and interpersonal skills [*Prince*, 2004]. This type of learning also takes the emphasis away from competition between learners, and more toward a greater understanding from colearners [*Prince*, 2004]. Studies have shown that "students who would score at the 53rd percentile level when learning individualistically will score at the 70th percentile when learning cooperatively" [*Johnson et al.*, 1998, 33]. This idea relates to *Lave and Wenger* [1991] and *Vygotsky* [1978], that participation and interaction with other learners, peers, and experts can lead to a greater level of understanding. Finally, problem-based learning fosters a deeper approach to learning, positive student attitudes, and knowledge retention [*Prince*, 2004]. Problem-based learning uses specifically built exercises for learning that build on previous knowledge and lead to a greater level of understanding [*Prince*, 2004]. This form of learning is greatly related to the work of Vygotsky [1978] and Anderson [1978]. Students are building on existing knowledge using materials specifically designed to lead to that greater level of comprehension, and through iterative interaction are achieving a higher level of understanding.

Freire [1970] theorized that students must be aware of their environment, engage in critical listening, thinking, and reflection, and purposefully use their experiences to be active and engaged participants with course content and material, as opposed to being mere repositories for information. This undergirding principle, the engaged and vested involvement of the student in his or her own learning and development, embodies the idea of active participation presented in this chapter.

12.5. Engaging Instructional Activities

Scholars note the importance of educational context in influencing learning and student outcomes. Cleveland-Innes and Emes [2005] posit that social interaction and the resulting socialization process influence learning and learning outcomes. The authors further impress that the context in which the social interactions occur influences both the perceptions of the learners, and the resulting pattern of socialization. In a study of the impact of social and academic interaction, the authors found that "context is a critical factor in the determination of approach to learning" [*Cleveland-Innes and Emes*, 2005, 257]. Cleveland-Innes and Emes align their findings with the learning process complex proposed by Biggs [1987], who theorizes that there are three main approaches to learning: surface, achieving, and deep (p. 10). The motive behind surface learning is to meet the academic requirements necessary to understand the bare essentials and have the ability to regurgitate facts and lessons [*Biggs*, 1987]. The underlying motive of the achieving approach is to gain content mastery merely for the purpose of attaining a higher grade and maintaining the status of model student [*Biggs*, 1987]. Last, the premise of the deep approach is motivated by an intrinsic interest in the topical matter and a genuine desire to achieve competence through meaningful experiences and connections with previous knowledge [*Biggs*, 1987]. According to Cleveland-Innes and Emes, "Deep Approaches lead to comprehensive understanding and positive emotional outcomes" [2005, 257]. These authors make a clear case for "adjusting context and [social] interaction factors to encourage most effective approaches to learning for higher education," to achieve better and more meaningful outcomes for the learners" (p. 258).

In encouraging a deep approach to learning, context is critical for fostering a connection between the learner's experience and the subject matter at hand. *Mumford et al.* [2008] argues that contextualization allows people greater flexibility in applying their knowledge to other problems and settings. Thus, to enable learners to develop a deep sense of connection to the topic of ethics, the combined educational theories behind active participation and contextualized experiences suggest that the content of the course, and the activities that support it, be engaging, meaningful, and relevant to the participants.

In describing an approach to ethics training for scientists, *Mumford et al.* [2008] recommends that instructors incorporate the following criteria into the design and curriculum of courses and classes. First, in order to allow students to make connections with the learning material, it is important for instructors to feature actual cases and real people to better situate examples within a student's reality. Second, it is also necessary to incorporate both positive and negative examples of these real-life situations, to prepare students for a wider range of possibilities, responses, and outcomes. Finally, it is also crucial to use course material that will engender an emotional impact so that the students can make a personal and relevant connection with the material. Combined, these iterative elements will help students draw upon their existing educational foundation and experience to build knowledge and cultivate a deep approach to their learning. It will also lead to the learning goals of analysis, reflection, the development of strategies for response, and the heightened ability to anticipate the consequences of their actions.

In constructing instructional activities for the teaching of ethics, it is important to consider these criteria for selection and inclusion. In order for the content to be relevant and promote active learning, it makes sense that the materials be based on real people and real situations. *Thiel et al.* [2013] argues that "emotional case content stimulates retention of cases and facilitates transfer of ethical decision-making principles demonstrated in cases" (p. 1). Thus, a discussion of an actual case of malfeasance in a cancer study, like the Potti case at Duke, is more believable and relatable than a fictitious issue without familiar context or characters [*Gunsalus*, 2015a]. Although in many cases it is imperative to allow institutions and the people involved to remain anonymous, there are instances where using real names is applicable as the information is already in the public domain. Similarly, it is important to use positive example cases where an institution employed the correct ethical approach in a research study, as well as cases where an institution did not behave ethically. Finally, by using examples that render an emotional impact on students, such as anger or disgust, instructors are able to foster participants' connection with the horrible feelings a family may have undergone knowing their loved one's participation in a flawed cancer study was for naught [*Gunsalus*, 2015a]. It is a valuable learning experience to make a meaningful connection with the struggles a family may have undergone by participating in an experiment gone ethically awry.

Depending on the discipline, audience, or instructor, it is possible to draw on various types of instructional activities for ethics courses or classes with elements of ethics contained therein. The following section features numerous activities that highlight a variety of ways to teach ethical theory, behavior, and approaches to decision making. The instructional activities described below use aspects of the experiential learning method, incorporating active participation and the criteria of real situations, positive and negative examples, and emotional impact.

12.6. The Curious Case of Tania Head

During a pedagogical demonstration at the Association for Practical and Professional Ethics (APPE) in February 2015, Marilyn Dyrud of the Oregon Institute of Technology led the audience through a peculiar case study involving a survivor of the September 11 attacks on the World Trade Center in New York. Dyrud created an intimate atmosphere in the room, using her skill at storytelling paired with a lively PowerPoint presentation to paint a personal portrait of the heroine while expertly drawing the audience in and creating empathy for a person they had never met. Each new facet and angle of the tale Ms. Dyrud told brought the fully engaged audience closer to the main character, to the point where they felt fully vested in her well-being. It was during the height of the audience's outpouring of support for this survivor that Dyrud delivered the punchline: that Tania Head was a fraud who was not even in New York City during the time of the attacks, and that her entire story was a lie. The audience was totally shocked and dumbfounded, with audible gasps of incredulity echoing in the room. It was the effect that Dyrud was going for, an intense emotional impact that would resonate with the audience and cement its connection to the sense of betrayal felt when this breach of ethical behavior occurred. As Dyrud explained to the audience, it was this physical and emotional impact that she sought to use to fuse this case into the minds of her students. She followed her bombshell revelation with a debriefing of the case, asking the audience to describe how the case made them feel, and helping them to identify with the emotions and understand the experience. Dyrud explained that she normally covers this case over a couple of days, using discussion, journaling and a writing assignment to further incorporate the learning experience into the psyche of her students.

The approach modeled by Dyrud is a good example of the experiential learning method as it features a real person in an actual situation to demonstrate a negative case of unethical behavior. She uses emotional impact as a hook to grip the audience and engages them in the process of deconstructing the case afterwards, which encourages active participation and fully incorporates all of the aspects of the experiential ethics education model. The case study method is oft-employed in the field of ethics education and is well supported in the literature [*Macrina*, 2005; *Bebeau et al.*, 1995].

12.7. Ethics Case Instruction for Collaborative and Community-Based Researchers

Natalie Baloy of the University of California at Santa Cruz uses an approach she terms "ethical case deliberation," a well-known element and approach to ethics education. At APPE in 2015, she demonstrated this pedagogical strategy for preparing researchers and their partners within the community to navigate ethical dilemmas that occur in collaborative research projects. Baloy discussed ethics and the key principles of collaborative research before delving into a case study that exemplified some of the typical ethical issues faced by stakeholders in community-based research contexts. To engage the audience in the lesson, Baloy had several of the audience members role-play the case study, acting out the roles of university-sponsored researcher, community activist, and study participant, as well as other important players. Encouraging the audience to identify with their roles, the participants acted out their parts with enthusiasm, trying to embody the contextual and social essence of their characters. Upon completion of the performance, Baloy provided questions for discussion, reflection, and deliberation. She created a lively discourse, with the participants volunteering not only how they felt about the case and its issues but also how their characters reacted, and why. By identifying with another stakeholder in the case, the audience members were able to actively engage with another perspective that, prior to the role play, they may not have been able to consider. By putting the audience into the unique position of gaining this valuable insight, Baloy was able to highlight these alternative viewpoints and call attention to the needs of various stakeholders in a way that simply reading the case study out of a book would not convey.

The method employed by Baloy is a also a good illustration of the experiential learning method as the case presented was modeled on real events involving real people to showcase perceptions of both negative and positive instances of ethical behavior. Having participants assume the roles and identities of the stakeholders allowed the audience to associate better with the social and contextual nuances of ethical conduct in that particular situation. Also, the role play granted the participants entree into the emotional mindset of those in the community research setting. Furthermore, Baloy's approach of forming small groups to discuss and analyze questions specific to the case study gave participants the opportunity to further consider the contextual and social dynamics of the case, bringing the complexities of ethical decision making to life and engaging them in encountering ethical tensions firsthand. This process of case role play and analysis encourages active participation, highlighting yet another good example of the experiential ethics education model. Role plays are commonly used in the field of ethics education and are vaunted by the literature [*Brummel et al.*, 2010].

Other good examples of using the experiential instruction method abound. Edward Spence [2015] of Charles Sturt University uses philosophy plays and a neo-Socratic approach to teach ethics to university students. He solicits volunteers from the audience to act out a modern-day situation using Platonic dialogue to engage the audience both intellectually by using high-level philosophical diction, and emotionally through the execution of the drama. By using real people performing both the positive and negative aspects of true ethical dilemmas, Spence garners the active participation necessary to elicit the emotional impact required to deliver his lesson on the application of ethics.

The National Center for Professional and Research Ethics features "Two-Minute Challenges" (2MC), which are very short cases that illustrate an ethical issue that participants then discuss to determine what they would do in the situation. These cases have been used for high school ethics education as well as for training in the responsible conduct of research. 2MCs are also used in executive education and professional programs in medicine, business, and law. In our own experience, we have employed 2MCs with audiences ranging from faculty to students. The laughter that ensues as colleagues and classmates watch each other read and act out various roles in the 2MCs serves to heighten interest and engagement in the topic matter, and participants' connection with the cases is evidenced in their discussions with us after the class has ended. Building on the 2MC idea, one of our colleagues developed specific scenarios for the American Chemical Society and used cell phones and QR codes to link to a "what would you do" response that when input tabulates the answers, allowing for more discussion [*Schelble*, 2015]. Loui [2015] references the Ethics Core Website (https://nationalethicsresource center.net/index.php/home) to access role play videos, then employs the "think-pair-share" technique to allow a greater discussion of topics and how people think about ethics. "Ethics Unwrapped" is another website, created by the University of Texas at Austin, that provides videos and teaching materials targeting a multidisciplinary audience (http://ethicsunwrapped.utexas.edu/) [*Biasucci*, 2015].

There are other examples that demonstrate the use of emotion in the instructional method, recognizing the "the direct effects of affect on decision making and the indirect effects of affect on cognitive processing," as posited by Mumford *et al.* [2008, 318]. Bivens [2015] uses comic books to convey ethical issues, with topics covering consequential, social contract, and virtue ethics. He has written several comic books that convey specific ethical ideas styled and illustrated in a comedic manner. While comedy may seem dismissive as a mode of instruction, taking a cue from the famed playwright Moliere, "the duty of comedy is correcting men while amusing them," meaning, it is much easier to communicate the truths and corrections in life with humor. Stolick [2015] uses fear and disgust to educate, quoting excerpts from Al Gore's "An Inconvenient Truth" to get people to consider the ethical and geoethical implications of global warming. To echo a point made by Daniel Mahoney, a professor of leadership and administration at Gonzaga University, ethics must be taught experientially and experienced by students. "Ethics is rooted in theory, but historically has been taught in a hybrid manner," he writes, alluding to the mix of theory and applied ethics employed in many ethics courses [*Mahoney*, 2015]. The authors concur with Dr. Mahoney's observation that theory-based ethics courses lack the exposure and intimacy necessary to allow students to gain familiarity with the material, and do not therefore instill the intrinsic value that active participation and emotional connection can provide.

An area that we predict will have an effect in the coming years will be the integration of games. Sher [2015] discusses his use of games in teaching ethical conundrums. There are a plethora of ethical issues implicit in traditional games like Risk or Monopoly that can also be used to get students to actively participate and engage with the instructional activity. With the ubiquity of digital games, an app or a game specifically focused on ethics is also a future possibility. The goal of any of these activities is to use them to engage and create a community culture where ethics is not an infrequent topic.

Kalichman [2015] discusses the importance of enculturating ethical behavior in faculty, postdocs, graduate students, and undergraduates in academic settings. He recommends working with faculty to talk with their students about ethical situations, and then working with the postdoctoral researchers and graduate students to talk with the undergrads and the faculty. This work attempts to create an ongoing conversation about ethical considerations within the workplace, rather than making it a hoop to jump through every four years, like online IRB training. Certainly within a discipline such as computer science, genetics, or the geosciences there are different issues that arise, as well as different students, faculty, and university cultures. Thus, it is important to leverage the knowledge of everyone within a department or university to help maximize the experiential learning method. For example, was there a new Marvel film that covered something related to ethics? A new episode of Doctor Who or NCIS? Some new game? Perhaps regularly scheduling a film or video viewing session followed by a debriefing and discussion will help create an environment more conducive to making ethical issues part of the water cooler culture, rather than something discussed for eight hours every four years.

12.8. Learning Goals and Outcomes

The learning goals of analysis and reflection follow the experiential instructional method of using engaging instructional activities with real people, real cases, and positive and negative outcomes to prompt active, participatory learning and emotional impact. By conducting an analysis of a problem, students are better enabled to consider how they would handle an issue, and more thoroughly plan how to react. Thinking about these responses and verbalizing them ahead of time makes it easier when we actually encounter like situations. Gunsalus [2012] terms the idea of envisioning situations and preparing responses as creating scripts. By thinking about potential predicaments, anticipating consequences, and practicing a script ahead of time, it is easier to react appropriately when actually in the situation. Thus, when a situation catches you off-guard, these memorized scripts allow you to respond accordingly while deflecting the need for action until you have time to consider the issue at hand. For example, when in an awkward situation, one might say, "You know I'm a better second-day responder, let me think about it overnight, and I'll email/call/talk to you about it tomorrow with a clear answer." This prerehearsed response gives you time to separate yourself from the situation and think about what needs to be done.

The process of analysis and response preparation also allows for reflection and contemplation not only on the situation but on the potential actions to follow as well. The idea of reflection dates back as far as Socrates and Plato, who espoused that the unexamined life is not worth living. More recent philosophers as William James and John Dewey saw reflection as essential to learning. More recent educators such as Freire also saw critical reflection as integral to understanding. Mezirow [1990] built his ideas of learning and the importance of critical reflection on the philosophies of these thinkers. When we reflect on an act, we make connections to our content knowledge, the knowledge of our own processes, and our presuppositions. Then when we act in the future, this reflection influences those actions. Thus, similar to Gunsalus' [2012] idea of scripts, thinking about what you might say helps you say it when necessary, just as reflecting upon how you thought and acted in the past can influence how you act in the future.

Thus, the learning goals of any ethics training and creation of an ethical culture should include analysis and reflection. It is a culture that needs to be created and perpetuated. As Mumford [2005] has shown through analyzing obituaries of scholars, it is through early and continual exposure to training that leads to renown. By using the above model as an outline, one can begin to train graduate students early in the research process. However, advisors and faculty need to continue this discussion in the classroom, in the laboratory, and in the field as ethical issues arise at other institutions or in the media. It is through these ongoing conversations that the culture is created, and our novice graduate students become master scholars and ethicists.

As a closing thought, the authors still cannot recall much about the issues or questions featured in our rote IRB training. Yet we readily remember the characters in the sessions featured at APPE and in our own cases, as well as the quandaries these characters faced. Much like the moral stories we hear as children, these anecdotes stay with us in our subconscious, teaching ethical lessons from a deeper place of connection than just a pat lecture or a static computer screen telling us what to do. The engrained nature of these teachings are bolstered by our weekly work meetings in which we discuss articles or events related to ethics in research, from which we gain an ongoing understanding of the issues and the atmosphere, something that cannot be covered by training every three years. Ethics is part of our mindset, practice, and bearing, and if we do not encourage this culture in our institutions every day, we will not achieve the desired results.

References

Anderson, R.C., R.J. Spiro, and C. Anderson (1978), Schemata as scaffolding for the representation of information in connected discourse, *American Educational Research Journal*, 15(3), 433–440.

Baloy, N. (2015, February), Ethics case instruction for collaborative and community-based researchers. Paper presented at APPE, Costa Mesa, CA.

Bebeau, M.J., K.D Pimple, K.M.T. Muskavitch, S.L. Borden, and D.H. Smith (1995), Moral reasoning in scientific research: Cases for teaching and assessment, Indiana University, retrieved from http://poynter.indiana.edu/mr/mr-main.shtml

Berger, P.L., and T. Luckmann (1967), *The social construction of reality: A treatise in the sociology of knowledge*, Anchor, New York.

Biasucci, C. (2015, February), Ethics unwrapped educational program and video series. Paper presented at APPE, Costa Mesa, CA.

Biggs, J.B. (1987), *Student approaches to learning and studying*, Australian Council for Educational Research, Hawthorn, Victoria.

Bivins, T.H. (2015, February), Designing online comic books for media ethics. Poster presented at APPE, Costa Mesa, CA.

Brummel, B.J., C.K. Gunsalus, K.L. Anderson, and M.C. Loui (2010), Development of role-play scenarios for teaching responsible conduct of research, *Science and Engineering Ethics*, 16(3), 573–589.

Cleveland-Innes, M.F., and C. Emes (2005), Social and academic interaction in higher education contexts and the effect on deep learning, *Journal of Student Affairs Research and Practice*, 42(2), 387–408.

Dewey, J. (1938), *Experience and education*, Collier Books, New York.

Dyrud, M. (2015, February), The curious case of Tania Head. Paper presented at APPE, Costa Mesa, CA.

Freire, P. (1970), *Pedagogy of the oppressed*, Continuum, New York.

Gunsalus, C.K. (2012), *The young professional's survival guide*, Harvard University Press, Cambridge, MA.

Gunsalus, C.K. (2015a), Misconduct expert dissects Duke scandal, *The Cancer Letter*, January 9. Retrieved from http://www.cancerletter.com/articles/20150123_3

Gunsalus, C.K. (2015b, April), Responsible conduct of research training. Lecture presented at Tufts University, Boston, MA.

Johnson, D.W., R.T. Johnson, and K.A. Smith (1998), Cooperative learning returns to college: What evidence is there that it works? *Change*, July/August, 27–35.

Kalichman, M. (2015, February), Moving RCR education outside the classroom. Paper presented at APPE, Costa Mesa, CA.

Lave, J., and E. Wenger (1991), *Situated learning: Legitimate peripheral participation*. Cambridge University Press, ISBN 0-521-42374-0.

Loui, M. (2015, February), Using short videos to teach research ethics. Poster presented at APPE, Costa Mesa, CA.

Macrina, F.L. (2005), *Scientific integrity: Text and cases in scientific integrity*, ASM Press, Birmingham, AL.

Mahoney, D.J. (2015, February), Ethics and educational leadership. Paper presented at APPE, Costa Mesa, CA.

Martin, V. (2011), Using wikis to experience history (Unpublished doctoral dissertation), University of Illinois, Urbana-Champaign. Retrieved from Ideals http://hdl.handle.net/2142/24084.

Martinson, B.C., C.R. Thrush, and A.L. Crain, (2013), Development and validation of the Survey of Organizational Research Climate (SORC), *Sci Eng Ethics*, 19(3), 813–834. doi: 10.1007/s11948-012-9410-7.

McDowell, G.S., K.T.W. Gunsalus, D.C. MacKellar, S.A. Mazilli, V.P. Pai, P.R. Goodwin, et al. (2015), Shaping the future of research: A perspective from junior scientists, *F1000Research*, 3(291), 1–20. doi:10.12688/f1000research.5878.2.

Mezirow, J. (1990), How critical reflection triggers transformative learning, in *Fostering critical reflection in adulthood* (1–20), edited by J. Mezirow et al., Jossey Bass.

Mumford, M.D., S. Connelly, R.P. Brown, S.T. Murphy, J.H. Hill, A.L. Antes, E.P. Waples, and L.D. Devenport (2008), A sensemaking approach to ethics training for scientists: Preliminary evidence of training effectiveness, *Ethics & Behavior*, 18(4), 315–339, doi:10.1080/10508420802487815.

Mumford, M.D., S. Connelly, G. Scott, J. Espejo, L.M. Sohl, S.T. Hunter, and K.E. Bedell (2005), Career experiences, and scientific performance: A study of social, physical, life and health sciences, *Creativity Research Journal*, 17, 105–129.

Mumford, M.D., S.T. Murphy, S. Connelly, J.H. Hill, A.L. Antes, R.P. Brown, and L.D. Devenport (2007), Environmental influences on ethical decision making: Climate and environmental predictors of research integrity, *Ethics & Behavior*, 17(4), 337–366, doi:10.1080/10508420701519510.

Niemi, P. (2015), Six challenges for ethical conduct in science, *Science and Engineering Ethics*, doi:10.1007/s11948-015-9676-7.

Prince, M. (2004), Does active learning work? A review of the research, *Journal of Engineering Education*, 93(3), 223–231.

Roland, M. (2007), Who is responsible? *EMBO Reports, European Molecular Biology Organization*, 8(8), 706–711.

Schelble, S. (2015, February), Using interactive case studies to create an ethical culture in the chemistry profession. Poster presented at APPE, Costa Mesa, CA.

Sher, S. (2015, February), Ethics games: Challenges, structural divisions, and keeping score. Poster presented at APPE, Costa Mesa, CA.

Spence, E.H. (2015, February), Philosophy plays: A neo-Socratic model for teaching ethics. Paper presented at APPE, Costa Mesa, CA.

Stolick, M. (2015), Scared straight: Teaching environmental ethics. Poster presented at APPE, February 20, Costa Mesa, CA.

Thiel, C.E., S. Connelly, L. Harkrider, L.D. Devenport, Z. Bagdasarov, J.F. Johnson, and M.D. Mumford (2013), Case-based knowledge and ethics education: Improving learning and transfer through emotionally rich cases, *Sci Eng Ethics*, 19, 265–286.

Vygotsky, L.S. (1978), Interaction between learning and development, in *Mind in society: The development of higher psychological processes* (pp. 79–91), edited by M. Cole, V.J. Steiner, S. Scribner, and E. Souberman, Harvard University Press, Cambridge, MA.

13. TEACHING GEOETHICS ACROSS THE GEOSCIENCE CURRICULUM: WHY, WHEN, WHAT, HOW, AND WHERE?

David W. Mogk[1], John W. Geissman[2], and Monica Z. Bruckner[3]

Abstract

The grand challenges that face humanity about living on Earth in the twenty-first century present ethical issues that will engage geoscientists personally, and collectively as a profession. Thus, there is strong reason to ensure that the geoscience workforce of the immediate future is cognizant of the need and is equipped to recognize, mitigate, and act upon geoethical issues as they arise. This will require geoethics instruction at the undergraduate and graduate levels across the curriculum. This chapter provides the rationale for, and summarizes key elements of, geoethics instruction. We emphasize that the domain of geoethics lies at the intersection of several human endeavors, involving understanding the Earth system, monitoring and addressing changing social and cultural value systems, responding to economic realities, communicating with the public, protecting human health, engaging responsible stewardship of the Earth, and philosophically assessing regrettable and unsustainable circumstances that lead to irreversible impacts on planet Earth and humanity. As such, the teaching of geoethics should be explicitly embedded into the geoscience curriculum; teaching faculty should accept this responsibility as a core component of all geoscience classes, at all instructional levels, for all students.

[1] Department of Earth Sciences, Montana State University, Bozeman, Montana, USA
[2] Department of Geosciences, University of Texas at Dallas, Richardson, Texas, USA
[3] Science Education Resource Center, Carleton College, Northfield, Minnesota, USA

Scientific Integrity and Ethics in the Geosciences, Special Publications 73,
First Edition. Edited by Linda C. Gundersen.
© 2018 American Geophysical Union. Published 2018 by John Wiley & Sons, Inc.

13.1. Introduction

As the world's population is expected to hurtle well past 10 billion by the end of the century [*United Nations News Centre*, 2011], humanity is facing numerous grand challenges associated with how to live responsibly and sustainably on planet Earth. The Earth Science Literacy principles [*Earth Science Literacy Initiative*, 2009, Big Ideas 7, 8, and 9] affirm that humans depend on Earth for resources, that natural hazards pose risks to humanity, and that humans significantly alter the Earth. Each of these principles encompasses ethical considerations about (1) how geoscientists conduct their work; (2) how the product of their science is communicated and applied for the health, safety, and economic security of humanity; and (3) the responsibilities of geoscientists for stewardship of Earth. Addressing these grand challenges requires the development of a geoscience workforce that is cognizant of those needs as well as well equipped to reduce vulnerability and nurture life by "fostering a sustainable future through a better understanding of our complex and changing planet" [*NSF*, 2009].

Solutions to the grand challenges that will face humanity and our relationship with Earth through this century and into the next are found at the intersection of four basic endeavors: (1) knowledge of how the Earth system operates, (2) understanding of changing social and cultural value systems (3) economic realities, and (4) philosophical approaches that address human actions that can catastrophically and irreversibly impact human existence, and indeed, all life on Earth [e.g., *Barnosky et al.*, 2011; *Kolbert*, 2014]. It is this intersection in which the domain of geoethics lies [*Moores*, 1997; *IAPG*, 2012; *Peppoloni and di Capua*, 2012, 2015] (Figure 13.1). Scientific and social imperatives converge in an Earth system approach [*NASA*, 2003] that focuses on the interactions between Earth's "spheres" (e.g., atmosphere, hydrosphere, geosphere, biosphere) and increasingly emphasizes the human dimensions that impact, and are impacted by, Earth [*Ireton et al.*, 1997]. Because Earth system science and geoethics are integrally connected at the interface between natural and social systems, comprehensive, coherent, and intentional instruction in geoethics is essential for all students (Figure 13.1).

If geoethics is an integral aspect of Earth system science, instruction in geoethics is an essential component of the professional development of geoscientists. Geoethics education addresses common themes informing how geoscience should be carried out in an ethical manner, and how the products of geoscience-related work and research are appropriately applied to societal issues. As demands for Earth's resources as well as pressures stemming from the consequences of natural hazards continue to place personal and communal values and priorities in conflict, geoscientists will be confronted more frequently with ethical issues requiring a balance between geoscience knowledge and societal issues [e.g., *Grunwald*, 2015; *Potthast*, 2015]. Consequently, there is a compelling need for geoscientists to have preprofessional (and continuing!) training in geoethics

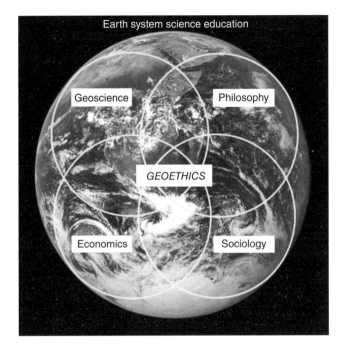

Figure 13.1 Geoethics is found at the intersection of geoscience, philosophy, economics, and sociology, and is an essential component of Earth system education.

[*Mayer*, 2015] in order to both recognize emerging ethical issues and be prepared to take appropriate actions when ethical issues materialize. We assert that the geoscience workforce of the future must be adequately trained to protect the integrity of their science, to act responsibly as part of a professional community, to serve society, and to foster stewardship of the planet [*AGI*, 2015; *Boland and Mogk*, 2018, this volume]. A robust course of instruction in geoethics should be included in geoscience curricula for undergraduate majors as well as graduate students. In addition, geoethics can and should readily be incorporated into "introductory-level" geoscience courses or possibly in cross-disciplinary courses team taught with colleagues in the humanities, arts, or engineering departments (e.g., "first-year experience" courses that address topical issues). These experiences may be the only exposure to the Earth system that end users in civil capacities (e.g., public policy makers, civic planners, businesspersons, and ordinary citizens who consume goods and hopefully vote regularly) may receive in their higher education experience.

Currently, there is no formalized course of study on ethics in the geoscience curriculum that is on par with, or comparable to, those required for other disciplines (e.g., ABET requirements in engineering; see their website [n.d.]). Although ethics education in the geosciences shares some commonalities with sister STEM disciplines, ethics programs "must be tailored to meet the needs of researchers in specific fields" [*National Academy of Engineering*, 2009]. This point is further emphasized by Potthast [2015], who suggests that an "ethics in science" approach is best suited to reveal the intricate connections between the facts (of a discipline) and values ("epistemic moral hybrids"). Potthast [2015] writes, "Scientists can no longer be viewed as value-neutral providers of knowledge and certain means, leaving only the user and society to think about the moral dimension of their 'application.'" Thus, there is a need for a dedicated effort to develop a geoscience-specific ethics curriculum that addresses the types of questions, methods, and motivations that animate the geosciences as a discipline. The preprofessional training of geoscience students is, we assert, unfortunately marginalized to the extent that ethical training is not adequately provided. Mastery of ethical principles and practice is too important to be left to mimicry, emulation, or chance circumstance; instruction and training in ethics must be explicitly and intentionally embedded across the geoscience curriculum by design. Regardless of how compelling our case may be, the reality is that few geoscience faculty recognize the need or have the interest in or the training required to provide instruction in geoethics to undergraduate and graduate students in either formal courses or through mentoring in research programs. In addition, there are currently few published or web-based resources available that address issues in the Earth and environmental sciences that can support a course of instruction in geoethics [e.g., *National Center for Professional & Research Ethics*, n.d.). To address this need, we have organized this chapter to provide an overview of "why," "when," "what," "how," and "where" to teach geoethics in the geoscience curriculum. The information presented in this chapter derives from a 2014 National Science Foundation–sponsored workshop titled "Teaching Geoethics across the Geoscience Curriculum" [*SERC*, n.d.-i].

13.2. What Is Geoethics?

The International Association for Promoting Geoethics [*IAPG*, 2012] defines geoethics in this manner:
- "Geoethics consists of the research and reflection on those values upon which to base appropriate behaviours and practices where human activities intersect the Geosphere.
- Geoethics deals with the ethical, social and cultural implications of geological research and practice, providing a point of intersection for Geosciences, Sociology and Philosophy.
- Geoethics represents an opportunity for Geoscientists to become more conscious of their social role and responsibilities in conducting their activity.
- Geoethics is a tool to influence the awareness of society regarding problems related to geo-resources and geo-environment."

Corollaries to these overarching principles indicate that geoethics encompasses numerous endeavors, including providing reference and guidelines for behaviors in conducting science; interacting with the public; building respect for the environment and Earth; defining the responsibilities of geoscientists; providing for the analysis of management of natural resources; addressing risk management, perception, and mitigation; requiring proper dissemination of scientific studies and solidifying the relationships between science and community; and illustrating the value of geoscience in addressing societal issues. The American Geosciences Institute has recently developed a comprehensive set of Guidelines for Ethical Professional Conduct that addresses the personal responsibilities of geoscientists as well as practicing members of a scientific community [*Boland and Mogk*, 2018, this volume; *AGI*, 2015].

There are many dimensions of geoethics. Typically, the focus of ethics training has emphasized responsible conduct in all forms and aspects of research. As important as this topic is, it falls far short of encompassing the totality of ethical issues that may confront geoscientists on personal, professional, societal, and global levels. In this chapter, in the context of endorsing a more formalized approach to incorporating geoethics in the classroom, we address four important dimensions of geoethics:

- *Geoethics and self*: What are the personal attributes of a geoscientist that establish the ethical values required to successfully prepare for and contribute to a career in the geosciences?
- *Geoethics and the geoscience profession*: What are the ethical standards expected of geoscientists if they are to contribute responsibly to the community of practice expected of the profession?
- *Geoethics and society*: What are the responsibilities of geoscientists in effectively and responsibly communicating the results of geoscience research to inform society about issues ranging from geo-hazards to natural resource utilization in order to protect the health, safety, and economic security of humanity?
- *Geoethics and Earth*: What are the responsibilities of geoscientists in providing good stewardship of the Earth based on their knowledge of Earth's composition, architecture, history, dynamic processes, and complex systems?

13.3. Causality, Uncertainty, and Analogous Thinking in the Geosciences

Why focus on "geo" in geoethics? The geosciences provide a global perspective of the complex relationships between the solid Earth, the hydrosphere, the atmosphere, the biosphere, and humanity as they interact in the Earth system. The Earth system is inherently open, dynamic, complex, and often chaotic. Earth processes (physical, chemical, biological) control the transport and storage of energy and mass between components of the Earth system on spatial scales that range from atomic to planetary and over the expanse of geologic "deep time." The geosciences study the geologic past in part to provide insight into what may occur in the future. However, the ancient geologic record is typically incomplete, and this occasionally leads to differences in the interpretation of field geologic relations. In addition, many Earth processes occur in spatially inaccessible environments, and temporally over time spans that are beyond direct human observation [*Ault*, 1998]. The result is often considerable, but inevitable, uncertainty in understanding the magnitude and frequency of geologic events that impact humanity (e.g., geohazards such as earthquakes, floods, extreme weather events) [*Pollack*, 2003] and the heterogeneous distribution of resources needed to sustain human life on a global scale. It is extremely difficult to assign causality to geologic phenomena. Hill's Criteria for Causality (strength of evidence, consistency of evidence) addresses many challenges in the geosciences: specificity, as events may have multiple causes and multiple effects; temporality (order of events); plausibility (does not violate known laws of Nature); coherence (should not conflict with generally known facts of Nature); experiment; and analogy [see *Crislip*, 2010]. Geoscientists must often rely on inference and analogy to explain the nature of their work and their interpretations [*Manduca and Kastens*, 2012]. Uncertainty and analogous thinking have been at the foundation of geologic thinking for centuries, perhaps as early as the time of Herodotus (484 to 425 BC), as he contemplated the age of the Nile River Delta in *The Histories*. More recently, the concepts of uncertainty and analogous thinking have been articulated by *Gilbert* (Inculcation of the Scientific Method [1886]) and *Chamberlain* (Method of Multiple Working Hypotheses [1890]), and further elaborated upon by contemporary philosophers of the geosciences such as Frodeman [1995]. The factors that are intrinsic to the nature of geoscience all conspire to set up potential conflicts and ethical dilemmas, both in the conduct of geoscience investigations and in the manner in which the results are presented, with or without uncertainties, to the geoscience community and to the public [e.g., *Lelliott et al.*, 2009; *MacCormack*, 2015; *Lark*, 2015; *Lachner and Kirchengast*, 2015].

13.4. Why Teach Geoethics?

13.4.1. Motivation

Certainly, a key reason for teaching geoethics centers on professional expectations. Ethics education is an increasingly important component of the preprofessional training of (geo)scientists, who are expected to conduct their work in an ethical manner so that colleagues will have confidence and trust in the integrity of their results [e.g., *McNutt*, 2015]. This expectation extends to application and communication of scientific results to the public, community planners, and policy makers who rely on this information to make the best informed decisions. Because considerable geologic work is focused on the economics of resource extraction, geoethics plays an important role in corporate decision making that may have significant impacts on the financial standing of a corporation or economic returns to investors and even local environments (e.g., the BreX gold scandal [*Schneider*, 1997; *Ro*, 2013], recent allegations that ExxonMobil misled the public about climate change risks [*Gillis and Schwartz*, 2015], and the decision by Tiffany and Company to decline participation in the Pebble Mine (gold) project in Bristol Bay, Alaska, which eventually led to the project being put on indefinite hold [*Kowalski*, 2015]).

External drivers have established a mandate to provide preprofessional training in ethics for all scientists. Funding agencies (e.g., the U.S. National Science Foundation (NSF)) require training of graduate students in the responsible conduct of research [e.g., NSF, n.d.-b]. Employers, to a greater extent, are expecting their workers to have basic training in ethics as businesses seek to minimize and avoid exposure to liabilities [e.g., *University of Texas*, 2014; cf. *Center for Energy Workforce Development*, n.d., "Workplace competencies"]. Furthermore, the public demands the highest standards of ethical conduct by scientists [e.g., *Center of Science in the Public Interest*, n.d.; *U.S. Department of the Interior*, n.d.; *U.S. Department of Agriculture*, 2013]. Society trusts that scientific research results are an honest and accurate reflection of a researcher's work [*NAS-NAE-IOM*, 2009; *Resnik*, 2011]. Trust in the process and products of the scientific enterprise are both particularly germane in the contemporary political environment, as the Pew Research Center [2015] reports: "One of the key trends in public opinion over the past few decades has been a growing divide among (United States) Republicans and Democrats into ideologically uniform 'silos.'" All of these mandates, among many others, establish the expectation of trust within the scientific community, and confidence by the larger community, that the products of scientific investigations have been conducted and reported according to the highest ethical standards.

There are also many pedagogic reasons for including geoethics in both undergraduate and graduate geoscience curricula. Formalized study of geoethics will likely accomplish several goals and will enhance student experiences in higher education. Geoethics promotes critical thinking and review. It provides students with opportunities to formulate responsible solutions to socioeconomic problems. It requires thoughtful and effective communication about scientific findings to peers and the public [e.g., *Geller*, 2015]. It also provides a systems-thinking context for dealing with risks, such that discussions of issues related to hazards and resources can integrate sciences with social policies as well as highlight environmental, social, and economic stewardship. Geoethics stimulates social awareness of nature and natural history. Finally, geoethics provides context and relevance of geologic content and skills outside the classroom. Geoethics can be considered a part of a continuum of learning objectives that are often referred to as the "cocurriculum," which is not necessarily "owned" by any specific course or discipline. Among these are communication skills (oral, written, graphical), quantitative skills, interpersonal skills, information skills (discovery, review, prioritization, and appropriate application), and an appreciation, and thus an understanding, of the history and philosophy of our science. Critical thinking, problem solving, and development of "geoscience habits of the mind" [*SERC*, 2012] are related learning objectives. Inclusion of geoethics instruction in the geoscience curriculum supplements and complements these essential components of the geoscience cocurriculum. The issues that are typically encompassed by geoethics provide currency and relevancy to the geosciences and may help attract and recruit students to the geoscience professions and motivate continued learning. Geoethics may be particularly important in recruiting students from underrepresented groups, as it addresses issues of environmental justice, sustainability, urban issues, and civic engagement in societal issues [SERC, n.d.-a].

Geoethics is a growing field of scholarship that has yet to gain a strong presence in the geoscience curriculum. However, geoethics should be an important component of the preprofessional training of geoscience undergraduate majors and graduate students as well as for a scientifically literate public. For example, how can a student/citizen begin to answer the question "Should we rebuild the 9th Ward in New Orleans in the aftermath of Hurricane Katrina?" without understanding both the scientific fundamentals (frequency and magnitude of hurricanes, topography of flood plains, coastal processes) and the societal and economic implications that result from either (1) allowing people to rebuild in areas that are imminently dangerous or (2) requiring whole neighborhoods to be permanently displaced? The U.S. National Science Foundation Directorate for Social, Behavioral, and Economic Sciences (SBE) has placed an emphasis on "partitioning the collection" of SBE resources across topics such as public health, complexity, crisis and disaster prevention and management, sustainability, global climate change, urbanization, ethics, and education in its research agenda [*NSF*, 2011] (see Figure 13.2). Corollary priorities are focused on applications to energy, environment, and human dynamics by addressing several key issues, such as "How can society become resilient in the face of both natural and human-made disasters?" "How will changing patterns of human activity alter energy consumption and interaction with the natural environment?" and "What influences how people perceive, value, and use energy and the natural environment?" [*SBE*, 2009].

The ultimate rationale for formal instruction in geoethics is this: The geosciences have a key place at the table of public discourse about topics such as geohazards, resources, and questions centering on the ability of humanity to live responsibly and sustainably on Earth. Humanity has been able to harness the energy of the planet in such a way that "Agent Man" is now capable of altering global processes in a new geologic epoch, referred to as the "Anthropocene" [*Crutzen and Stoermer*, 2000; *Crutzen*, 2002; *Zalasiewicz, et al.*, 2008; *Ellis and*

Intersections of
Geoscience and Sustainability

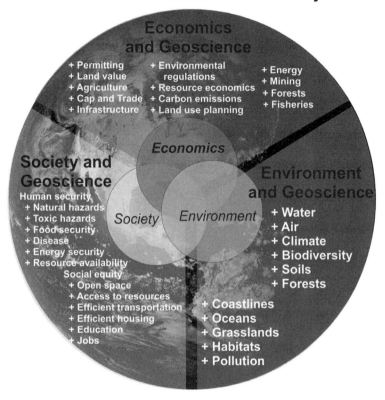

Figure 13.2 Issues related to sustainability map directly onto the same landscape that is encompassed by geoethics.

Haff, 2009], although the International Commission on Stratigraphy has not officially adopted the term for Earth's youngest epoch. A current example is the "Grand Inter-Oceanic Canal" planned across Nicaragua by the HK Nicaragua Canal Investment Co. Ltd., which will accommodate mega container ships that cannot pass through the Panama Canal and will require moving more soil and rock than the volume of 2 million Olympic-size swimming pools. Traditionally, we have trained geoscientists who have sought to understand Earth processes and, in doing so, have often said, "Yes, we can control and/or exploit Earth." We assert that it is now time to begin training geoscientists who also have the wisdom to ask, "But should we?"

13.4.2. Goals of Teaching Geoethics

Instruction in geoethics should prepare students to recognize ethical dilemmas as they emerge, to develop strategies and skills needed to responsibly participate in the profession, and to gain experience in ethical decision making. The learning goals for geoethics should be student centered, should articulate what students should know and be able to do, should promote development of higher order thinking skills (interpretation, analysis, synthesis, evaluation), and should be sufficiently concrete to yield measurable outcomes [SERC, n.d.-d]. The results of a workshop sponsored by the National Academy of Engineering [2009] described student learning goals for ethics education and identified the skill and knowledge sets that should be developed in ethics education. Participants recognized that students should be capable of recognizing and defining ethical issues; identifying relevant stakeholders and socio-technical systems; collecting relevant data about stakeholders and systems; understanding relevant stakeholder perspectives; identifying value conflicts; constructing viable alternative courses of action or solutions and identifying constraints; assessing alternatives in terms of consequences, public defensibility, and institutional barriers; engaging in reasoned dialogue or negotiations; and revising options, plans, or actions.

Workshop participants at the 2014 Teaching Geoethics across the Geoscience Curriculum workshop expanded on these goals by including the demonstration of a moral commitment and responsibility of action. More specifically, this demonstration and responsibility would involve moral sensitivity in being able to recognize ethical issues/concerns; moral understanding and recognition of moral concepts; moral reasoning involving deliberating about ethical conduct, interpreting and applying rules, balancing competing values, and resolving moral/ethical dilemmas; and moral courage to ensure that the right thing is done, even if it may involve some personal cost. These goals for geoethics instruction as an essential component of undergraduate and graduate curricula must then be well aligned with course content and assessments. In the following sections, we provide some suggestions on how to best promote learning about ethics for students.

13.5. What to Teach about Geoethics?

Many overarching concepts and themes that can be incorporated into geoethics instruction are easily accessible to students and faculty. Having explored overarching concepts and discussed general themes, students will be well positioned to evaluate and discuss specific examples of geoethical situations, which of course can be tailored to the level of the course and the geographic/geosocial setting. Here we present some of the central principles of geoethics that can be used as the foundation for more detailed explorations of geoethical issues and case studies.

13.5.1. Microethics and Macroethics

The concepts of microethics and macroethics should be introduced early in the sequence of instruction in geoethics. As described by Herkert [2001], microethics deals with personal and professional ethics and can be tied to responsibilities at the personal and intraprofessional level (e.g., an environmental consultant's ethical responsibility to provide a client with reliable data). Macroethics, on the other hand, deals with the ethics of a society or culture and can be tied to personal and professional responsibilities toward society (e.g., environmental consultants' responsibilities as a profession to ensure environmental stewardship in their professional conduct) [Herkert, 2004]. These concepts can help establish a foundation to assist students in analyzing ethical dilemmas at multiple levels.

13.5.2. Beneficence

Beneficence is the concept that scientific research should have as a goal: the welfare of society. This concept is rooted in medical research, where the central tenet is "do no harm" (and the corollaries remove harm, prevent harm, optimize benefits, and "do good"). The Belmont Report [1979] was written with a focus on "ethical principles and guidelines for the protection of human subjects of research" but also has important applications to geoethics. Beneficence is understood as an obligation to minimize risks, optimize benefits, and make sure there are adequate provisions to protect the privacy of subjects and to maintain the confidentiality of data. In applying the concept of beneficence to ethical dilemmas, the questions of who benefits, who are the stakeholders, who are the decision makers, who is impacted, and what are the risks all should be taken into consideration and should serve as the basis for extended discussion involving numerous useful examples.

13.5.3. Ethical Decision Making

Another essential concept is that of ethical decision making, which includes review and integration of several elements, including the context/facts of the situation, the stakeholders, and the decision makers. These elements factor into the number of alternate choices, which are mediated through the evaluation of impacts and negotiations among the parties and which should lead to the selection of an optimal choice [*Taylor*, 2014]. Vallero [2014] discussed ethical decision making and formulated a seven-step approach to ethical decision making (see also *Davis* [1999]):
1. State or define the problem/issue.
2. Gather information ("facts") from all sides.
3. Delineate all possible resolutions.
4. Apply different values, rules, principles, and regulations to the different options.

5. Resolve conflicts among values, rules, etc.
6. Make a decision and act.
7. Review steps 1–6; determine action and mitigation to prevent future situations.

These basic steps of ethical decision making can be universally applied to a very broad range of ethical learning opportunities in coursework. The systematic discussion of the adherence to these steps using specific and pertinent real life examples will serve students well as they engage in their professional careers.

13.5.4. Domains of Geoethics

For most working scientists, consideration of ethics is largely understood to involve responsible conduct of research, lab practices, and issues related to publication [e.g., *American Geophysical Union*, n.d.]. However, we take a much broader view of ethics and advocate an exploration of ethics from personal to planetary perspectives. For each of the four domains of geoethics that we initially defined, there are several topics and considerations that can readily be addressed in virtually any geoscience course.

13.5.4.1. Geoethics and Self

"Geoethics and Self" centers on how students learn right from wrong in their preprofessional training and how students embody the practices and standards required of their profession. Well-designed class activities should have students examine the personal attributes of a geoscientist that establish the ethical values required to successfully prepare for and contribute to a career in the geosciences; develop self-monitoring and self-regulating behaviors that adhere to standards of the profession; develop the ability to identify ethical dilemmas and their implications; and build a toolkit of strategies and practices to employ ethical decision making. For example, consider these scenarios: (1) While working alone late at night in an analytical laboratory, you notice significant instrument drift. Do you stop and take the time to run an internal standard, go back and rerun the whole calibration routine, decide to apply an "empirical" correction to the data at a later time, or shut down the operation and plan to identify the problem and rectify the situation tomorrow? (2) You are close to completing a difficult traverse in the field, the day is getting late, a thunderstorm is blowing in, and this is a location where you will not have the opportunity to revisit. Do you continue the traverse to acquire needed field data and critical samples despite the personal discomfort and potential danger, or do you retreat to safety and comfort? What would you do? There are no easy answers. But a geoscientist in these situations must make personal decisions about how to proceed, and these will ultimately impact the quality and integrity of the scientific product. A notable recent example of the need for geoscientists to be self-aware about potential ethical dilemmas can be found in the case of a geoscientist who was hired to compile a database of oil and gas wells in China and who was subsequently arrested and sentenced to eight years in prison for selling state secrets. The geologist reported, "I was doing honorable work in bridging the international oil industry with a country that needed investment, and in helping my firm revamp its products. The profession of geologist seemed the safest in the world….in my mind I never did anything wrong. I had such peace of mind" [*Matacic*, 2015]. Be forewarned, be prepared, and be aware of the potential consequences of your actions.

13.5.4.2. Geoethics and the Profession

13.5.4.2.1. Ethics of Members of a Professional Community

Students cannot be expected to fully understand the accepted norms and practices of the geoscience profession without explicit guidance. A good place to start is by assigning students to systematically review and to compare and contrast the codes of ethical conduct that have been developed by geoscience professional societies (a partial listing can be found on the Teaching Geoethics across the Geoscience Curriculum website [SERC, n.d.-i]). The American Institute of Professional Geologists [n.d.] publishes a monthly column by David Abbott on professional ethics and practices in *The Professional Geologist*; the 30-year archive of these columns provides rich and accessible readings to support geoethics instruction. Similarly, the American Association for the Advancement of Science [n.d.] has published a quarterly *Professional Ethics Review* since 1995 that highlights a wide array of cases largely focused on responsible conduct of research. A systematic review of responsible conduct of research can be found in *On Being a Scientist* [*NAS-NAE-IOM*, 2009]. Students should also be formally introduced to the Singapore Statement on Research Integrity [2010], which provides 4 fundamental principles and 14 responsibilities of researchers, and the Montreal Statement [2013], which deals with issues and problems that may arise in cross-boundary research collaboration and which recognizes responsibilities related to the increasing trend of collaborative and interdisciplinary research [*Mayer*, 2015]. Other related resources on responsible conduct of research can be found at the Teaching Geoethics across the Geoscience Curriculum website [SERC, n.d.-l].

Schienke *et al.* [2011] have identified three ethical dimensions of scientific research that should be articulated for students: (1) procedural ethics, which encompasses ethical aspects of conducting scientific research and disseminating results; (2) extrinsic ethics, which includes ethical factors to consider regarding the impact of scientific research on society; and (3) intrinsic ethics, which examines ethical issues and values that are embedded in the scientific process, such as choice of models, data reduction procedures, data acceptance criteria, and reporting errors and uncertainties. These ethical considerations may be constitutive (e.g., related to accepted professional practice) or contextual (accurate and complete reporting of background, assumptions, and external factors that direct what questions are asked and how the research is to be done). Other geoethics topics may be embedded into class activities for discussion or further exploration as time, opportunities, and circumstances warrant and/or allow. One example of these topics is the identification of ethical standards expected of geoscientists if they are to contribute responsibly to the community of practice (e.g., SERC, n.d.-k).

The assumption of trust in scientific research between all contributors (authors) is essential to the success of the scientific enterprise, and based on our experience, is a topic of considerable student interest. This relies on scientists' willingness to self-police, with a clear obligation to understand the implications of scientific misconduct. Related attributes include *accountability* (the obligation to justify one's actions, to report or answer to others, and to be evaluated upon one's choices) and *transparency* ("shedding light on rules, plans, processes, and actions"). Transparency requires that all procedures used in data manipulation be documented and reported so that others can understand and replicate the manipulations [*Kelman*, 2015]. Trust in the geosciences also includes transparent data collection and management practices. Another aspect of trust centers on the question of authorship and authorship criteria. The Committee on Publication Ethics and the International Committee of Medical Journal Editors have weighed in heavily on the criteria that should be used to establish authorship and discusses numerous steps to take during the course of a research enterprise to ensure that proper credit is given to all appropriate authors as well as steps to take to limit the possibility of unethical actions by one or several members of the research team. The advancement of science depends on the integrity of all aspects of data management, including the collection, use, and sharing of data, as recently discussed by Alberts *et al.* [2015] and Nosek *et al.* [2015] in the June 26, 2015 issue of *Science*. If the research is to be presented in a useful and significant way, critical decisions about the selection and analysis of data must be made before research commences if at all possible.

Data should be shared. Since 2011, the U.S. National Science Foundation has required all funded projects to implement a Data Management Plan for sharing and dissmination of research results [*NSF*, n.d.-a]. An open data policy reflects positively on those who share and benefits science by increasing the likelihood for new insights, collaboration, and reciprocal sharing. The new Guidelines for Authors published by *Science* [*Nosek et al.*, 2015] serve as an excellent template for student discussion of these key topics. A corollary to data management is the use of information in a digital world; from this, several questions arise, including Who controls the information? What are the expectations of making information universally accessible? How does one cite or indicate credit for information that's "in the ether"? Who can use the information and under what circumstances? What responsibility do researchers have if they think their data are being used inappropriately by other parties?

The nature of the geoscience work environment can create potential personal and/or interpersonal ethical and/or moral conflicts in the field, office, or laboratory. For example, is a professional stratigrapher/sedimentologist who is employed in the carbon extraction industry essentially constrained from acting as a "free moral agent" in advising that society take a more sustainable path that leads us away from the extraction and burning of fossil fuels? Such dilemmas potentially involve job security and may conflict with other personal and professional values. Geoscientists may find themselves in positions where their work assignments come in conflict with governmental regulations, policies, and laws; consequently, geoscientists should have a basic understanding of their rights and responsibilities related to whistleblowing [e.g., *NAS-NAE-IOM*, 2009; 19–23].

Are there, or should there be, limits to the extent or scope of geoscience research? Just because geoscientists have the ability to collect data (sometimes invasively or in a disruptive manner), do we have an inherent right to do so when individuals, populations, or natural systems may be negatively impacted? *Kelman* [2015] provides a warning related to the ethics of disaster research, as there is the potential that research projects may interfere with disaster-related recovery activities. Who decides when or if geoscience research goes "out of bounds"? Do the geosciences need the equivalent of an Institutional Review Board to certify that certain types of geoscience research can be conducted (e.g., issues related to geoengineering approaches to mitigate climate change; *Borgomeo* [2012], *Bohle* [2013])? In this context, are there instances when the acquisition of geoscience data may actually do harm to natural systems or have negative impacts on humanity (e.g., geotechnical recommendations that a dam could be built or a mine developed that could ultimately result in the displacement of human populations and significantly affect the environment)? Much has been written about the preservation of "classical" geologic sites and/or the ethical issues surrounding the collection of material from rare and unique localities. Collection of vertebrate fossils in the United States is subject to the Omnibus Public Land Management Act of 2009, Title VI Subtitle D—Paleontological Resources Preservation, and the Society of Vertebrate Paleontologists [n.d.] has adopted a strong ethics policy that governs collection, record keeping, adherence to regulations and property rights, fossil preparation, curation of specimens, and commercial sale or trade of specimens. Examples of global cooperation to govern research in environmentally sensitive areas include the Antarctic Treaty [*U.S. Dept. of State*, n.d.; *Berkman*, 2002; *Antarctic Treaty Summit*, 2009] and establishment of World Heritage Sites [*UNESCO World Heritage Sites*, n.d.] and Global Geoparks Network [n.d.]. Finally, what should geoscientists know about conducting their work while showing full respect for indigenous people and cultural norms [e.g., *Semken*, 2005; *Skandari*, 2015]?

13.5.4.2.2. Interpersonal Ethics

Interpersonal conflicts often reflect the distribution of power in social structures [*French and Raven*, 1959]. Social power is the ability to influence or control behaviors of people. Recognition of who controls power, and how it is administered, is important for conflict prevention, management, and resolution.

Most organizational structures are hierarchical in nature, and consequently, power is unequally distributed between dominant and submissive actors. The exercise of power may range from enabling to achieve common goals to coercive to extract behaviors that are contrary to personal will. Ultimately, professional conduct requires responsibility (i.e., moral agency) in personal conduct and respect for people, community, science, and our environment [e.g., see *American Geosciences Institute*, 2015].

The nature of the geoscience work environment can create potential personal and/or interpersonal ethical situations in the field, office, and the laboratory. We return to the question raised above, concerning whether a a professional stratigrapher/sedimentologist employed in the carbon extraction industry is limited in her/his ability to be a "free moral agent" in advocating that society takes a more sustainable path away from the extraction and burning of fossil fuels. Geologists may confront ethical dilemmas such as this, where job security is in conflict with other personal and professional values.

Interpersonal and work relations are particularly germane to geoscience education and subject to extensive discussion, as many class projects require collaborative and cooperative learning, and the workforce requires progressively more interdisciplinary teamwork. Collaborative work has the potential to result in ethical issues related to increased pressure for recognition, promotion, and funding and generate possible conflicts about who gets credit (appropriate authorship, as discussed above). In their preprofessional training, students should be prepared for potential ethical dilemmas arising through different types of interprersonal professional relations (e.g., client-consultant, supervisor-employee, instructor-student, mentor-student).

Interpersonal relations may involve gender issues. Particularly concerning for the geosciences are issues of sexual harassment and assult in field settings. In their recent survey, Clancy *et al.* [2014] report that 64% of the 666 respondents reported that they had personally experienced some form of sexual harassmant, and over 20% of respondents reported that they had personally experienced sexual assault. This issue was further articulated in a *New York Times* editorial [*Jahren*, 2014]. Additionally, institutions of higher education must take a critical look at policies and codes of conduct for faculty, including tenured faculty, in order to prevent abuses and to protect students from harrassment [*Flaherty*, 2015]. Discussions should center on those steps that can be taken to engender respect and to ensure a safe and productive work environment for all participants. Zero tolerance must be a universal standard in the scientific community [*Wood*, 2015].

13.5.4.3. Geoethics and Society

The association between geoethics and society is multifaceted. The geosciences have a central role in addressing the grand challenges facing humanity, particularly as they relate to geo-hazards and natural resources. Students must be given the opportunity to realize the connections between their future work in the geosciences and policies that ultimately impact human health, security, and economic well-being. As such, students should consider the responsibilities of geoscientists in service to society. An effective way to begin to address this issue is to give students opportunities to practice communicating to the public about complex Earth processes, replete with all the inherent uncertainties of the Earth system, in a responsible, accurate, and accessible manner. For example, a service-learning project in an environmental geology class was used to present a public forum to alert citizens about seismic hazards in the northern Rocky Mountains [*Mogk and King*, 1995]. Another example is the InTeGrate teaching module "Map your Hazards," which integrates interdisciplinary geoscience and social science methodologies to understand societal impacts that result from natural hazards. Students engage in place-based exploration of natural hazards, social vulnerability, risk, and the factors that shape their community's perception of natural hazards and risk and present on these findings [*Brand et al.*, 2014].

Other related topics concerning geoethics and society deserve discussion in the classroom. One is social responsibility, which can be considered as the role of geoscientists in addressing issues of global resources, sustainability, human health, and peace and security [e.g., see *Resources for Research Ethics Education*, n.d.-b; and Figure 13.2]. The American Association for the Advancement of Science [1998] affirms,

> If the U.S. is to respond effectively to the challenges of the 21st century, we must find ways to reorganize our science and technology enterprise to address tomorrow's needs and aspirations, by maintaining global sustainability, improving human health, addressing economic disparities, understanding our place in the universe, promoting peace and security, and directing the products of technology toward the betterment of society, nationally and worldwide.

Related to this issue is the question of whether scientists should be "above" policy and implications, avoiding involvement in activities that are not directly related to their scientific expertise or interests. Alternatively, is there a growing duty for scientists to become more and more involved in the societal implications of their science, going well beyond publishing in peer-reviewed academic venues [*Kelman*, 2015; *Dolce and Di Bucci*, 2015]? If scientists recognize that a proposed endeavor (e.g., development of a new water-intensive metal extraction facility in a highly arid environment) is fraught with technical and environmental challenges and complications, should they use their knowledge and speak out, publicly in one form or another, about the issue?

Public perception of, and trust in, geoscience research is highly varied. With this in mind, how can we best facilitate cooperative social interactions, risk taking (the expectation that skills and sound judgment will be applied), establishing trustworthiness (competence, experience, goodwill), and adhering to ethical and legal duties (i.e., the obligation to do what is expected [*Resnik*, 2011])? What do we consider, or assume, as trust in terms of relations between or among individuals or groups, that define a profession? Hopefully, society trusts researchers in the use of public resources (e.g., monetary support for scientific research) to provide knowledge and expertise that can inform public policy as well as provide knowledge that will yield beneficial applications in medicine, industry, engineering, technology, agriculture, transportation, communication, and other domains [*Resnik*, 2011]. Yet technology is a two-edged sword. Who, and in what capacity, has the authority to decide what applications are ethical? As Adlai E. Stevenson stated in his 1952 "The Atomic Future" speech, "Nature is neutral. Man has wrested from nature the power to make the world a desert or make the deserts bloom. There is no evil in the atom; only in men's souls." Furthermore, how can and do we convey the clear and unavoidable uncertainties associated with scientific and technologic advances to the public and planners [e.g., *Pollack*, 2003; *Bilham*, 2015; *Tinti et al.*, 2015]? With this in mind, how can we best prepare future geoscientists to responsibly inform the public and work with journalists, civil planners, and policymakers [*Dolce and Di Bucci*, 2015]?

In terms of public policy and litigation, the role of a geoscientist in informing (and possibly advocating for) public policy issues is certainly a topic deserving considerable attention in geoethics instruction. The preparations required for a geoscientist to serve as a court-appointed scientific expert are important considerations in these discussions. A facet of this discussion centers on the geosciences and underrepresented populations. What can the geosciences contribute to issues related to environmental justice [e.g., *Anand*, 2004; *Grunwald*, 2015]? The U.S. Environmental Protection Agency (EPA) defines environmental justice as "the fair treatment and meaningful involvement of all people regardless of race, color, national origin, or income with respect to the development, implementation, and enforcement of environmental laws, regulations, and policies." In areas such as the American West, for example, where water-related issues and prolonged drought have resulted in crop shortages, massive forest destruction, and an attendant increase in wildfires [*Field and Michalak*, 2015], and where the perspectives and rights of Native Americans are critical considerations [*Cozzetto et al.*, 2013], issues of environmental justice have become center stage. In addition, discussions of environmental justice issues tied to mining on Native lands can unearth rich geoethical discussions (e.g., case studies from *Klauk* [n.d.]). Matters of justice and fairness include ethical issues related to generational justice and responsibilities toward future generations. These concerns can obviously generate great discussion and interest among students and should involve stewardship and wise use of limited or nonrenewable resources and the knowledge that actions taken today may have irreversible impacts on the Earth system in the future (e.g., nuclear waste disposal, *Hocke* [2015]). This topic is not new. The "seventh generation principle" dates back to the formulation of "The Great Law of the Iroquois Confederacy," prepared sometime between 1142 and 1500 AD, and is based on the Iroquois philosophy that any decision made at a particular point in time should result in a sustainable world seven generations afterward.

There are also ethical concerns about the equitable distribution of resources among people living today and the justice associated with that distribution.

The issue of the distribution of global resources between developed and developing countries and its impacts on indigenous people grows more and more acute [e.g., *Hostetler*, 2015; *Zuniga et al.*, 2015; *Das*, 2015; *Limaye*, 2015]. A current example is the series of widespread electrical blackouts imposed by South Africa's state utility to avoid a collapse of the entire national electricity grid during the 2015 southern winter months [*Public Finance International*, n.d.; *Areff*, 2015]. Currently, sub-Saharan Africa accounts for 13% of the global population but only 4% of global energy demand, leaving more than 620 million people (two-thirds of the population) without access to electricity [*International Energy Agency*, 2014].

Grunwald [2015] enumerates several considerations that relate sustainability and geoethics. Dangers to and intolerable risks for human health must be avoided. Provisions must be made for minimum basic services (nutrition, housing, clothing, etc.) and protection against life risks (illness, disability) for all members of society. The rate of consumption of renewable resources must be stabilized at or below the replenishment rate without endangering the efficiency and reliability of the respective ecosystem. The demonstrated reserves of nonrenewable resources must be preserved over time. The release of any foreign substance into an ecosystem must not exceed the absorption capacity of that ecosystem. Technical risks associated with any enterprise that have potentially disastrous impacts must be avoided. Cultural and natural landscapes must be conserved. All members of society should be provided with the opportunity to participate in decision-making processes. Finally, all members of society must be granted equal access to information and educational and occupational opportunities. All of these considerations can readily be discussed in geoethics instruction, with pertinent examples and, at least in some cases, solutions.

13.5.4.4. Geoethics and Earth

The topic geoethics and Earth provides many opportunities for discussion of issues related to stewardship of the planet and sustainability, most of which are associated with rich examples of ethical dilemmas. Students should be aware of the fact that geoscientists have a special responsibility to foster stewardship of Earth based on their knowledge of Earth's composition, architecture, history, dynamic processes, and complex systems. These responsibilities clearly address the issues of the day, which to a large degree are the breadth of topics encompassed by environmental ethics, climate change, geoengineering, and preservation of geodiversity, to name just a few [*Osborne*, 2000]. There is potential here for elaborating on several big questions that students likely will confront in the course of their careers. Perhaps the biggest one is "What does it mean to live in the Anthropocene" [*Zalasiewicz et al.*, 2008]? How do geoscientists relate to "deep ecology" (i.e., "biospherical egalitarianism": all living things have value independent of their usefulness [*Naese*, 1973])? What is the intrinsic value of natural places (e.g., aesthetics, spiritual)? What is meant by "sustainability" and "stewardship"? What are the responsibilities of geoscientists toward sustainability and stewardship? In describing a new sustainability initiative in the Geosciences Directorate of the National Science Foundation, the directorate head wrote, "In a sustainable world, human needs would be met without chronic harm to the environment and without sacrificing the ability of future generations to meet their needs. This incorporates four topical themes: environment, energy, materials and resilience" [*Killeen*, 2012]. Each of these topical themes merit considerable student discussion.

An equally important question is how geoscientists should address preservation versus conservation. Discussions could focus on the reflections of H. D. Thoreau, "In wildness is the preservation of the world," from the essay "Walking" [1862] or "A man is rich in proportion to the number of things he can afford to let alone" from the book *Walden; or, Life in the Woods* [*Thoreau*, 1854]. The concept of Aldo Leopold's "The Land Ethic," published posthumously in 1949 as the key essay in *A Sand Country Almanac, and Sketches Here and There* includes several key considerations. One is, "The land ethic simply enlarges the boundaries of the community to include soils, waters, plants, and animals, or collectively: the land." Another is, "That land is a community is the basic concept of ecology, but that land is to be loved and respected is an extension of ethics." The final point is, "We can be ethical only in relation to something we can see, feel, understand, love, or otherwise have faith in." And a final example question is whether a river has rights [*Phillips*, 2014]. Under what circumstances might it be acceptable to dump waste in a river? Would any of these be acceptable: from a single individual (e.g., tossing food scraps, relieving oneself) on a hike, a couple washing dishes on a camping expedition, a single boat (dumping untreated sewage) on a fishing expedition, a single home with a faulty septic system, or a factory or city dumping its waste (treated or untreated)?

13.6. How to Teach Geoethics

A class studying geoethics should be dedicated to question asking, critical thinking, and problem solving. Students should be expected to develop questioning skills that result in series of logical, cascading thought processes, including recognizing the ethical issues; finding, evaluating, and prioritizing evidence (thus developing an ethical toolkit); and finally, formulating appropriate responses and actions (thus becoming moral agents).

13.6.1. Best Practices

Ethical issues are complex and are not necessarily associated with a right or wrong answer, so it is usually imperative that students practice and reflect on the actual *process* of ethical decision making rather than a specific, certain, and absolute product. Geoethics requires students to ask deep and eternal questions, as Samuel Becket [1957] presages in *Endgame*:

CLOV (*wearily*): You've asked me these questions millions of times.
HAMM: I love the old questions. (*With fervour.*) Ah the old questions, the old answers, there's nothing like them!

Active learning has been well demonstrated to be a key element in student learning [e.g., *Bransford et al.*, 1999; *Handelsman et al.*, 2004; *Freeman et al.*, 2014], and geoethics instruction provides rich opportunities for students to engage their peers, as well as their instructor(s), in classroom activities, questions, and discussions. Best practices in STEM education emphasize more and more the use of inquiry, discovery, and the prioritization and use of evidence, all essential components of geoethics instruction. We encourage instructors to explicitly include geoethics-related learning goals in any of their course syllabi to emphasize the importance and universality of this topic in the course and curriculum and to demonstrate coherency among the goals, activities, and assessments.

Traditionally, case studies have been widely used in ethics instruction. Several examples are provided at the Ethics Core website [*National Center for Professional & Research Ethics*, n.d.] and at the growing collection of case studies at the Teaching Geoethics across the Geoscience Curriculum website [SERC, n.d.-j]. Ethics Core also provides a large collection of "Two-Minute Challenges" that can easily be embedded into regular course instruction. Numerous well-documented classroom strategies create dynamic and interactive learning environments: think-pair-share, role-playing, Socratic questioning, critical review of videos and other animations, and gallery walks (see the Starting Point and Pedagogy in Action projects for a comprehensive review of "what," "why" and "how" to use active learning in the classroom in *SERC* [n.d.-g, n.d.-h]). More comprehensive or term-long class projects emphasizing geoethics may utilize problem-based learning or service learning.

Effective instruction in geoethics must recognize that many of the topics addressed may be somewhat contrary to or simply confront students' personal moral or societal values and thus may be exceedingly controversial. Such instruction moves into the realm of the *affective domain*, which encompasses the internal and external factors that affect a student's ability to learn [*SERC*, n.d.-c]. The affective domain includes factors such as student motivation, curiosity, fear, attitudes, perceptions, social barriers, and values. Instructors should be aware of "best practices" in teaching controversial (perceived or not) subjects [*SERC*, n.d.-f]. Instruction in geoethics may push students out of their comfort zone, but not too far, in what *Vygotsky* [1978] calls the "zone of proximal development." Student mastery, including willingness to address issues outside of their comfort zone, is achieved through scaffolded exercises with guidance and encouragement by the instructor as well as formative interaction with their peers.

Another important aspect of learning that is closely related to teaching geoethics is *metacognition*. Metacognition is described as being aware of one's own learning processes or "thinking about thinking" [*SERC*, n.d.-e]. Students must develop skills for self-monitoring their learning, which leads to self-regulation of their behaviors. With respect to geoethics, students must first be aware of the ethical dimensions of a given situation and then possess the tools to appropriately respond to the ethical situation. A metacognitive approach to teaching geoethics also strongly supports critical-thinking and problem-solving skills.

In teaching geoethics, instructors must be prepared to be guides in class exploration of ethical dilemmas. This task is, of course, challenging, in that it is human nature to invoke one's own bias or prior experience. Instructors should prepare themselves with a strong grounding in the specifics of the situation and the confidence to engage open-ended discussions with the class (balancing the need to keep the discussion on target but without constraining exploration and reflection toward a prescribed or personally biased outcome). Students should be encouraged to ask questions and be cognizant of the fact that often the questions they pose have no right or wrong answers. Because the topics they address will often confront personal and communal values, it is important to establish ground rules that prohibit *ad hominem* attacks on the students themselves but that encourage a rigorous scholarly critique of the evidence and ideas that are presented. *Taylor* [2014] discusses several components that are important for a productive ethical dilemma discussion, including trust, respect, and the ability to disagree without personal attacks; the difference between being judgmental and making a judgment; the importance of process over conclusion; the recognition that uncertainty is acceptable; and the adherence to the notion of "description, then prescription."

We believe that all faculty have a responsibility to introduce some measure of geoethics instruction into their courses at all levels. Simply because few faculty have formal training (or interest) in teaching ethics does not relieve them of this responsibility. Faculty must often teach "out of field" in their many courses, but they do so with confidence based on wide personal and professional experience and as they have developed the ability to be lifelong learners. In creating the Teaching Geoethics across the Geoscience Curriculum website, we have provided resources, references, tutorials, strategies, and case studies to help lower the threshold for all faculty to engage teaching geoethics. We think it safe to assume that individuals who are interested in or willing to teach geoethics will be grounded enough in ethical and moral considerations in science that they will be able to lead and guide in-class discussions that center around numerous questions pertaining to ethical issues. Such questions include:

- Who is responsible for his/her actions? (moral agent)
- Who/what can be acted upon in a way that may be right or wrong? (moral subject)
- Does something have intrinsic value? (because it exists)
- Does something have instrumental value? (useful to someone or something)
- What are the consequences of an action? (as broadly as possible)

- What is the *value* of the Earth and its physical, chemical, and biological components?
- What are the different ethical approaches that exist? Are there rules that apply universally to everyone all the time, or are issues and answers contextual?

More specific examples include the following:

- Is it acceptable to develop a strip mine in a very arid part of the American Southwest to extract over 100 million tons of magnesium from dolomite? What will happen if we continue to inject wastewater from hydrofracking "experiments" into shallow levels of Earth's crust? Who will be affected if sea level continues to rise, and how?
- What are the reasons we should, or should not, address the warnings in Pope Francis' [2015] papal encyclical concerning global warming, air and water pollution, the destruction of forests, and wasteful use of resources, to name just a few?

It does not take much to get a discussion of ethics started in class, and certainly one measure of an effective and provocative interaction in class is when such discussions spill over beyond class time, across campus, and into nearby coffee houses and other watering holes after class. Another measure is whether students are willing and/or excited about bringing up these issues with their parents or peers.

13.6.2. Assessment of Student Mastery of Geoethical Principles

Assessments of student learning must be closely aligned with the stated learning goals. Measures of student mastery of principles and applications of geoethics can be obtained using traditional course requirements such as exams, essays, and recitation participation and can even be embedded into lab assignments. Davis [1999] suggests that students can carry out an ethical analysis of class projects using criteria such as identification of stakeholders and their interests; identification of the standards or norms that are used in decision making; economic impacts; safety, public health, and environmental impacts; and adherence to professional guidelines [cf. *Online Ethics Center*, 2010]. This type of ethical analysis provides a mechanism for students to externalize their understanding of the many facets that must be considered when confronted with an ethical issue.

The Pittsburgh-Mines Engineering Ethics Assessment Rubric can be generalized to guide the evaluation of class assignments according to five attributes (1 being the lowest and 5 being the highest) [*Online Ethics Center*, n.d.] These five attributes provide a sufficiently thorough means for the instructor to objectively judge student mastery of the material related to a specific assignment. The first is the recognition of dilemmas; the range of responses is from students failing to see problem (1) to the student clearly identifying key ethical issues (5). The second bears on information, with responses ranging from students ignoring important facts (1) to students identifying unknown facts and using their own expertise to add appropriate information (5). The third considers analysis of the topic at hand, and the range of responses can run from students providing no analysis (1) to students citing analogous cases, offering more than one alternative solution, and identifying risks for each solution (5). Fourth, the breadth of perspective taken ranges from students showing nothing but a wondering perspective (1) to a single perspective (2) to a much broader, global perspective of the issue (5). Finally, the actual resolution is a key component of student input. Here the range of responses, for example, might run from the absence of a resolution or the resolution lacking integrity (1) to a situation where the student resolves case thoroughly, through clear argumentation, and understands consequences of various actions (5). Other rubrics for assessing students' ethical thinking can be found in *Rudnicka* [2005], which presents a model of a grading rubric for evaluating students' understanding of ethics case studies in an operations and management course, and the Ethical Reasoning VALUE rubric developed by the Association of American Colleges & Universities [n.d.].

Although the focus in the above assessment rubric is on measures of student mastery of ethical principles and applications, coursework in geoethics must also require students to demonstrate mastery of underlying scientific principles, to use evidence, and to apply critical-thinking and problem-solving skills.

13.7. Where to Teach Geoethics in a Geoscience Curriculum

We assert that geoethics should be taught early and often, for all students at all levels in a geoscience curriculum. Training in ethics, like practicing a musical instrument or training for an athletic event, needs to be practiced repeatedly and in many contexts, so that practitioners are fully prepared to engage geoethics in their future careers, whatever they may be. We believe that multiple exposures to geoethics across the geoscience curriculum should be an important component in training the next generation of geoscientists as well as for the education of the general public.

Introductory geoscience and geoscience-related courses (e.g., Physical Geography, Physical Geology, Environmental Geology, Earth System Science, Oceanography, Earth History) provide ample opportunities to embed geoethical topics, particularly related to societally relevant and/or topical issues such as geohazards, natural resources, climate change, and paleoclimatology. Introductory courses provide an excellent platform to introduce the following:

- Principles of sound science and the need for ethics in scientific research (e.g., understanding scientific uncertainty, ability to replicate experimental observations, data integrity and availability, etc.).
- Discussions of scientific uncertainty and the fact that too often society perceives science as a set of unchanging facts. Thus, when interpretations change

as more evidence is obtained, the public perception is that scientists don't know what they're doing. The public must understand that science advances through the give and take of testing, interpretation, and reinterpretation as new facts and evidence emerge.

- A reference and guidelines for behavior in addressing concrete problems of human existence and sustainability by exploring socioeconomic solutions compatible with a respect for the environment and the protection of all eco-systems and land.
- The societal role played by geoscientists and their responsibilities in high-lighting the ethical, cultural, and economic repercussions that their recom-mendations and/or decisions may have on society.
- Critical analysis of the use and management of geo-resources.
- Problems related to risk management and the mitigation of geohazards, including uncertainties related to geohazards.

Introductory-level courses are of particular importance in the geoscience curriculum, as they are the incubators that often generate unanticipated student interest and thus attract and recruit students to careers in the geosciences. Introductory courses provide a gateway for many such students who might never otherwise consider such possibilities but eventually may consider pursuing a degree and then a career in geoscience. In addition, for the general student population, a class in geosciences, possibly with a laboratory, may be the only sci-ence-based class that they will ever take. Consequently, such a course may be the sole basis of their understanding of fundamental Earth processes and the nature of science and may well inform their future personal actions, including voting and consuming practices, as responsible citizens. We are reminded of a passage in Frank H.T. Rhodes' [2001] *The Creation of the Future*, where he writes in chapter 5, "Teaching as a Moral Vocation,"

> After thirty-five years of teaching American history, the most striking thing about Professor Walter LaFeber [Cornell University] is that he has not lost a glimmer of his love for his subject, and still finds the birth of a similar passion in his students a cause for celebration. "It's the best thing about teaching," he said, "You see them livening up in class. You see their interests take off. And you sit there thinking, 'Is this going to be the next Secretary of State?'"

So, knowledge of the impacts of earth processes on and by humanity is important for the salvation of humanity; now, more than ever, we need a scientif-ically literate public. Introductory geoscience courses are also often required in the degree programs of preservice teachers, and the ethical dimensions related to the geosciences should be part of their professional training as well.

Geoethics principles and case studies can readily be embedded across the "core" courses of the geoscience curriculum. Every geoscience course has the opportunity (and we would even advocate the responsibility) to embed geoethical issues into standard course work. The extent of this presentation could be as small as periodic two-minute writing reflections to full class projects. In addition to explicit instruction in the responsible conduct of research (e.g., issues related to use and representations of data, lab or workplace safety, plagiarism, conflicts of interest), topical issues with ethical dimensions can also be used as class examples and case studies. Some examples, from an almost endless list, include the uses and medical impacts of asbestos in a mineralogy course; the extraction and burning of coal to provide electricity and/or heat in environmental geology courses; volcanic hazards at Mount St. Helens, Mount Pinatubo, Montserrat, Yellowstone, and responsible communications with the public in an igneous petrology course [e.g., *De Lucia*, 2015]; earthquake preparedness, prediction, and response (e.g., the L'Aquila, Italy, Case [*Mucciarelli*, 2015]) in environmental geology, geophysics, or tectonics courses; impacts of Hurricanes Katrina or Sandy in oceanography, geomorphology, or meteorology courses; and impacts of water diversion projects (e.g., Salton Sea, Aral Sea) or dealing with any Superfund site in the United States (e.g., Love Canal; Butte, Montana) in geohydrology and geomorphology courses. The InTeGrate project offers a variety of undergraduate-level teaching modules that address several of these issues in the context of teaching about the Earth and sustainability issues [*SERC*, n.d.-b]. We suggest that any course in the geoscience curriculum can easily embed geoethical topics into its regular coursework and that this expectation should be standard for course credit on par with development of communication and quantitative skills.

Even more notable is the need to fill an important niche in the geoscience curriculum: a dedicated course on geoethics, for which this contribution provides the rationale. Such a course could be offered as a capstone course for senior geoscience majors or an entry course in a graduate curriculum. Initial plans for such a course at some institutions are in development [*Cronin et al.*, 2014, 2015]. This course, and others like it, would be intended to provide fundamental concepts and practices in geoethics and will provide the forum for discussion of a series of ethical issues that students might expect to encounter in their professional careers.

Ethics training should not be confined to formal classroom instruction, however. One-on-one mentoring between faculty and students is the most common mode of training during supervision of independent study, as a growing number of undergraduate students become involved in research experiences or graduate thesis/dissertation research projects. This type of training typically focuses on laboratory protocols such as data collection and reporting, safety in the lab, and authorship. Further, this type of training usually concentrates on the immediate priorities of the research group and thus may not be comprehensive in its coverage of the breadth of ethical topics outlined in this chapter. Also, the quality of mentoring may be quite variable as determined by mentor-student interactions. Mentoring and career counseling of students requires particular attention to ethical consequences of career decisions. Students should recognize that life decisions should anticipate the ethical implications of pursuing specific career paths (e.g., extractive resources, environmental consulting, public policy.) Advice on effective mentoring can be found at the Resources for Research Ethics Education website [nd.-a] and in *NAS-NAE-IOM* [2009].

Other extracurricular opportunities for ethics training can be found at the department level, through departmental seminar series (e.g., invite experts in ethics education from sister departments or industry), journal clubs, student-initiated geoclubs, and in interactions with practicing alumni and other working geoscientists who are willing to discuss the importance of ethics and their own personal experiences in their own professional settings.

13.8. Conclusions

"It cannot be stressed enough that the key to maintaining RI (research integrity) has to be through education and training at an early stage in a researcher's career plus learning by good example and mentoring by one's seniors plus constant vigilance on infringements" [*Mayer*, 2015]. In this chapter, we have attempted to make the case that geoethics should be a central part of the geoscience curriculum from introductory courses through the core of the geoscience major into graduate education and, eventually, beyond into the broad geoscience workforce. The principles of geoethics, and of ethical decision making, should be an integral part of the fabric of geology training for all students. Ample resources are available to support instruction in geoethics, and we encourage all faculty to explicitly embed geoethics topics in their formal coursework and in informal advising and mentoring of students. Recent geologic (e.g., during the proposed Anthropocene epoch) and human history have both shown clearly that there will be continuing and likely more complicated human interactions with the Earth system, interactions that will cause more global issues. Will Durant [1946] famously wrote, "Civilization exists by geological consent, subject to change without notice," and his sentiments have become even more true today. Geoethics education is essential to provide next-generation geoscientists with the tools they will need to address (avoid, prepare for, and mitigate) ethical issues that they will surely encounter in their geoscience careers and to provide the ethical professional guidance needed to help inform public policy and planning about the geoscience-related issues that will certainly impact humanity.

Acknowledgments

We would like to thank the participants of the 2014 Teaching Geoethics across the Geoscience Curriculum workshop for generously sharing their ideas and experiences in teaching geoethics. Much of the work reported in this chapter derives from the discussions and contributions of this group. We would also like to thank the editor of this volume, Linda Gundersen, and two anonymous reviewers for their comments, which substantially improved the quality of this chapter.

This work was supported by NSF Ethics Education in Science and Engineering (EESE) program grant number NSF 1338741. Any opinions, findings, and conclusions or recommendations expressed in this material are those of the author(s) and do not necessarily reflect the views of the National Science Foundation.

References

Accreditation Board in Engineering and Technology (ABET) (n.d.), Website, http://www. abet.org/. Accessed June 30, 2015.

Alberts, B., R.J. Cicerone, S.E. Fienberg, A. Kamb, M. McNutt, R.M. Nerem, R. Scheckman, R. Shiffrin, V. Stodden, S. Suresh, M.T. Zuber, B.K. Pope, K.H. Jamieson (2015), Self-correction in science at work, *Science*, 348(6242), 1420–1422.

American Association for the Advancement of Science (n.d.), Professional Ethics Review Archives website. http://www.aaas.org/page/professional-ethics-report-archives. Accessed on June 30, 2015.

American Association for the Advancement of Science, Board of Directors (1998). A framework for federal science policy (cited at Resources for Research Ethics Education http://research-ethics.net/topics/social-responsibility/). Accessed on June 30, 2015.

American Geophysical Union (n.d.), Scientific ethics for authors and reviewers, http://publications.agu.org/author-resource-center/scientific-ethics-authors/. Accessed November 1, 2015.

American Geosciences Institute (AGI) (2015), Guidelines for ethical professional conduct, http://www.americangeosciences.org/community/agi-guidelines-ethical-professional-conduct. Accessed June 30, 2015.

American Institute of Professional Geologists (n.d.), Website, http://www.aipg.org/publications/TPGPublioc.htm. Accessed on June 30, 2015.

Anand, R. (2004), *International environmental justice: A north-south dimension*, Ashgate, Hampshire, UK.

Antarctic Treaty Summit (2009), Website, http://www.atsummit50.org/session/book.html. Accessed November 7, 2015.

Areff, A. (2015), Over 50% chance of total blackout, Nersa hears, *FIN24*, http://www. fin24.com/Economy/Eskom/Over-50-chance-of-grid-meltdown-energy-analyst-20150624. Accessed November 15, 2015.

Association of American Colleges & Universities (n.d.), Ethical reasoning VALUE rubric, http://www.aacu.org/ethical-reasoning-value-rubric. Accessed on November 7 2015.

Ault, C.R. (1998), Criteria of excellence for geological inquiry: The necessity of ambiguity, *Journal of Research Science Teaching*, 35, 189–212.

Barnosky, A.D., et al. (2011), Has the Earth's sixth mass extinction already arrived? *Nature*, 471, 51–57.

Becket, S. (1957), *Endgame: A play in one act*, http://samuel-beckett.net/endgame.html. Accessed on June 30, 2015.

Belmont Report (1979), Ethical principles and guidelines for the protection of human subjects research, United States Department of Health and Human Services, http://www.hhs.gov/ohrp/humansubjects/guidance/belmont.html. Accessed on June 30, 2015.

Berkman, P.A. (2002), *Science into policy: Global lessons from Antarctica*, Academic Press, San Diego.

Bilham, R. (2015), M_{max}: Ethics of the maximum credible earthquake, in *Geoethics ethical challenges and case studies in Earth sciences*, edited by M. Wyss and S. Peppoloni, 120–140, Elsevier, Amsterdam.

Bohle, M. (2013), To play the geoengineering puzzle? OIAPG—The International Association for Promoting Geoethics [blog], http://iapgeoethics.blogspot.it/2013/12/to-play-geoengineering-puzzle-by-martin.html.

Boland, M.A., and D.W. Mogk (2018), The American Geosciences Institute's guidelines for ethical professional conduct, in *Scientific integrity and ethics in the geosciences*, edited by L. Gundersen, AGU, Washington D.C.

Borgomeo, E. (2012), Why I study Earth sciences. *Ann. Geophys.*, 55, 361–363, 10.4401/ag-5635, ISSN:2037-416X.

Brand, B., P. McMullin-Messier, and M. Schlegel (2014), Map your hazards! Assessing hazards, vulnerability and risk, in InTeGrate's Earth-focused Modules and Courses Peer Reviewed Collection, edited by D. Gosselin, http://serc.carleton.edu/92962. Accessed November 18, 2015.

Bransford, J.D., A.L. Brown, and R.R. Cocking (1999), *How people learn: Brain, mind, experience and school*, National Academy Press, Washington, D.C.

Center for Energy Workforce Development (n.d.), Energy: Generation, transmission and distribution competency model, http://www.careeronestop.org/competencymodel/competency-models/energy.aspx. Accessed June 30, 2015.

Center of Science in the Public Interest (n.d.), Website, http://www.cspinet.org/index.html. Accessed June 30, 2015.

Chamberlain, T.C. (1890), The method of multiple working hypotheses, *Science*, 15, 92–96; reprinted 1965, vol. 148, 754–759.

Clancy, K.B.H., R.G. Nelson, J.N. Rutherford, and K. Hinde (2014), Survey of Academic Field Experiences (SAFE): Trainees report harassment and assault, *PLoS ONE* 9(7), e102172, doi:10.1371/journal.pone.0102172.

Cozzetto, K., K. Chief, K. Dittmer, M. Brubaker, R. Gough, K. Souza, F. Ettawageshik, S. Wotkyns, S. Opitz-Stapleton, S. Duren, P. Chavan (2013), Climate change impacts on the water resources of American Indians and Alaska natives in the U.S., *Climatic Change* 123, 569–584.

Crislip, M. (2010), Causation and Hill's Criteria, Science-Based Medicine, http://www.sciencebasedmedicine.org/causation-and-hills-criteria/. Accessed June 30, 2015.

Cronin, V., G. Di Capua, C. Palinkas, C. Maenz, S. Peppoloni, and A-M. Ryan (2014), A collaborative effort to build a modular course on geoethics, Amer. Geophys. Union Fall Meeting, San Francisco, ED23D-3499.

Cronin, V., A-M. Ryan, G. Di Capua, N. McHugh, C. Palinkas, C. Maenz, and S. Peppoloni (2015), Seeking community input to a modular course on geoethics, Amer. Geophys. Union Joint Assembly, Montreal, ED 42A-04.

Crutzen, P.J. (2002), Geology and mankind, *Nature*, 415, 23.

Crutzen, P.J., and E.F. Stoermer (2000), The Anthropocene, *Global Change Newsletter*, 41, 17–18.

Das, M. (2015), Ethics and mining—moving beyond the evident: A case study of manganese mining from Keonjhar District, India, in *Geoethics ethical challenges and case studies in Earth sciences*, edited by M. Wyss and S. Peppoloni, 393–407, Elsevier, Amsterdam.

Davis, M. (1999), *Ethics and the university*, Routledge, New York, 143–174.

De Lucia, M. (2015), "When will Vesuvius erupt?" Why research institutes must maintain a dialogue with the public in a high-risk volcanic area: The Vesuvius Museum Observatory, in *Geoethics ethical challenges and case studies in Earth sciences*, edited by M. Wyss and S. Peppoloni, 335–349, Elsevier, Amsterdam.

Dolce, M., and D. Di Bucci (2015), Risk management: Roles and responsibilities in the decision-making process, in *Geoethics ethical challenges and case studies in Earth sciences*, edited by M. Wyss and S. Peppoloni, 211–221, Elsevier, Amsterdam.

Durant, W. (1946), What is civilization? *Ladies Home Journal*, January issue.

Earth Science Literacy Initiative (2009), Earth science literacy principles, http://www.earthscienceliteracy.org/. Accessed June 30, 2015.

Ellis, E.C., and P.K. Haff (2009), Earth science in the Anthropocene: New epoch, new paradigm, new responsibilities, *EOS Trans.*, 90(49), 473.

Field, C.B., and A.M. Michalak (2015), Water, climate, energy, food: Inseparable & indispensable, *Daedalus Journal of the American Academy of Arts & Sciences*, Summer 2015: On Water, 7–17, doi:10.1162/DAED_x_00337.

Flaherty, C. (2015), Enough time for justice, *Inside Higher Ed*, https://www.insidehighered.com/news/2015/10/19/u-california-examine-deadlines-disciplining-professors-harassment-cases. Accessed November 18, 2015.

Freeman, S., S.L. Eddy, M. McDonough, M.K. Smith, N. Okoroafor, H. Jordt, and M.P. Wenderoth (2014), Active learning increases student performance in science, engineering, and mathematics, *Proceedings of the National Academy of Sciences*, 111, 8410–8415.

French, J.R.P., and B. Raven (1959), The bases of social power, in *Studies in social power*, edited by D. Cartwright, 259–269, University of Michigan Press, Ann Arbor.

Frodeman, R. (1995), Geological reasoning: Geology as an interpretive and historical science, *Geological Society of America Bulletin*, 107(8), 960–968.

Geller, R.J. (2015), Geoethics, risk-communication, and scientific issues in earthquake science, in *Geoethics ethical challenges and case studies in Earth sciences*, edited by M. Wyss and S. Peppoloni, 263–272, Elsevier, Amsterdam.

Gilbert, G.K. (1886), The inculcation of scientific method by example, with an illustration drawn from the Quaternary geology of Utah, *Amer. Journal Sci.*, 31, 284–299.

Gillis, J., and J. Schwartz (2015), Exxon Mobil accused of misleading public on climate change risks, *International New York Times*, October 30, 2015, http://nyti.ms/1Hi1bOJ.

Global Network of National Geoparks (n.d.), Website, http://www.globalgeopark.org/index.htm. Accessed November 7, 2015.

Grunwald, A. (2015), The imperative of sustainable development: Elements of an ethics of using georesources responsibly, in *Geoethics ethical challenges and case studies in Earth sciences*, edited by M. Wyss and S. Peppoloni, 25–35, Elsevier, Amsterdam.

Handelsman, J., D. Ebert-May, R. Beichner, P. Bruns, A. Chang, R. DeHaan, J. Gentile, S. Lauffer, J. Stewart, S.M. Tilghman, and W. Wood (2004), Scientific teaching, *Science*, 304, 521–522.

Herkert, J.R. (2001), Future directions in engineering ethics research: Microethics, macroethics and the role of professional societies, *Science and Engineering Ethics*, 7, 403–414.

Hocke, P. (2015), Nuclear waste repositories and ethical challenges, in *Geoethics ethical challenges and case studies in Earth sciences*, edited by M. Wyss and S. Peppoloni, 359–367, Elsevier, Amsterdam.

Hostetler, D. (2015), Mining in indigenous regions: The case of Tampakan, Philippines, in *Geoethics ethical challenges and case studies in Earth sciences*, edited by M. Wyss and S. Peppoloni, 371–380, Elsevier, Amsterdam.

IAPG (2012), What is geoethics? http://www.iapg.geoethics.org/home/what. Accessed June 30, 2015.

International Energy Agency (2014), World Energy Outlook 2014 Fact Sheet, http://www.worldenergyoutlook.org/media/weowebsite/2014/141112_weo_factsheets.pdf. Accessed November 17, 2015.

Ireton, M.F., C.A. Manduca, and D.W. Mogk (1997). Shaping the future of undergraduate Earth science education: Innovation and change using an Earth system approach, Workshop report from the American Geophysical Union: 61 pp.

Jahren, H. (2014), Science's sexual assault problem, *New York Times*, 18 September, Opinion Pages, p. A23.

Kelman, I. (2015), Ethics of disaster research, in *Geoethics ethical challenges and case studies in Earth sciences*, edited by M. Wyss and S. Peppoloni, 37–47, Elsevier, Amsterdam.

Killeen, T. (2012), A focus on science, engineering, and education for sustainability, *EOS*, 93(1), 1–3.

Klauk, E. (n.d.), Impacts of Resource Development on American Indian Lands (part of Integrating Research and Education project). http://serc.carleton.edu/research_education/nativelands/index.html. Accessed November 18, 2015.

Kolbert, E. (2014), *The sixth extinction: An unnatural history*, H. Holt and Company, New York.

Kowalski, M.J. (2015), When gold isn't worth the price, *New York Times*, Opinion Pages, 6 November.

Lachner, B.C., and G. Kirchengast (2015), Communication and perception of uncertainty via graphics in disciplinary and interdisciplinary climate change research, *Geophysical Research Abstracts*, 17, EGU2015-2240, 2015 EGU General Assembly.

Lark, M. (2015), Map scale and the communication of uncertainty, *Geophysical Research Abstracts*, 17, EGU2015-1953, 2015 EGU General Assembly.

Lelliott, M., M. Cave, and G. Wealthall (2009) A structured approach to the measurement of uncertainty in 3D geological models, *Quarterly Journal of Engineering Geology and Hydrogeology*, 42(1), 95–105, doi:10.1144/1470-9236/07-081 DOI:10.1144/1470-9236/07-081#_blank.

Leopold, A. (1949). *A Sand County almanac and Sketches here and there*, Oxford University Press, New York.

Limaye, S.D. (2015), Geoethics and geohazards: A perspective from low-income countries, an Indian experience, in *Geoethics ethical challenges and case studies in Earth sciences*, edited by M. Wyss and S. Peppoloni, 409–417, Elsevier, Amsterdam.

MacCormack, K. (2015), The importance of communicating uncertainty to the public 3D geological framework model of Alberta, *Geophysical Research Abstracts*, 17, EGU2015-1880, 2015 EGU General Assembly.

Manduca, C.A., and K. Kastens (2012), Geoscience and geoscientists: Uniquely equipped to study Earth, in *Earth and mind II: A synthesis of research on thinking and learning in the geosciences*, edited by K. Kastens and C. Manduca, Geological Society of America Special Paper 486, 1–12, doi:10.1130/2012.2486(01).

Matacic, C. (2015), Geologist reflects on life behind bars in China, *Science*, 349(6243), 13–14, doi:10.1126/science.349.6243.13.

Mayer, T. (2015), Research integrity: The bedrock of the geosciences, in *Geoethics ethical challenges and case studies in Earth sciences*, edited by M. Wyss and S. Peppoloni, 71–81, Elsevier, Amsterdam.

McNutt, M. (2015), Integrity, not just a federal issue, *Science*, 357(6229), 1397.

Mogk, D., and J. King (1995), Service learning in geology classes, *Journal of Geological Education*, 43(5), 461–465.

Montreal Statement on Research Integrity (2013), 3rd World Conference on Research Integrity, Montreal, 5-8 May 2013. http://www.cehd.umn.edu/olpd/Montreal Statementpedf. Accessed June 30, 2015.

Moores, E.M. (1997), Geology and culture: A call for action, *GSA Today*, 7(1), 7–11.

Mucciarelli, M. (2015), Some comments on the first degree sentence of the "L'Aquila trial," in *Geoethics ethical challenges and case studies in Earth sciences*, edited by M. Wyss and S. Peppoloni, 205–210, Elsevier, Amsterdam.

Naese, A. (1973), The shallow and the deep, long-range ecology movement: A summary, *Inquiry*, 16(1), 95–100. doi: 10.1080/00201747308601682.

NASA (2003), Earth Science Enterprise Strategy http://science.nasa.gov/media/medialibrary/2010/03/31ESE_Strategy2003.pdf. Accessed June 30, 2015.

NAS-NAE-IOM (National Academy of Sciences, National Academy of Engineering, and Institite of Medicine) (2009), *On being a scientist: A guide to responsible conduct in research* (third edition). National Academy Press, Washington, D.C.

National Academy of Engineering (2009), *Ethics education and scientific and engineering research: What's been learned? What should be done? Summary of a workshop*, National Academies Press, Washington, D.C., 10.17226/12695.

National Center for Professional & Research Ethics (n.d.), Ethics Core Collaborative Online Resource Environment, https://nationalethicscenter.org/. Accessed on June 30, 2015.

National Science Foundation (2009), Geovision Report, NSF Advisory Committee for Geosciences, http://www.nsf.gov/geo/acgeo/geovision/start.jsp. Accessed on June 30, 2015.

National Science Foundation (2011), Rebuilding the Mosaic, Research in the Social, Behavioral and Economic Sciences at the National Science Foundation in the Next Decade, NSF 11-086, http://www.nsf.gov/pubs/2011/nsf11086/nsf11086.pdf

National Science Foundation (n.d.-a), NSF Data Management Plan Requirements, https://www.nsf.gov/eng/general/dmp.jsp. Accessed November 7, 2015.

National Science Foundation (n.d.-b), Responsible Conduct of Research, http://www.nsf.gov/bfa/dias/policy/rcr.jsp). Accessed June 30, 2015.

Nosek, B.A., Alter, G., Banks, G.C., Borsboom, D., Bowman, S.D., Breckler, S.J., Buck, S., Chambers, C.D., Chin, G.,Christensen, G., Contestabile, M., Dafoe, A., Eich, E., Fresse. J., Glennnerster, R., Goroff, D., Green, D.P., Hesse, B., Humphreys, M., Ishiyama, J, Karlan, D., Kraut, A., Lupia, A., Mabry, P., Madon, T.A., Malhotra, N., Mayo-Wilson, E., McNutt, M., Miguel, E., Paluck, E. L., Siminsohn, U., Soderberg, C., Spellman, B.A., Turitto, J., Vandenbos, G., Vazoire, S., Wagenmakers, E.J., Wilson, R., and Uarkoni, T. (2015), Promoting an open research culture, *Science*, 348(6242), 1422–1425.

Omnibus Public Land Management Act of 2009 website, http://www.gpo.gov/fdsys/pkg/ PLAW-111publ11/pdf/PLAW-111publ11.pdf. Accessed November 7, 2015.

Online Ethics Center (n.d.), http://www.onlineethics.org/. Accessed on June 30, 2015.

Online Ethics Center (2010), Evaluation & assessment bibliography, http://www.online ethics.org/Resources/Bibliographies/evaluationbiblio.aspx. Accessed on June 30, 2015.

Osborne, R.A.L. (2000), Geodiversity: "Green" geology in action. *Proc. Linn. Soc. N.S. W.*, 122, 149–176.

Peppoloni, S., and G. Di Capua (2012), Geoethics and geological culture: Awareness, responsibility and challenges, *Ann. Geophys*, 55, 335–341, 10.440/ag-6099. ISSN: 2037-416X.

Peppoloni, S., and G. Di Capua (2015), The meaning of geoethics, in *Geoethics ethical challenges and case studies in Earth sciences*, edited by M. Wyss and S. Peppoloni, 3–14, Elsevier, Amsterdam.

Pew Research Center (2015), Americans, Politics, and Science Issues, http://www.pewinternet. org/2015/07/01/americans-politics-and-science-issues/. Accessed on June 30, 2015.

Phillips, M. (2014), Geoethics: An exploration of perspectives. Presentation made at the Teaching Geoethics across the Geoscience Curriculum workshop, http://serc.carleton. edu/geoethics/program.html. Accessed on June 30, 2015.

Pollack, H.N. (2003), *Uncertain science…uncertain world*, Cambridge University Press, Cambridge.

Pope Francis (2015), Encyclical Letter Laudato Si' of the Holy Father Francis On Care for Our Common Home, http://w2.vatican.va/content/francesco/en/encyclicals/documents/ papa-francesco_20150524_enciclica-laudato-si.html. Accessed on June 30, 2015.

Potthast, T. (2015), Toward an inclusive geoethics: Commonalities of ethics in technology, science, business and environment, in *Geoethics ethical challenges and case studies in Earth sciences*, edited by M. Wyss and S. Peppoloni, 49–56, Elsevier, Amsterdam.

Public Finance International (n.d.), http://www.publicfinanceinternational.org/feature/ 2015/06/economic-impact-south-africa%E2%80%99s-energy-crisis. Accessed November 15, 2015.

Resnik, D.B. (2011), Scientific research and the public trust, *Sci. Eng. Ethics*, 17(3), 399–409 doi:10.1007/s11948-010-9210-x DOI:10.1007%2Fs11948-010-9210-x#pmc_ext.

Resources for Research Ethics Education (n.d.-a), Mentoring, http://research-ethics.net/ topics/mentoring/#resources. Accessed June 30, 2015.

Resources for Research Ethics Education (n.d.-b), Social Responsibility, http://research-ethics.net/topics/social-responsibility/. Accessed June 30, 2015.

Rhodes, F.H.T. (2001), *The creation of the future*, Cornell University Press, Ithaca, NY.

Ro, S. (2013), BRE-X: Inside the $6 billion gold fraud that shocked the mining industry, *Business Insider*, 3 Oct 2013 http://www.businessinsider.com/bre-x-6-billion-gold-fraud-indonesia-2014-10. Accessed March 27, 2015.

Rudnicka, E.A. (2005), Ethics in an operations management course, *Science and Engineering Ethics*, 11(4), 645–654.

SBE (National Science and Technology Council Subcommittee on Social, Behavioral and Economic Sciences) (2009), Social, Behavioral and Economic Research in the Federal Context, http://www.nsf.gov/sbe/prospectus_v10_3_17_09.pdf. Accessed on June 30, 2015.

Schienke, E.W., S.D. Baum, N. Tuana, K.J. Davis, and K. Keller (2011), Intrinsic ethics regarding integrated assessment models for climate management, *Sci. Eng. Ethics*, 17(3), 503–523. doi:10:1007/s11948-010-9209-3.

Schneider, H. (1997), A lode of lies: How Bre-X fooled everyone, *Washington Post Foreign Service*, 18 May 1997, http://www.washingtonpost.com/wp-srv/inatl/longterm/canada/stories/brex051897.htm. Accessed 27 March 27, 2015.

Science-Based Medicine (n.d.), http://www.sciencebasedmedicine.org/causation-and-hills-criteria/. Accessed June 30, 2015.

Semken, S. (2005), Sense of place and place-based introductory geoscience teaching for American Indian and Alaska Native undergraduates, *Journal of Geoscience Education*, 53(2), 149–157.

SERC (2012), Teaching the Methods of Geoscience, *InTeGrate*, http://serc.carleton.edu/integrate/workshops/methods2012/index.html. Accessed June 30, 2015.

SERC (n.d.-a), Attract Diverse Students to STEM, *InTeGrate*, http://serc.carleton.edu/integrate/programs/diversity/attract.html. Accessed June 30, 2015.

SERC (n.d.-b), Authored Modules and Courses, *InTeGrate*, http://serc.carleton.edu/integrate/teaching_materials/modules_courses.html. Accessed November 18, 2015.

SERC (n.d.-c), On the Cutting Edge, Affective Domain in the Classroom, http://serc.carleton.edu/NAGTWorkshops/affective/index.html. Accessed June 30, 2015.

SERC (n.d.-d), On the Cutting Edge, Course Design, http://serc.carleton.edu/NAGTWorkshops/coursedesign/tutorial/goals.html. Accessed June 30, 2015.

SERC (n.d.-e), On the Cutting Edge, The Role of Metacognition in Learning, http://serc.carleton.edu/NAGTWorkshops/metacognition/index.html. Accessed June 30, 2015.

SERC (n.d.-f), On the Cutting Edge, Teaching Controversial Subjects, http://serc.carleton.edu/NAGTWorkshops/affective/controversial.html. Accessed on June 30, 2015.

SERC (n.d.-g), Pedagogy in Action. http://serc.carleton.edu/sp/index.html. Accessed November 18, 2015.

SERC (n.d.-h), Starting Point, Teaching Introductory Level Geoscience, http://serc.carleton.edu/introgeo/index.html. Accessed June 30, 2015.

SERC (n.d.-i), Teaching Geoethics across the Geoscience Curriculum, http://serc.carleton.edu/geoethics/index.html). Accessed on June 30, 2015.

SERC (n.d.-j), Teaching Geoethics across the Geoscience Curriculum, Case Study Collection, http://serc.carleton.edu/geoethics/case_studies.html. Accessed on June 30, 2015.

SERC (n.d.-k), Teaching Geoethics across the Geoscience Curriculum, Professional Societies Mission Statements and Codes of Ethics. http://serc.carleton.edu/geoethics/prof_soc.html. Accessed on June 30, 2015.

SERC (n.d.-l), Teaching Geoethics across the Geoscience Curriculum, Responsible Conduct of Research, http://serc.carleton.edu/geoethics/rcr_responsible.html. Accessed on June 30, 2015.

Singapore Statement on Research Integrity (2010), 2nd World Conference on Research Integrity, Singapore, 21–24 July 2010, www.singaporestatment.org. Accessed June 30, 2015.

Skandari, M. (2015), What foreign rescue teams should do and must not do in Muslim countries, in *Geoethics ethical challenges and case studies in Earth sciences*, edited by M. Wyss and S. Peppoloni, 194–201, Elsevier, Amsterdam.

Society of Vertebrate Paleontologists (n.d.), Member Ethics, http://vertpaleo.org/Membership/Member-Ethics.aspx. Accessed November 7, 2015.

Taylor, S. (2014), Geoethics Forums, Presentation made at the Teaching Geoethics Across the Geoscience Curriculum workshop, http://serc.carleton.edu/geoethics/program.html. Accessed on June 30, 2015.

Thoreau, H.D. (1854), *Walden; or, life in the woods*, Ticknor and Fields, Boston.

Thoreau, H.D. (1862), Walking, published posthumously in the Atlantic Monthly.

Tinti, S., Armigliato, A., Pagnoni, G., and Zaniboni, F. (2015), Geoethical and social aspects of warning for low-frequency and large-impact events like tsunamis, in *Geoethics ethical challenges and case studies in Earth sciences*, edited by M. Wyss and S. Peppoloni, 176–192, Elsevier, Amsterdam.

UNESCO World Heritage Sites (n.d.). http://www.state.gov/t/avc/trty/193967.htm. Accessed November 7, 2015.

United Nations News Centre (2011), Global population to pass 10 billion by 2100, UN projections indicate. http://www.un.org/apps/news/story.asp?NewsID=38253#. UP1ehuik0Xw. Accessed on June 30, 2015.

University of Texas at Austin, Jackson School of Geosciences (2014), Summit on the future of undergraduate geoscience education, http://www.jsg.utexas.edu/events/future-of-geoscience-undergraduate-education/. Accessed June 30, 2015.

U.S. Department of Agriculture (2013), Scientific Integrity Policy Handbook, DR 1074-001, http://www.usda.gov/documents/usda-scientific-integrity-policy-handbook. pdf. Accessed on June 30, 2015.

U.S. Department of the Interior (n.d.), Integrity of Scientific and Scholarly Activities, http://www.doi.gov/scientificintegrity/index.cfm. Accessed on June 30, 2015.

U.S. Department of State (n.d.), Antarctic Treaty, http://www.state.gov/t/avc/trty/193967. htm. Accessed November 7, 2015.

Vallero, D. (2014), Engineering: Lessons from the study of RCR for emerging engineering issues, Presentation made at the Teaching Geoethics Across the Geoscience Curriculum workshop, http://serc.carleton.edu/geoethics/program.html. Accessed on June 30, 2015.

Vygotsky, L. S. (1978), *Mind in society: The development of higher psychological processes*, Harvard University Press, Cambridge, MA.

Wood, B. (2015), Zero tolerance. Period, *Science*, 350 (6260), 487.

Zalasiewicz et al. (2008), Are we now living in the Anthropocene? *GSA Today*, 18(2), doi:10.1130/GSAT01802A.1

Zuniga, F.R., J. Merlo, and M. Wyss (2015), On the vulnerability of the indigenous and low-income population of Mexico to natural hazards. A case study: The Guerrero Region, in *Geoethics ethical challenges and case studies in Earth sciences*, edited by M. Wyss and S. Peppoloni, 381–391, Elsevier, Amsterdam.

14. FACILITATING A GEOSCIENCE STUDENT'S ETHICAL DEVELOPMENT

Vincent S. Cronin

Abstract

It is the responsibility of each generation of mature geoscientists to help subsequent generations develop as ethical geoscientists. Society depends on professional geoscientists to provide reliable information and unbiased expert advice about some of the most vexing challenges involving our interactions with the natural world. If the results of geoscience are not reliable, if our results are driven by financial, political, or social agendas rather than solely by good science, the professional work of geoscientists will not be sought or valued by society. Ethics is central to science. Professional geologists have a commitment to advancing scientific knowledge and serving society, and we must develop and propagate a shared sense of geoethics to help guide our work.

While mature geoscientists have a responsibility to act ethically and help in the development of novice geoscientists, the final responsibility for our ethical development rests with each of us alone. Geoscientists must accept, adopt, and internalize the imperative to act in an ethical manner every day of their academic and professional careers. The process of facilitating the ethical development of undergraduate geoscience students needs to be tailored to individual institutions and ranges from an *interior* strategy of infusing ethics education throughout an undergraduate curriculum to an *exterior* strategy that primarily uses online resources and short-course

Department of Geosciences, Baylor University, Waco, Texas, USA
International Association for Promoting Geoethics, (www.geoethics.org)

Scientific Integrity and Ethics in the Geosciences, Special Publications 73,
First Edition. Edited by Linda C. Gundersen.

opportunities organized by professional societies to provide ethical content. Well-articulated case studies with an ethical dimension are particularly effective for ethics instruction. Major geoscience organizations should support and participate in the development of resources for in-class and online learning about geoethics.

14.1. Introduction

It is possible for people who are not professionals to give no formal thought to personal ethics and yet still lead a good life. Under routine circumstances, they simply move from one task to the next, making small decisions along the way until the day is done. Occasionally, circumstances require a more consequential decision to be made, and they do what seems best at the time without much prior thought or subsequent examination. And while entire departments of philosophy professors might object by reiterating Socrates' dictum that the unexamined life is not worth living, my sense is that many lives are led simply without intentional reference to any particular ethical theory or system. The motivation is supplied by the need to feed, clothe, shelter, and care for yourself and help do the same for the ones you love, while humbly seeking happiness and fulfillment within that small local community.

The situation is different for professional geoscientists. To be called a *professional* means, obviously enough, that you are part of a profession, which is a very special type of job. Professions require specialized knowledge and extensive advanced academic training to develop the skills and background knowledge to undertake certain tasks and responsibilities on behalf of society. Professions are generally recognized formally by the legal system and are often self-governed, at least in part, by professional societies. The most visible examples of professions include medicine, law, and engineering. An important and perhaps essential feature of professions is that their principle purpose is service to the public [e.g., *Andrews*, 2014].

As geoscientists, we are part of a large professional community of scientists that has its own customs and standards as it seeks to fulfill its local and global responsibilities. Today the global community faces major problems in which geoscience will supply essential insights: climate change, water supply, waste management, energy resources, acquisition and reclamation of mineral or chemical resources, and soil conservation, to name a few [e.g., *Geological Society of London*, 2014]. Each of these challenges involves issues of government, public economics and private wealth, labor, national sovereignty, and political will. As geoscientists are deployed to help manage these vital problems in a sustainable manner, it will be necessary to develop a coherent and shared sense of geoethics within the geoscience community. Perhaps the term *geoscience community* means nothing more than "the set of all geoscientists," but a true community functions more effectively to achieve mutually beneficial outcomes. Geoethics is a fundamental tie that can bind the geoscience community together.

We have more than two millennia of source material to draw upon in studying ethics. If our goal is to facilitate a geoscience student's ethical development, we need to take care not to bury him or her in literature or arcane jargon. Achieving that goal while the student is an undergraduate is much more like planting a few seeds or saplings that will grow in time, rather than trying to transplant a mature forest. The prerequisite for learning about geoethics is not a full and functional grasp of the history and contemporary understanding of moral philosophy. We require only that a geoscience student be willing to do the work of learning every day, and to put that newly acquired knowledge into practice throughout their professional life.

The core challenge in facilitating student learning about geoethics is to help each student accept, adopt, and internalize the need to act as an ethical geoscientist all the time. The struggle is not with others who act in unethical ways, although we have a responsibility to oppose unethical behavior as part of our professional community. Rather, the more important struggle is within each of us. Striving for professional integrity is a struggle that is not won once and for all time but continues every day throughout an entire career. How do we help each novice geoscientist to fully understand the central importance of that struggle? How do we encourage each to engage in that struggle, using intellectual tools developed over two millennia of moral philosophy?

The purpose of this essay is to touch on some if the ideas and issues associated with applied-ethics education, and to provide a few thoughts on geoethics education for geoscience students. This is part of an active international conversation, and perhaps an evolving understanding, of how to help geoscience students become ethical geoscientists.

14.2. Background

14.2.1. Basic Ideas and Definitions About General Ethics

Humans are capable of thoughts and actions that are of such amazing grace, beauty, and complexity that they scarcely seem possible. We can develop bonds of love and trust with one another that can withstand the most searing trials, and our courage in the face of existential threats can be breathtaking. Humans are also capable of thoughts and actions that debase themselves and harm others, rendering them virtually unworthy of coexistence with the rest of humanity. This range exists as a potential in all of us. Our intellect allows us the opportunity to condition our thoughts, words, and actions to reflect the most positive, most constructive characteristics of our society.

A very basic definition of *ethics* is that it involves the discrimination of "good" and "bad" behavior. In law, the perceived ability to understand the difference between "right" and "wrong" is fundamental to the determination of whether a defendant is competent to stand trial for their actions. Stephen Carter's definition of *integrity* involves "(1) *discerning* what is right and what is wrong, (2) *acting* on what you have discerned, even at personal cost, and (3) *saying openly* that you are acting on your understanding of right from wrong" [*Carter*, 1996; *Benjamin*, 1990]. One might say that our *character* is the summation of all the ethical choices we make about how we interact with each other, and how we treat ourselves.

Development of our character is a responsibility that we all bear as individuals. There is no escaping that we are responsible for ourselves. And yet there is a significant role to be played by the constellation of people around us. We learn from each other. We are influenced by each other. We are responsible to each other. To some extent, we are responsible *for* each other.

Humans are social animals. Within the tribes that we inhabit, there are mores, customs, and standards of behavior and belief [*Greene*, 2013]. The meaning of the word *moral* reflects the manners or customs of the community, an understood standard of behavior that is expected of any member of the tribe. As moral philosopher Bernard Gert noted, "Every feature of the moral system must be one that is known to and could be chosen by any rational person" [*Gert*, 2005]. Behaviors that are outside of the moral envelope of a group mark a person as an unreliable member of the tribe, or perhaps not a member at all. Throughout much of human history, being ostracized from your native tribe has led to dire consequences for you and your family. Survival was most easily assured through alignment with the moral expectations of your group.

Bernard Gert recognized a universal sense of morality among humans, in contrast to the notion that every group is defined by its own morality that might or might not have significant overlap with other groups [*Gert*, 2005]. "No one discusses the question, 'Should you lie in order to cause problems for a person you do not like?' because everyone agrees on the answer" [*Gert*, 2005, 4]. In a series of accessible, well-reasoned books, Gert articulated a moral approach whose goals are to reduce the harm or evil, and to increase the amount of good in the world [*Gert*, 2004, 2005]. He recognized five principal evils to be avoided (death, pain, disability, loss of freedom, loss of pleasure) and four fundamental goods to be sought or protected (pleasure, freedom, ability, consciousness). His moral rules, required for people to inflict less harm on each other, can be summarized by two dicta: *do no harm* (do not kill, do not cause pain, do not disable, do not deprive of freedom, do not deprive of pleasure) and *do not violate trust* (do not deceive, keep your promises, do not cheat, obey the law, do your duty). Gert admitted exceptions to his moral rules, and suggested a two-step procedure to evaluate whether an exception is justified: collect all of the relevant facts "to provide a complete morally relevant description of the action," and consider what would happen if other people knew that they could also violate that rule under similar circumstances.

Gert's moral insights join with older ethical theories or systems as a worthy attempt to structure our understanding of how to live a moral life. Other major systems that are commonly described in any brief summary of applied ethics include various forms of the utilitarian (or consequential) approach originally described by John Stuart Mill, virtue ethics described by Aristotle, the natural law ethics of Thomas Aquinas, deontological ethics of Kant, duty ethics and the theory of justice described by John Rawls, and human rights ethics exemplified by the work of John Locke and A. I. Melden, to name a few [*Martin and Schinzinger*, 2005; *Starr*, 1991; *Shafer-Landau*, 2012, 2015; *MacIntyre*, 1998].

Anthony Weston is a moral philosopher who has attempted to provide an accessible "21st century toolbox" filled with useful ideas about how to navigate our world [*Weston*, 2013]. He offers two clues to the nature of ethics: first, "ethics asks us to pay attention to something beyond ourselves"; and second, "we work from the inside of [ethics]." He asks us to recognize that we are part of an ethical continuum that began long before our time here and that will continue long after we are gone. We stand in the middle of this ethical continuum and have an effect on it just as it affects us. He suggests, "To think or act ethically is to take care for the basic needs and legitimate expectations of others as well as our own." To him, a moral person might simply follow the customs and traditions of the community, but an *ethical* person is engaged in questioning and challenging those moral values, "systemizing and criticizing and possibly even revising our moral values, as well as more consciously embracing them" [*Weston*, 2013].

Robert Barger encourages us to frame our understanding of how ethical decisions are made by considering how such judgments are affected (or perhaps determined) by an ideology or worldview [*Barger*, 2008]. Barger conceives of metaethics as an attempt to understand the source or foundation of ethical judgments and their justification, and "how a particular worldview…underlies and determines the formulation of such ethical judgments."

Hugh Gauch, Jr., also considers worldviews in his treatise on the scientific method, stating, "A worldview is a person's beliefs about the basic makeup of the world and life" [*Gauch*, 2012]. Some people have a worldview that makes it impossible for them to conceive of an Earth that is older than a few thousand years, or that life on Earth evolved over time. An ideology effectively blocks their ability to understand facts that are inconsistent with their views, or to fully value opinions that differ from theirs. To Gauch, the key question with respect to worldviews is, "How much does worldview pluralism affect science's claims and fortunes?" The worldview of a skeptical purist does not presuppose that the physical world is real, orderly, and generally comprehensible by humans. Gauch concluded that, with the exception of skeptical purists, a unified expression of science is equally comprehensible and useful to all cultures and worldviews.

I am going to use the word *geoethics* in this chapter in a rather informal manner to mean, approximately, "ethics applied to the geosciences." Geoethics is also sometimes used in a more macroethical sense to capture the global responsibilities and values that are related to our stewardship of Earth, including the need to adjust our actions and economies so that we can achieve a more just and sustainable future.

Ethics asks us to pay attention to something beyond ourselves, to act with beneficence and avoid causing needless harm, to be worthy of trust. Science seeks an understanding of the physical world that is consistent with our most careful observations, and geoscience focuses our attention on understanding our planetary home. Professionals are experts in command of hard-won technical knowledge, whose work is useful and beneficial to society. Our shared goal is to help each geoscience student become an ethical professional geoscientist.

14.2.2. Young Adults, Old Adolescents

I focus on undergraduates in this essay because a significant number of novice geoscientists enter the workforce after earning a bachelor's degree. Much of what I write in this essay also applies to the ethical education of graduate students, but instruction about geoethics is too important to wait until graduate school.

The average college student in the United States who has attended school continuously since childhood is between the ages of ~18 and 23 years: right in the middle of a time of almost unimaginable change physically, emotionally, socially, and intellectually. The typical undergraduate student is still in the process of becoming functionally independent of the family that was the foundation of their childhood world. They are old enough to vote, marry, and volunteer for military service, but they might be described as older adolescents as accurately as young adults. They have worked their way through an educational system that has promoted certain virtues and a sense of "right" and "wrong" behavior, whether or not that message has also been promoted in their homes, churches, teams, media, and other associations. First-year college students are not ethical blank slates; however, neither are they mature, informed ethical agents. Recent studies in neuroscience indicate that our brains do not fully mature until around age 25 [e.g., *Aamodt and Wang*, 2011], so the undergraduate college years are enormously important to the ethical development of any person who is fortunate enough to attend college.

My experience is that it is easy to engage undergraduate students in discussions about ethical issues because they have an innate interest in these matters. What they lack is life experience and a full set of intellectual tools that are useful in the process of ethical reasoning. On average, their formal knowledge of ethics is largely limited to general descriptions of a small set of virtues along with a handful of aphorisms. A few geoscience students have taken a course involving ethics in the university's philosophy department, where theory and dense reasoning might take precedence over practical applicability. Recently, students at my university have been required to take an online ethics tutorial prior to registration for courses. That experience seems to have focused on the need to avoid plagiarism and other forms of cheating.

The ethical value that undergraduate students seem to talk about most often is fairness, a fundamental sensitivity they have had since earliest childhood. In conversations I have had with undergraduate students about their ethical concerns, they typically focus on the ethical issues they perceived within their immediate world as a college student in a geoscience department: cheating, teachers they thought were unfair, teachers who are too busy on research to spend time with them, and various injustices they had endured. They tend to focus on ethical failures rather than examples of positive ethical behavior. Their concerns tend not (yet) to be global in scope, but rather they are just trying to survive their courses in chemistry and calculus. They are focused on microethical concerns affecting their local world but are far less attuned to macroethical concerns that affect the geosciences and its relationship to society as a whole [*Herkert*, 2005].

In working with adolescents and young adults, we should understand that many of them have a strong tendency to be quick to judge, label, criticize, and shun a person for words or actions they consider to be inconsiderate, unjust, untrue or unfair. These tendencies are not unique to the young, of course. It is important to consider whether a single lapse of judgment is sufficient to characterize a person's character as intrinsically poor. I would argue that it does not, or else we would all be in trouble. We cannot begin to learn deeply about ethics until we abandon the practice of branding other people as unethical simply because they have made a mistake. As David Brooks reminds us, we are all flawed and should face our inadequacies openly, with honesty and humility [*Brooks*, 2015]. We are each "crooked timbers," to borrow a phrase from *Kant* [1784]. We need to approach our studies of ethics with kindness, charity, empathy, and humility. Our own flawed character should lead us to know, or maybe just hope, that forgiveness and redemption are requisite parts of healing and growth.

The undergraduate years can be essential to the ethical development of a novice geoscientist. While undergraduate geoscience students are not ethical blank slates, they are effectively blank slates with respect to the ethical conduct of science in general and geoethics in particular. My sense is that these students are generally unaware of the existence of geoethics, and have not given much or any thought to the role that ethics plays in science. Geoscience faculty members have a responsibility to help students develop their functional knowledge of applied ethics. The most direct ways to do so are to teach through the personal example of being an ethical professional, and to include geoethics as an intentional and integral part of the geoscience curriculum.

14.2.3. Why Do We Need to Spend Time Teaching Geoethics within a Geoscience Curriculum?

An undergraduate geoscience curriculum is already crowded with material that well-educated geologists or geophysicists need to master. Time is not elastic, so if something is added to the undergraduate curriculum, something else must be eliminated. Hence, there is often concern (if not outright hostility) about the idea of including nonscience material in the undergraduate geoscience curriculum. The same concerns and hostility have accompanied the inclusion of ethics in engineering curricula [e.g., *Herkert*, 1999; *Sigma Xi*, 1993]. Bernard Lo described a commonly held opinion in the science community, "Only unethical persons have ethical problems. Ethics is a matter of common sense and experience. Therefore, studying ethics isn't useful" [*Sigma Xi*, 1993]. Many university faculty members in STEM fields observe that they have no formal training in applied ethics, so it would be improper and perhaps even unethical for them to teach applied ethics. The faculty of many philosophy departments would probably agree.

I would argue, however, that ethics is actually a central imperative in science because the scientific enterprise is concerned with discovering reliable information about the materials, properties, processes, and history of the physical world. Rather than "reliable information," I am tempted to use the controversial word *truth* here in the sense suggested by Albert Einstein: "Truth is what stands the test of experience" [*Einstein*, 1950]. Truth to a scientist is not an ultimate and immutable explanation but rather is a provisional approximation within some measure of uncertainty [e.g., *Rutherford and Ahlgren*, 1991; *Gauch*, 2012]. The idea of "truth" is central to ethics, as expressed through virtues such as "honesty," "trust," "reliability," and "understanding" [e.g., *Ahearne*, 2011; *Gert*, 2004].

One of the recommendations that emerged from a forum sponsored by Sigma Xi titled "Ethics, Values, and the Promise of Science" [*Sigma Xi*, 1993] involves ethics education:

> Appropriate ethical behavior needs to be communicated to and practiced at all levels of academic, governmental, industrial and other research organizations associated with science and engineering. Ethics should be taught as an integral component of formal scientific education, in cooperation with technical professionals and scholars in the humanities.

This sentiment is echoed by Stephanie Bird, who wrote that "it has become apparent" that the practice of simply *hoping* students will learn about responsible research conduct and ethical behavior by observing exemplary behavior in the lab "is inadequate and serves neither the needs of the research community nor those of society as a whole" [*Bird*, 2014].

Unfortunately, there has been a history of significant ethical lapses among scientists in general and geoscientists in particular. This history provides the motivation for sustained efforts concerning geoethics. A short and incomplete list of ethical problems follows: fabrication or falsification of data; conflicts of interest; failure to maintain confidentiality; practicing without a license in certain jurisdictions; inappropriate client advocacy; creation of biased or inaccurate reports that are favorable to clients' interests; selective data acquisition, analysis or disclosure; insufficient scope of work to recognize or effectively address problems; regulatory violations; failure to disclose regulatory violations; misrepresentation of professional qualifications; unethical treatment of geoscience students;

trespassing or conducting investigations without permission on private property; damaging key geological outcrops; sampling without requisite permission; plagiarism and self-plagiarism; insufficient or missing attribution of sources; practicing beyond competency; accepting or offering bribes (gifts) to influence decisions; submission of biased reviews that are tainted by self-interest; inadequate protection of human safety; and retaliation against whistle-blowers [e.g., *ASBOG*, 2015; *Goodstein*, 2010; *Oreskes and Conway*, 2010; *Swazey et al.*, 1993; *Andrews*, 2014].

The geoscience community demonstrates that it believes ethical practice to be important in several ways. Most professional and academic geoscience societies have some form of a code of ethics for their members, strongly indicating that the geoscience community as a whole feels that there is an essential need to define standards of ethical practice. Lists of links to the ethics codes of various professional societies are available in several locations on the web, several of which are provided in the references cited in this chapter. Canada requires successful completion of a Professional Practice Exam that emphasizes professional ethics to become a licensed geoscientist [e.g., *Andrews*, 2014]. The Association of State Boards of Geology (ASBOG) affirms the importance of geoscience ethics and includes questions about applied ethics on its licensing exam [*ASBOG*, 2015]. Many states require licensed geoscientists to participate in ethics short courses or workshops on a regular basis in order to retain their licenses. One of the central purposes of the American Institute for Professional Geology (AIPG) is to encourage the ethical practice of geology, including the recognition of geoscientists as professionals. AIPG "certifies geologists based on their competence, integrity, and ethics" and has strongly supported efforts to license the practice of geology in every state in the United States (http://www.aipg.org). The American Geosciences Institute (AGI) is a federation of member societies representing approximately 250,000 geoscientists across the globe and has compiled "Guidelines for Ethical Professional Conduct" through consultation with its members (http://www.americangeosciences.org/community/agi-guidelines-ethical-professional-conduct). The International Association for Promoting Geoethics has arisen in recent years in response to the need to develop, promulgate, and strongly encourage the ethical practice of geoscience worldwide (http://www.geoethics.org).

14.2.4. Personal Responsibility and Education

We educate ourselves. The purpose of our schools as they are conventionally structured is to collect materials and expertise in a convenient place, to bundle ideas in logical sets and categories, to facilitate understanding of existing ideas and the generation of new ideas, and to bring together diverse people who want to learn. The role of a teacher is to facilitate learning. With all of the resources that society has marshaled to help me learn, at substantial cost, the enduring truth is that I have to concentrate and work diligently to accomplish that learning. Without my personal commitment and effort, all of those resources are wasted on me. My learning is *my* responsibility.

Learning about anything requires a commitment by the student to do the work necessary to learn, to make the connections, and to internalize the hard-won knowledge by consolidating short-term memory into functional long-term memory [e.g., *Khan*, 2012]. This is as true of learning about geoethics as in gaining a functional knowledge of calculus or reflection seismology. In order to retain our expanding knowledge of geoethics, we must use it every day in different ways, making connections between how we live our personal and academic or professional lives. We choose to learn, accept, adopt, and practice living with integrity as a personal commitment.

On the inside of the ceremonial gates of Pomona College are the words of former college president James Blaisdell: "They only are loyal to this college who departing bear their added riches in trust for all mankind." The opportunity to engage in higher education carries with it responsibilities toward others, recognizing that only a small part of Earth's population will have the opportunity of a college education. Given that opportunity, we *ought* to—it is *ethical* to—take full advantage by working hard to master the core knowledge in our chosen field of study. For geoscientists, the ethical responsibility for mastering our core material relates to our responsibilities to be good scientists, to be stewards of Earth, and to serve society. Society depends on geoscientists for reliable information about our many geologic challenges. We hold our added riches in trust for all of humanity, across the globe and forward in time.

My development as an ethical person is *my* responsibility. Fulfilling that responsibility takes intentional effort, exerted daily. Fulfilling that responsibility requires me to listen to and learn from the diverse voices throughout history and across cultures: voices of people who have graced us with their insights about how we might choose to live our lives. My development as an ethical geoscientist is essential if I am to fulfill the three great responsibilities I have as a geoscientist: contributing to and sharing scientific knowledge, service to society, and stewardship of planet Earth.

14.2.5. Our Shared Responsibility to Develop the Character of Following Generations

It is the responsibility of each generation to help subsequent generations develop as ethical people. Within the geosciences, that responsibility to help novice geoscientists develop is shared by all mature geoscientists, but special responsibility rests with geoscience teachers and advisors, and those who hire novice geoscientists after graduation. The personal ethical norms for an entire career can be established early in a geoscientist's training and apprenticeship.

In engaging in the development of applied ethics, we are part of a continuum that began long before our time and that depends critically upon us for its propagation into the future.

14.3. Strategies

14.3.1. Do Not Reinvent the Wheel

We can learn from efforts in other professions to define their ethical concepts and standards as we continue the process of developing the field of geoethics. A vast literature exists about the nature and ethics of science. Thankfully, there are useful summaries and syntheses of this sea of knowledge that we can utilize. Hugh Gauch, Jr., has written an excellent, accessible book on the scientific method that includes his thoughts on the nature of truth and the scope of scientific ethics [*Gauch*, 2012]. It provides a nice introduction to the broad field of the responsible conduct of scientific research, which is the focus of many books and websites [e.g., *Macrina*, 2014; *D'Angelo*, 2012; *Goodstein*, 2010; National Center for Professional & Research Ethics, 2015]. The pamphlets *On Being a Scientist: A Guide to Responsible Conduct in Research* by the National Academies [2009] and *Honor in Science* [*Jackson*, 2000] are of an appropriate scope and scale to be required reading for any undergraduate science major.

Geoscientists have good models for applied ethics in several allied fields. Engineers have developed a culture that values applied ethics, and engineering is perhaps the most closely related discipline to the applied geosciences. A practical example of this consonance is the *Professional Practice Exam*, given to engineers and geoscientists alike who are seeking licenses to practice in Canada [*Andrews*, 2014]. The field of engineering ethics is more fully developed than geoethics at this time, in part because of the long history of professional engineering societies articulating their views of engineering ethics through codes and guidelines [*Luegenbiehl and Davis*, 1992]. The Accreditation Board of Engineering and Technology's (ABET) *Code of Ethics of Engineers* is still a widely referenced resource for engineering in the United States (https://engineering.purdue.edu/MSE/academics/undergraduate/ethics.pdf). Its most direct counterpart in the geosciences might be the Code of Ethics of the American Institute for Professional Geologists [*AIPG*, 2003].

Engineers have developed popular texts explaining successes and failures in their field, textbooks that are partly or entirely devoted to engineering ethics, and a variety of educational resources developed in support of courses in engineering ethics taught in many (but not all) engineering programs [e.g., *Petrosky*, 1985; *Andrews*, 2014; *Harris et al.*, 2014; *Martin and Schinzinger*, 2005; *Seebauer and Barry*, 2001; *Shuirman and Slosson*, 1992]. The journal *Science and Engineering*

Ethics focuses on engineering but has many articles of more general interest. Given that geoscientists typically spend a significant amount of time in front of a computer, the field of applied ethics for computer engineers/scientists also supplies resources and educational models that can inform the development of geoethics [e.g., *Barger*, 2008; *Stichler and Hauptman*, 1998]. John D'Angelo is a chemist and ethicist who wrote a brief, accessible text on ethics and misconduct in scientific research that is largely relevant to the geosciences [*D'Angelo*, 2012]. Environmental ethics also has a rapidly expanding literature that overlaps substantially with geoethics. Two early environmental works that are still essential reading are Rachel Carson's *Silent Spring* and Aldo Leopold's *A Sand County Almanac* [*Carson*, 1962; *Leopold*, 1949].

The textbooks developed for courses in science ethics or engineering ethics tend to have a broadly similar structure. This structure might be adapted effectively for a course on geoethics. The applied-ethics textbooks begin with a summary of general ethics and associated terminology, followed by ideas and terms that relate more directly to the practice of science or engineering. Engineering texts also tend to define and emphasize the special responsibilities that professionals have toward society, often citing the codes of ethics of relevant professional groups. In virtually all of the pedagogical texts I have reviewed, case studies follow the account of general ethics and are used to illustrate the importance of ethical reasoning while making engineering or scientific decisions. Case studies provide opportunities for students to acquire technical knowledge and skills while considering practical examples of applied ethics. Many case studies also illustrate the social impact of decisions made by scientists or engineers.

14.3.2. Content

What questions, ideas, and resources about geoethics should an undergraduate geoscience student be familiar with before graduating with a bachelor's degree? We will be optimistic and assume that the student has access to sufficient internal (within their department) and external (online and elsewhere) resources to gain familiarity with geoethics content.

We should recall and consider seriously that there are limits to the amount of material that a student can process, a fact that is commonly ignored by university teachers to the detriment of students. If we want students to internalize their growing understanding of geoethics, we will have to provide them with time to read, write, discuss, experiment, and reflect.

The following list is long, and yet it is certainly incomplete. It is offered as a starting point for reflection and discussion. Graduates with bachelors' degrees in geoscience should do the following:
- be able to give an informed answer to the question "Why be ethical?" [*GWU*, 1998]

- be honest and trustworthy in all professional and academic interactions [e.g., *Ahearne*, 2011]
- be concerned for the welfare of others
- strive daily to live a life of personal, professional/academic, and scientific integrity
- respect "the fundamental rights, dignity, and worth of all people" [*Fisher*, 2003]
- have "an understanding of professional and ethical responsibility" [*ABET*, 2015]
- have "the broad education necessary to understand the impact of [geoscience] in a global, economic, environmental, and societal context" [*ABET*, 2015]
- have "a knowledge of contemporary issues" that might relate to intersections of geoscience and society, or to geoethics [*ABET*, 2015]
- understand the basic vocabulary of moral philosophy (e.g., ethics, morals, values, dilemmas)
- have a basic functional knowledge of the half-dozen or so most prominent theories, systems, or models of general ethics (e.g., utilitarianism, values ethics)
- be able to apply ethical perspectives, theories, models or systems to help clarify or resolve an ethical problem or reach an ethical decision
- have a functional knowledge of the responsible conduct of science [*Gauch*, 2012; *Mogk*, 2014]
- be able to assess the uncertainty of a measurement or computational result involving observational data
- fully and accurately cite the source of data and ideas, in a manner sufficient to allow another person to locate and evaluate the primary source material
- be familiar with the general rules or customs of the science and geoscience communities with respect to authorship, as well as the customs associated with acknowledgments in papers prepared for public viewing
- be able to argue respectfully "from example, analogy, and counter example" [*GWU*, 1998]
- be able to identify stakeholders in a situation that requires ethical judgment [*GWU*, 1998]
- be able to identify and differentiate ethical issues from other types of issues [*GWU*, 1998]
- understand the different types or purposes of various codes of ethics: normative, educational, aspirational or regulatory
- be familiar with the general contents of codes of ethics related to general science and to various aspects of geosciences, as published by relevant professional and academic societies. It is particularly important that they notice the strong similarities in what different segments of the geoscience community consider to be ethical behavior.
- be able to apply "ethical codes to concrete situations" [*GWU*, 1998]

- be able to work toward resolution of ethical issues by "identifying and evaluating alternative courses of action" [*GWU*, 1998]
- place ultimate importance and value on the protection of human life, safety, health, and welfare
- accept and understand the responsibility to act as a steward of Earth, by virtue of a geoscientist's education, experience, and perspective
- support the sustainable use and development of Earth's resources, and understand the practical and ethical imperative to maintain a healthy biosphere
- inform the public or public authorities about the risks associated with geological hazards, including hazards that might be induced by human activities
- remember to consider the macroethics (the "big picture") of a situation in which the primary focus has been on microethics, and vice versa
- understand the role and responsibility of a geoscientist who is acting as an expert in governmental or judicial proceedings
- understand the role and responsibility of a geoscientist who is acting as an expert supplying members of the general public with information about geological events or processes
- understand and be able to articulate the role, rights, and responsibilities of a geoscientist to a client, and their limits
- understand and be able to articulate the role, rights, and responsibilities of a geoscientist to an employer, and their limits
- work toward resolution of ethical conflicts with courtesy, respect, openness, and attentiveness to alternate viewpoints
- understand the likelihood, if not the inevitability, of encountering a person in your professional life who does not share your commitment to personal or professional/academic ethics, and who might actually want to do you harm. Graduates should be able to recognize this situation and understand the need to take responsible steps to protect themselves.

14.3.3. Case Studies

Case studies fulfill the same purpose in ethics education for scientists and engineers as their more ancient counterparts of parables, myths, fables, and allegories fulfilled in classical ethics education, although case studies are more firmly tied to actual events. Case studies form an important core of engineering ethics and geoethics pedagogy. Bernard Lo stressed the importance of teaching about applied ethics using cases that have ethical dimensions, so that students can discuss, think through, and experiment with authentic ethical dilemmas in the safety of the classroom before they encounter them later in their careers [*Sigma Xi*, 1993].

Perhaps the best feature of case studies is the seamless manner that technical, social, and ethical issues are often woven together in a well-articulated case.

Sometimes case studies are direct descriptions of actual events based on public records, and sometimes they are synthetic cases that might or might not be based on actual events. Cases can describe instances where the decisions were appropriate and the outcome was positive, but it seems that many describe instances where problems arose and failure (if not outright disaster) ensued. The cases are always simplified to some extent, but not necessarily because we want to avoid the messiness of actual ethical situations. We simply do not have access to the same data that were available to the original participants in the case, and cannot fully recreate the same physical, social, professional, or psychological context in which the original participants made their ethical judgments.

When we work with case studies that portray or closely parallel actual events, we must be sensitive to two facts: real cases involve real people whose families are probably still coping with the aftermath of the problem, so we should display appropriate consideration for their privacy and well-being; and real cases might involve active or potential litigation. Here is a brief example of a case study.

After the disaster of the San Fernando earthquake in 1971, the State of California began to enact statutes and policies to better protect its people from earthquake-related hazards. The first of these, usually called the Alquist-Priolo Act, established special study zones (SSZ) around the ground-surface trace of faults that are known to have caused surface rupture during the Holocene, within the last ~11,000 years [*Bryant*, 2010; *Bryant and Hart*, 2007; *State of California*, 2007; *CGS*, 2004, 2008]. Development within an SSZ requires a detailed site investigation by a geologist licensed in the State of California, and a plan that would ensure that no habitable structures would be built across an active fault.

A developer wanted to build houses on a small undeveloped area within the SSZ of a fault that experienced ~4 m (13 feet) of right-lateral shear during a single M ~8 earthquake around a century ago. The developer's consulting geoscientist carefully mapped the property, logged trenches across suspected fault traces, and located the main trace of the active fault as well as some minor fault splays. A design was developed that established a ~15 m (50 foot) setback from the mapped main trace, and a ~9 m (30 foot) setback from the minor splays (Figure 14.1a) [*Borchardt*, 2010]. Using the average size of other houses in the area as a guide, the developer created a design that maximized the number of new houses that could be built on the property without intruding on the setback zones around the fault traces (Figure 14.1b). The access roads were all located along the faults, and all of the utility lines (gas, water, sewer, electricity, telecommunications) were buried under the roadways. The design could be implemented profitably, met all legal requirements, and was reviewed and approved by the appropriate regulatory agencies. The development was built as designed, and now people live in those houses.

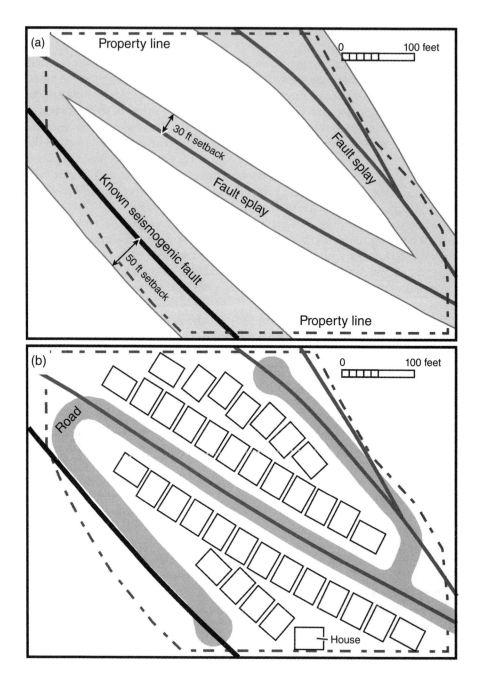

Figure 14.1 Maps to accompany the example of a case study in geoethics. (a) Pre-development site investigation identified the trace of a known-active fault and several splays within a property intended for residential development. Setback zones (gray) are measured from the fault traces to identify the area on which houses might be built (white). (b) Map of house locations that do not impinge on any of the faults or setback zones.

A case like this might generate a variety of responses and a lot of questions. Those questions might lead students to investigate the applicable laws and building codes, the potential danger associated with proximity to a seismogenic fault during an earthquake, the actual experiences of people whose houses are located near a seismogenic fault that generated a recent earthquake that was strong enough to cause damage, and so on. Perhaps the students will consider the question of whether we *ought* to build certain projects that are consistent with all applicable codes yet seem to expose innocent people to significant potential harm. Students typically wonder whether the developer has an inviolable right to realize a profit on a real estate investment through the legal construction of a development that meets all relevant code requirements but that might pose a danger to people and their property. In the context of ethics, we want students to learn how to set their initial emotional reactions aside to analyze the situation [after *GWU*, 1998].

- Who are the stakeholders in this case?
- What is the social context of this case? For example, how would building this development impact society (a) if no serious damage, injury, or death ever occurs due to its proximity to an active fault, and (b) if serious damage, injury, or death occurs due to its proximity to an active fault? What level of loss related to an earthquake would it take to negate the positive impact of this development?
- What are the ethical issues involved in this case? [e.g., *Cronin and Sverdrup*, 1998]
- What pertinent evidence do we have, and what evidence is missing that we would like to have? (In a real situation, recognition that you do not have all the facts would lead to additional research.)
- What governmental restrictions (laws, statutes, codes, policies) apply?
- What rights are endangered, if any, (a) if the development is built, and (b) if the permits for the development are denied and the land remains undeveloped?
- What part(s) of relevant professional codes of ethics might apply to this case?
- What alternative actions could be (or could have been) taken to protect the safety and well-being of the public?

Geoscientists have a deep literature of case studies to draw upon. Early in my career as a faculty member, I taught a course called *Case Studies in Engineering Geoscience*, which included a substantial component of applied ethics. The two texts that I used for that course were dominated by case studies: *Forensic Engineering: Environmental Case Histories for Civil Engineers and Geologists* [*Shuirman and Slosson*, 1992] and *Engineering Ethics* [*Martin and Schinzinger*, 2005]. Civil engineering cases such as dam failures or landslide problems often have a substantial amount of geoscience content and can be readily adapted for use in geoscience courses. Symposia devoted to geoscience ethics occasionally generate volumes that contain case studies [e.g., *Hoose*, 1993]. Recent texts by *Wyss and Peppoloni* [2015], *Peppoloni and Di Capua* [2015], and *Lollino et al.* [2014] continue this tradition of publishing case studies that involve ethical judgments.

The periodical literature is another rich source of useful case studies. The Engineering Geology Division (now the Environmental and Engineering Geology Division) of the Geological Society of America produced for many years a series called *Case Studies in Engineering Geology*, and continues to produce a parallel series called *Reviews in Engineering Geology*. Both series feature a wide variety of case studies. The *Association of Engineering Geologists Bulletin* was also a source of brief case studies, and its current manifestation is called *Environmental and Engineering Geoscience*, a joint publication of the AEG and the GSA. The International Association for Engineering Geology and the Environment (IAEG) publishes the *Bulletin of Engineering Geology and the Environment*, which also is a source of case studies that could be adapted for a geoethics course. Since June of 1987, David Abbott has written a regular column titled "Professional Ethics and Practices for the Professional Geologist," a publication of the American Institute of Professional Geologists (AIPG). Many of these columns include brief case descriptions, followed by Abbott's comments that he bases on the AIPG Code of Ethics and his own remarkably deep well of experience and common sense.

14.3.4. Structuring Geoethics Education: Internal and External Solutions

If our goal is to encourage geoscientists to conduct their professional careers ethically, we should infuse ethics throughout everything we do in a geoscience curriculum. And beyond course content, ethical behavior should be the expected norm throughout academic geoscience departments: in teaching, in research, in administration. Efforts to facilitate the ethical development of undergraduate geoscience students should be tailored to each geoscience department, based on the level of awareness and acceptance of the need to include geoethics training in the curriculum of the department.

The advantage that our friends in engineering have over geoscientists, with respect to applied-ethics education, is that familiarity with relevant aspects of applied ethics is an explicit point in the review criteria specified by their primary accreditation agency [*ABET*, 2015]. Hence, there is a tangible risk to their program if they neglect ethics instruction. The same cannot be said of geosciences.

I would like to frame the following suggestions by imagining two scenarios that bound a range of situations that might be encountered within a geoscience department. In the first scenario, the departmental faculty and relevant administrators are genuinely interested in promoting ethical behavior and in facilitating their students' growth as ethical geoscientists. In this scenario, the approach will be a set of actions taken *internally* while utilizing external resources made available by professional societies, ethics centers, and other groups working to support professional/scientific ethics. In the second scenario, at least one faculty member or administrator in a department is interested in promoting geoethics, while others are disinterested and will not contribute. In this scenario, the few who are interested in geoethics will do what they can while helping students connect with *external* resources developed to facilitate their growth as ethical geoscientists. As reportedly observed by the American humorist Garrison Keillor, "Sometimes you can only do so much. But you do have to do that much."

Even in departments in which the faculty and administrators are initially open or even enthusiastic about infusing geoethics throughout the curriculum, there might be a problem in achieving persistence over the course of years as departmental faculty and priorities change. What is undeniably persistent, however, is the need for geoethics education.

14.3.4.1. Internal Ethics Education

An ideal way to facilitate student learning and to highlight the central importance of ethical practice is to disseminate the principles of geoethics throughout every course in a department. Tasks as small as collecting data in a reproducible manner, properly defining the uncertainty of quantitative measurements and analytic results, properly attributing the source of data or an idea, taking care of field and laboratory equipment, and being careful to leave outcrops unchanged by your presence are all fundamentally ethical. Faculty members who understand that ethics content is important can highlight the ethical nature of these routine tasks for students. Routinely providing each student who declares a geoscience major or minor with a free copy of *On Being a Scientist* [*NAS-NAE-IOM*, 2009] or *Honor in Science* [*Jackson*, 2000] as soon as they declare their degree plans would be a significant statement in favor of ethics by the department. Another positive step would be to include a book like *Scientific Method in Brief* [*Gauch*, 2012] as a required text in a low-level required core course for majors, and make it a required text in all subsequent core courses so that the students retain it as a reference.

A benefit of this approach of spreading geoethics across the curriculum is that the faculty will send a clear message to students that ethical behavior is important in all aspects of the geosciences. Another potential benefit is that this effort might, over time, have a sustained positive effect on the ethical climate within the department as a whole.

There are a number of challenges to the long-term success of this approach that would have to be addressed. Time must be allocated to develop resources for each course and implement the plan. The faculty members who teach undergraduate courses will have to agree to make the necessary changes to the way they teach their courses, and to communicate and coordinate their actions with each other. They will need to be willing to listen and learn from each other. Training might be necessary to accomplish this plan. Implementation will take time and might require funding. As participating faculty members leave and are replaced over time, it might be challenging to sustain this model with new members whose focus will be on research and tenure.

A single course might be devoted entirely or in large part to geoethics, as was my Case Studies in Engineering Geoscience course. Ideally, the course would augment other efforts to emphasize geoscience ethics, rather than replace them. It would be best if the course could be developed or taught in collaboration with a willing moral philosopher from the school's philosophy department, and perhaps even a sociologist. Perhaps the course could combine study of general science ethics (i.e., the responsible conduct of research) with geoethics, or an account of the roles of geoscience in society coupled with geoethics. The course could be a required capstone experience. A potential challenge might involve the administrative process of having a new undergraduate geoscience course approved by a university or college course-and-curriculum committee, because the course content would include a substantial amount of material about ethics that is normally taught through a department of philosophy.

Many departments have a weekly colloquium or brown-bag lunch at which speakers are invited to talk about their work. Part of this time can be used to talk about topics with an ethical dimension, or about professional responsibilities, or about issues in which society and geoscience are joined. Whether the opportunity is a weekly colloquium or a class period devoted to ethics, resources are available on the web in support of brief presentations about applied ethics, science ethics, or geoethics. The Northeast Ethics Education Partnership at Brown University has developed presentations on a number of topics that are available for download. The Center for the Study of Ethics in the Professions at Illinois Institute of Technology also has modules on various topics in applied ethics that are available for download, as does the Ethics CORE and other groups. The *Teaching Geoethics Across the Geoscience Curriculum* website, supported by NSF and the Science Education Resource Center at Carleton College, is a growing source of useful materials and teaching strategies.

14.3.4.2. External Ethics Education

Not all geoscience departments will support the idea of spreading geoethics throughout the undergraduate curriculum. I challenge the geoscience-ethics community to recognize this issue and to collaborate on effectively mitigating the problem.

I became aware of the Khan Academy several years ago while developing a mathematically challenging module for GPS geodesy. Khan Academy (KA) is the inadvertent creation of Salman Khan, a graduate of MIT and Harvard whose free, high-quality, online educational resources reach many more unique students in a typical week than Harvard University has taught in its entire history since 1636 [*Khan*, 2012]. Khan's short videos about math provide a mechanism for people to explore and learn mathematics on their own, without worrying about tuition or high-pressure high-stakes exams. KA provides a model for how some parts of geoethics education might be accomplished through web-based technology.

Sal Khan has taken learning off the clock, so we are not constrained by class days or times, or by the number of weeks in a semester or quarter. He frees us from complete dependence on what he calls the Prussian model of teaching, in which arbitrary boundaries are placed around knowledge so that linear teaching of discreet chunks of material can be conducted to large audiences of students during fixed times and terms. The Prussian model is fundamentally an industrial-production model for education. In Khan's model, progress is not defined along a straight line from one topic to the logical next, but rather is defined by *mastery* of information and a developing interest in the connections between one bit of information and all the rest. Learning is nonlinear, asynchronous, and largely self-directed. The material is organized into coherent, short strands that can be examined within the limitations of a typical person's attention span. If you are learning about one thing and realize that you no longer remember how to do a required step, you simply go back and refresh your memory without apology or penalty. You move back and forth, this way and that, and wherever else your interests take you as you develop a more profound mastery of the material. The KA resource is always there (unlike your teachers), and enhances rather than necessarily replaces traditional coursework [*Khan*, 2012].

There are many types of resources that are currently available online for helping people learn about ethics (e.g., Ethics Unwrapped from the University of Texas, the Wikipedia Portal for Ethics, the *Stanford Encyclopedia of Philosophy* and the *Internet Encyclopedia of Philosophy*, ethics codes from various organizations, videos, and educational modules). I would like to imagine that the geoscience ethics community could develop its own online ethics resources, in coordination with applied philosophers, sociologists, and others with content knowledge, and with necessary assistance from web artists, information technologists, and computer science experts. The KA model serves as an encouragement that self-directed learning facilitated by a well-designed online resource is possible and can be effective. And just as KA supports students in traditional math (and other) courses, online geoethics resources can support any internal geoethics educational effort that is undertaken within a geoscience department.

An undergraduate geoscience student or postgraduate working anywhere in the world in the coming decade might be able to go online and work through the basics of general and applied ethics, access provocative and informative case studies, and begin to understand the many codependencies of geoscience and society. With appropriate sponsorship for that online resource, a novice geoscientist in the future could learn about geoethics for free, whenever and wherever they choose, and perhaps demonstrate mastery through some online mechanism whose current manifestation is the digital badge. Perhaps earning or maintaining a digital badge in geoscience ethics will become a requirement for academic degrees, professional license acquisition and renewal, and geoscience jobs in the future.

While web-based opportunities should be developed for self-guided education about geoethics, digital delivery of information is not a fully adequate substitute for direct human interaction. It is difficult to imagine that a web-based curriculum could impart the same benefit as watching experienced geoscientists acting ethically in all of their interactions and investigations, every day. There is no technological substitute for an ethical mentor who can talk with a novice and discuss issues and problems as they arise in the real world. There is no escaping the reality that experienced geoscientists teach by example, and thereby demonstrate whether we actually believe that ethical practice is essential. We demonstrate the strength of our commitment through our actions.

If some in the academic community do not want to participate in geoscience ethics education, the broader geoscience community worldwide should step in to offer students an alternative. Professional and academic geoscience organizations should provide backbone services in support of geoethics education. Ethics education should be a topic represented by sessions at every regional or national meeting of all professional or academic geoscience organizations. Papers about geoethics should be accepted for publication in our professional and academic geoscience journals. Workshops and short courses about geoethics should be held frequently, independently, and as a part of local or national geoscience meetings. Education in geoethics is the shared responsibility of the entire geoscience community, just as ethical behavior is (or should be) the expectation *for* the entire geoscience community.

Acknowledgments

My education in geoethics began as an apprentice engineering geologist under Dr. James E. Slosson. Through the workshop "Teaching Geoethics across the Geoscience Curriculum," organized in 2014 by Dave Mogk, Monica Bruckner, Cathy Manduca, and the staff of the Science Education Resource Center at Carleton College, I began fruitful and productive interactions with Silvia Peppoloni, Giuseppe Di Capua, Cindy Palinkas, Catherine Pappas-Maenz, Anne Marie Ryan, John Geissman, and Susan Kiefer, and became affiliated with the International Association for Promoting Geoethics. I adapted the case study about the housing development along an active fault from a story told by Bill Bryant. Anonymous reviewers offered insightful suggestions that I have adopted, with thanks. Of course, I bear sole responsibility for the shortcomings of this chapter.

References

Aamodt, S., and S. Wang (2011), *Welcome to your child's brain: How the mind grows from conception to college*, Bloomsbury Press, New York.

ABET (2015), Criteria for Accrediting Applied Science Programs, 2015–2016, Accreditation Board for Engineering and Technology, http://www.abet.org/accreditation/accreditation-criteria/criteria-for-accrediting-applied-science-programs-2015-2016/.

Ahearne, J.F. (2011), Honesty, *Am. Sci., March-April*, http://www.americanscientist.org/issues/pub/2011/2/honesty

AIPG (2003), Code of Ethics, American Institute of Professional Geologists, http://www.aipg.org/about/ethics.htm.

Andrews, G.C. (2014), *Canadian professional engineering and geoscience practice and ethics* (5th edition), Nelson Education, Toronto, Canada.

ASBOG (2015), Task analysis survey 2015: A study of the practice of geology in the United States and Canada, National Association of State Boards of Geology, http://www.asbog.org.

Barger, R.N. (2008), *Computer ethics: A case-based approach*, Cambridge University Press, Cambridge.

Benjamin, M. (1990), *Splitting the difference: Compromise and integrity in ethics and politics*, University Press of Kansas, Lawrence.

Bird, S.J. (2014), Social responsibility and research ethics: Not either/or but both, *Professional Ethics Report*, 27(2), http://www.aaas.org/news/social-responsibility-and-research-ethics-not-eitheror-both.

Borchardt, G. (2010), Establishing appropriate setback widths for active faults, *Env. & Eng. Geosci.*, 16(1), 47–53.

Brooks, D. (2015), *The road to character*, Random House, New York.

Bryant, W.A. (2010), History of the Alquist-Priolo Earthquake Fault Zoning Act, California, USA, *Env. & Eng. Geosci.*, 16(1), 7–18, http://s3.documentcloud.org/documents/83738/10-mapping-history-of-alquist-priolo-act.pdf.

Bryant, W.A., and E.W. Hart (2007), Fault-rupture hazard zones in California, *California Geological Survey, Special Publication*, 42, 42 p., ftp://ftp.consrv.ca.gov/pub/dmg/pubs/sp/Sp42.pdf.

CGS (2004), Recommended criteria for delineating seismic hazard zones in California, *California Geological Survey, Special Publication*, 118, http://www.conservation.ca.gov/cgs/shzp/webdocs/Documents/SP118_Revised.pdf.

CGS (2008), Guidelines for evaluating and mitigating seismic hazards in California, *California Geological Survey, Special Publication*, 117A, http://www.conservation.ca.gov/cgs/shzp/webdocs/Documents/SP117.pdf.

Carson, R. (1962), *Silent spring*, Houghton Mifflin, Boston.

Carter, S.L. (1996), *Integrity*, Harper Perennial, New York.

Cronin, V.S., and K.A. Sverdrup (1998), Preliminary assessment of the seismicity of the Malibu Coast Fault Zone, Southern California, and related issues of philosophy and practice, in *A paradox of power: Voices of warning and reason in the geosciences, Reviews in Engineering Geology*, edited by C.W. Welby and M.E. Gowan, 123–155, Geological Society of America, Boulder, Colorado.

D'Angelo, J. (2012), *Ethics in science: Ethical misconduct in scientific research*, CRC Press, Boca Raton, Florida.

Einstein, A. (1950), The laws of science and the laws of ethics, in *Albert Einstein, Out of my later years, Estate of Albert Einstein*, 114–115, Wings Books, Englewood Cliffs, New Jersey.

Fisher, C.B. (2003), Developing a code of ethics for academics: Commentary on "Ethics for all: Differences across scientific society codes," *Science and Engineering Ethics*, 9, 171–179.

Gauch, H.G., Jr. (2012), *Scientific method in brief*, Cambridge University Press, Cambridge.

Geological Society of London (2014), Geology for Society, http://www.geolsoc.org.uk/geology-for-society.

Gert, B. (2004), *Common morality: Deciding what to do*, Oxford University Press, Oxford.

Gert, B. (2005), *Morality: Its nature and justification* (revised edition), Oxford University Press, Oxford.

Goodstein, D. (2010), *On fact and fraud: Cautionary tales from the front lines of science*, Princeton University Press, Princeton, N.J.

Greene, J. (2013), *Moral tribes: Emotion, reason, and the gap between us and them*, The Penguin Press, New York.

GWU (1998), ImpactCS website, George Washington University, http://www.seas.gwu.edu/%7Eimpactcs//index.html

Harris, C.E., M.S. Pritchard, M.J. Rabins, R. James, and E. Englehardt (2014), *Engineering ethics: Concepts and cases* (5th edition), Wadsworth, Boston.

Herkert, J. (1999), ABET's Engineering Criteria 2000 and Engineering Ethics: Where Do We Go From Here? Online Ethics Center for Engineering and Research, http://www.onlineethics.org/cms/8944.aspx.

Herkert, J.R. (2005), Ways of thinking about and teaching ethical problem solving: Microethics and macroethics in engineering, *Science and Engineering Ethics*, 11(3), 373–385.

Hoose, S. (editor) (1993), Symposium on Ethical Considerations in the Environmental Practice of Engineering Geology and Hydrogeology: Association of Engineering Geologists, 36th Annual Meeting, San Antonio, Texas.

Jackson, C.I. (2000), Honor in Science, Sigma Xi, Research Triangle Park, N.C., https://www.sigmaxi.org/docs/default-source/Programs-Documents/Ethics-and-Research/free-pdf.pdf?sfvrsn=2.

Kant, I. (1784), Idea for a universal history from a cosmopolitan perspective, in *Toward perpetual peace and other writings on politics, peace, and history: Immanuel Kant*, edited by P. Kleingold (2007), Yale University Press, New Haven, Connecticut.

Khan, S. (2012), *The one world school house: Education reimagined*, Twelve, New York.

Leopold, A. (1949), *A Sand County almanac*, Oxford University Press, Oxford.

Lollino, G., M. Arattano, M. Giardino, R. Oliveira, and S. Peppoloni (editors) (2014), *Education, professional ethics and public recognition of engineering geology*, Proceedings of the XII International IAEG Congress, Torino, Engineering Geology for Society and Territory, volume 7, Springer, Cham, Switzerland.

Luegenbiehl, H.C., and M. Davis (1992), Engineering codes of ethics: Analysis and applications, Unpublished paper originally written to be part of the Exxon Modules for Applied Ethics Series, http://ethics.iit.edu/publication/CODE--ExxonModule.pdf or http://hdl.handle.net/10560/2228.

MacIntyre, A. (1998), *A short history of ethics: A history of moral philosophy from the Homeric age to the twentieth century*, University of Notre Dame Press, Notre Dame, Indiana.

Macrina, F.L. (2014), *Scientific integrity: Text and cases in responsible conduct of research* (4th edition), ASM Press, Washington, D.C.

Martin, M.W., and R. Schinzinger (2005), *Ethics in engineering* (4th edition), McGraw-Hill Science/Engineering/Math, New York.

Mogk, D. (2014), Responsible conduct of research, *Teaching Geoethics across the Geoscience Curriculum*, Science Education Resource Center, http://serc.carleton.edu/geoethics/rcr_responsible.html.

National Academy of Sciences, National Academy of Engineering, and Institute of Medicine (NAS-NAE-IOM) (2009), *On being a scientist: A guide to responsible research conduct*, National Academies Press, Washington, D.C.

National Center for Professional & Research Ethics (2015), Ethics CORE: National Center for Professional & Research Ethics, University of Illinois at Urbana-Champaign, https://www.nationalethicscenter.org.

Oreskes, N., and E.M. Conway (2010), *Merchants of doubt*, Bloomsbury Press, New York.

Peppoloni, S., and G. Di Capua (2015), *Geoethics: The role and responsibility of geoscientists*, Lyell Collection, Geological Society of London, Special Publication 419.

Petrosky, H. (1985), *To engineer is human: The role of failure in successful design*, St. Martin's Press, New York.

Rutherford, F.J., and A. Ahlgren (1991), *Science for all Americans: A Project 2061 report on literacy foals in science, mathematics, and technology*, Oxford University Press, Oxford, http://www.project2061.org/publications/sfaa/online/sfaatoc.htm

Seebauer, E.G., and R.L. Barry (2001), *Fundamentals of ethics for scientists and engineers*, Oxford University Press, Oxford.

Shafer-Landau, R. (2012), *The fundamentals of ethics* (2nd edition), Oxford University Press, Oxford.

Shafer-Landau, R. (2015), *The ethical life: Fundamental readings in ethics and moral problems* (3rd edition), Oxford University Press, Oxford.

Shuirman, G., and J.E. Slosson (1992), *Forensic engineering: Environmental case histories for civil engineers and geologists*, Academic Press, San Diego, California.

Sigma Xi (1993), A Summary of the 1993 Sigma Xi Forum *Ethics, Values and the Promise of Science*, https://www.sigmaxi.org/docs/default-source/Programs-Documents/Ethics-and-Research/1993ethicsforumsummary.pdf?sfvrsn=4.

Starr, W.C. (1991), Ethical theory and the teaching of ethics, in *Ethics across the curriculum: The Marquette experience*, edited by R.B. Ashmore and W.C. Starr, Marquette University Press, Milwaukee, Wisconsin.

State of California (2007), Seismic Hazards Mapping Act, California Public Resources Code, Division 2, Chapter 7.8, http://www.conservation.ca.gov/cgs/shzp/Pages/shmpact.aspx.

Stichler, R.N., and R. Hauptman (editors) (1998), *Ethics, information and technology readings*, McFarland & Company, Jefferson, N.C.

Swazey, J., M. Anderson, and K. Louis (1993), Ethical problems in academic research, *Am. Sci., November-December*, 542–553, http://www.americanscientist.org/issues/feature/ethical-problems-in-academic-research/1.

Weston, A. (2013), *A 21st century ethical toolbox* (3rd edition), Oxford University Press, Oxford.

Wyss, M., and S. Peppoloni (2015), *Geoethics: Ethical challenges and case studies in Earth sciences*, Elsevier, Amsterdam.

Appendix A

CASE STUDIES FOR SCIENTIFIC INTEGRITY AND GEOETHICS PRACTICE

Linda C. Gundersen

Introduction

This appendix contains 25 case studies and a brief lesson plan to be used in a classroom or workshop setting for teaching scientific integrity and geoethics. Chapters 13 and 14 of this book also provide guidance and resources for teaching geoethics. Many of the chapters in this book can be used in conjunction with this appendix to focus discussion on a subject such as scientific integrity, geoethics, publishing, data ethics, climate change, and others. The appendix concludes with a list of websites and books that have additional case studies, lesson plans, and resources for students and teachers.

Case studies are an important tool in teaching scientific integrity and ethics. They provide an active learning experience in a safe environment where ethical issues can be explored and discussed with peers, faculty, and/or institutional leadership. By grappling with problems that may arise in student and professional life in a nonthreatening setting, participants can think through problems and gain ethical skills and insights to better prepare themselves when an actual issue arises. The case studies below are predominantly set in a geoscience context, addressing situations that may occur in the field, classroom, laboratory, industry, and professional settings. Some of these case studies can be used in the context of classroom curricula or for a specific session on integrity and ethics. The case studies in this appendix are brief and based on real situations. Those inexperienced with integrity and ethics may want to work with their institution's integrity and ethics experts for help in designing a class using these case studies and the resources provided in this book.

Retired, U.S. Geological Survey, Ocean View, Delaware, USA

Scientific Integrity and Ethics in the Geosciences, Special Publications 73,
First Edition. Edited by Linda C. Gundersen.

An Example of How to Use These Case Studies in a Classroom or Workshop Setting

The goal of these exercises is to help participants reason through integrity and ethical issues, know what possible alternatives, codes, rules, and resources they have, and realize that such cases may not always have a "right" answer but a set of possible solutions. Before the class, instructors should provide copies of any relevant ethical codes as appropriate, such as the integrity and ethical codes in this book and the codes of your institution, and review them with the class. Instructors can also focus in on a subject such as climate change or publishing and augment one of the case studies with a chapter from this book or a journal paper on the subject.

Divide participants into 4–6 person teams and allow 10 minutes (depending on the complexity of the case study) for the teams to read through the case study, discuss the possible ethical issues, and provide solutions. Have participants consider the following questions for each case study:
1. What are the ethical issues in this situation? Is there research misconduct? Detrimental behavior? Has a crime been committed?
2. What is the evidence?
3. What are some of the codes, rules, or regulations that could be used to evaluate or resolve this dilemma? What are the requirements for reporting?
4. Who is potentially affected and what are the short- and long-term effects?
5. How is science affected?
6. What are some of the resources that you could utilize? (examples include institutional integrity officers, professional societies, positions of authority, peers, student help services)
7. What would you do to resolve the situation?
8. When is it appropriate to talk with a colleague about their unacceptable behavior and when is it not appropriate or possibly unsafe or detrimental?
9. What are the options for moving forward?
10. How could this situation be avoided or prevented?
11. What are the important lessons learned?

At the end of the ten minutes, have each of the teams present their results. Encourage participants to be respectful, open, pose alternate scenarios, and ask questions. End with the lessons learned in each study. The case studies provided below include both very short and longer, more complex studies. Shorter case studies can usually be done in five minutes.

A. Scientific Integrity Case Studies

A1. You have been given permission to use the equipment in the laboratory of a prominent geochemist. These analyses are critical to your thesis. While learning to standardize the instrument, the postdoctoral researcher who runs the lab suggests a way to calibrate the machine that is not in the manual, and to you, appears to bias the analyses. You operate the machine as instructed but then decide to also try the analyses using the procedure for calibration in the manual. When you compare the data, you find there is a real difference in the results. What do you do?

A2. You help a friend with a science project that is similar to yours. You provide some leads, references, and a draft of your paper "in press" that you think will be helpful. Your colleague is struggling; he does not have much data and is short on time. In a month, you head to a geoscience meeting, visit your friend's poster, and you recognize passages from your own paper and from the references you gave to him. The passages are nearly verbatim and there are no citations; the data graphs look perfect; your colleague is the lone author; and now here he is strolling up to you, smiling. What do you do?

A3. You have been formally asked to review a paper by an author that you recently met at a scientific meeting. The two of you hit it off and discuss in detail your respective research projects. You agree to do the review, but when you begin reviewing the manuscript, it appears to contain many of the ideas you discussed with the author at the meeting, including some new work you were proposing to the National Science Foundation. It is clear this paper is well written and could have far reaching impact, but it completely scoops your own ideas. What do you do?

A4. You are working with an old respected friend on a geochemical survey of a river and its tributaries that flow through a highly mineralized area. The survey will be the basis for research by several students and serve as the baseline information for the state in their environmental assessment of the river basin. As the field work progresses, you note that your friend often takes shortcuts, such as not waiting for instruments to be fully calibrated and not leaving analytical probes in the water for sufficient time. You mention this to him and he says that in his long experience, he sees little difference in measurements. You decide not to follow what he does, and whenever you take samples you follow protocol meticulously. When you start to analyze the data, you generally see a slight difference between your data and your friend's data, with your data being consistently higher. Overall the data trends appear to agree with known point sources of contaminants and ore deposits. In several instances, there are large departures where your friend's data is anomalous and makes no sense with surrounding data. He suggests that you throw those points out as obviously having some kind of problem. What do you do?

A5. You are a PhD student working with an internationally known adviser in a prominent laboratory. As your thesis develops, you realize there is an important story related to the origin of the rare earth elements (one of your advisor's specialties) in the rocks of your study. Additionally, there is potential for a significant ore deposit. You want to impress your adviser with developing this part of the thesis, so you work diligently on it alone during your first year whenever you have time. At the end of the year, you feel like you have a publishable piece of work focusing on the unique origin and chemistry of the rare earth elements. You bring the paper to your adviser and tell him about the work you have been doing, along with the conclusions. He is very excited and commends you for your great work. You send the digital document to him to review. Several weeks pass and when you ask about the status of the paper, he says he wants more time with it. More time passes and after a month you express concern that you would like to submit it and he says he will send it to you. When his e-mail finally arrives, not only is he first author on your manuscript, but the second author is a famous rare earth element scientist you have never met, but who owns a mining company. You are the third author. The paper has been largely rewritten to emphasize the economic geology. A short version of your origin story for the rare earth elements is included to support the main points of the paper, the discovery of a world-class ore deposit, and a new ore deposit-type for rare earth elements. Your adviser concludes his e-mail with an enthusiastic postscript that the paper is on its way to a prominent journal. What do you do?

A6. You are a graduate student and third author on a paper with your advisor and another student. You have contributed data and interpretation of results, and worked on editing and refining the thoughts and ideas in the other sections of the paper. The adviser sends the paper to a prominent journal, but the reviews indicate issues regarding your data and its interpretation. You feel that the issues are superficial and partly a matter of opinion related to the interpretation; however, your advisor disagrees. In order to meet the reviewer's requirements and run additional statistical tests, it will take a month. Your advisor, who is seeking tenure, feels that the paper can stand without your data; he takes out your data and interpretation and resubmits the paper without you as author. What do you do?

A7. Helen has spent a long summer sampling soils and stream water for radionuclides, including radon gas and dissolved oxygen. Weather, temperature, barometric pressure, and time of day are critical observations when sampling gases. Back in her office she starts entering the data from her notebook and notices she was not always consistent in recording all her parameters. She decides she can estimate time of day from her other sampling stations and the distance between them. For ambient temperature, barometric pressure, and weather conditions she can use the nearest weather station, and for the few water temperatures she forgot to record she decides to use an average of adjacent sites, particularly since water temperature varied in those places by only a few degrees. She enters these data without indicating they are estimated and utilizes them like any of her directly measured data. In the subsequent paper she publishes, she states that all the data were measured directly in the field, and when she deposits the data in a national repository of similar data, she does not include any information on the fact that these are estimates. Is there a scientific integrity problem?

A8. You and Shelly share an X-ray laboratory, examining clays and other minerals for your own research, and also as paid research assistants for the rest of the department. One evening when you are running some analyses, Shelly comes in to set up some analyses on the X-ray machine and appears to be drunk. You have smelled alcohol on Shelly's breath before, but not noticed any drunken behavior until now. You help Shelly get set up and advise her to go home, that you will finish the analyses. Several days pass and you walk into the laboratory and see Shelly about to turn on the X-ray machine; however, one of the safety shields is not in place. You stop Shelly from turning the machine on and notice the smell of alcohol on her breath. What do you do?

A9. You and Bob are describing sediment core from lakes for your master's thesis on past climate change. Bob is doing an isotope study and you are focusing on changes in sediment textures that reveal the wetness or dryness of the environment. These are expensive and important cores that will be used in a larger continental study of recent climate change events. Bob's preliminary analysis shows some sharp differences in the isotopes in several places that do not seem to make sense with your work. Bob asks you to go through the textures with him, and using photo enhancement software, he enhances some of the areas where he sees a change in the isotopes. You agree that there might be something there, but feel that the photo enhancement is too great and exaggerates the color and textural differences to the point of distortion. After inspecting the core, chemistry, and the photographs again, you confirm your conclusion with Bob that the sediment textures do not indicate the changes Bob suggests, but Bob continues to disagree with you. Six months pass and you notice on a colleague's desk a copy of a paper he is reviewing, written by Bob and containing the enhanced pictures of the core and the conclusions you disputed. What do you do?

A10. A research team, Bob, Bill, and their postdoctoral fellow Jean, have developed a model for the distribution of pollutants from a point source. The model predicts the spread of contaminant concentrations in three dimensions over time. The model is published in a peer-reviewed journal and well received. Almost immediately other scientists apply the model to numerous critical locations including drinking water and endangered species habitat in lakes and reservoirs. While working on the data for a new study, the research team discovers a programming glitch in the published model. The error had only a small impact on the numbers used in their earlier study, but it produces large errors with higher concentrations of contaminants over longer periods of time, underestimating the concentrations and distribution. Bob and Bill decide to remedy the situation by posting the corrected simulation software on the project website as version 2 without explanation. Jean disagrees with this approach and thinks that in addition, a note in the journal should be published discussing the error and that researchers who have used the model should be notified. Jean is outvoted and version 2 is posted to the website with no explanation. What should Jean do?

B. Harassment, Discrimination, and Bullying Case Studies

B1. You are an early career scientist on your first panel to hire an important new research scientist vacancy at your institution. Your panel is the second hurdle to determine the final three candidates. You interview six candidates using a common script the committee has put together. The questions include information about their special area of research, their funded projects and publication record, their teaching and mentoring, their professional service, and a general question asking the candidate to share anything they would like the committee to know about themselves personally. In this last question, all the candidates share information about their families, their desire to live or continue to live in the area, and some of their hobbies. During private discussion with the panel about the candidates, you realize that the two men who are married and already have homes and children in the area are looked on more favorably than the man who is married and needs to move his family. You also note that the woman candidate without children is discussed more positively than the two women candidates with children. Additionally, there is a broad ranging discussion on the fitness of the woman candidate who has a new baby at home. You feel that the candidates are all very close in terms of meeting the requirements of the job. It appears to you that the selection of the final three candidates is coming down to personal criteria. What do you do?

B2. You are on a three-week geophysics exploration cruise, mapping a small area near an important plate junction that you studied in your master's thesis. You are the only woman on the cruise with five male graduate students and your male advisor. You and your new doctoral thesis adviser have really hit it off and you are excited about the prospects of conducting this research. After a few days on the cruise, you feel uncomfortable with the sexual and other crude jokes and language of the students. The other students also express resentment toward you because you get more of the instrument time than they do. The other students seem to indicate that you are having an affair with your advisor, hence your preferential treatment. You feel isolated and intimidated by their actions. What should you do?

B3. You are with a group of graduate students and university faculty at a geosociety function celebrating a fellow student who is receiving an important geosociety honor. There are many important people in attendance, providing a great opportunity for you to network. You notice that two of the attendees are visibly intoxicated, are being rude, and are making sexually explicit comments to several other attendees who are clearly uncomfortable with the situation. One of the intoxicated attendees, a famous person you do not know personally, gets up, walks over to another attendee and forces a kiss on her. You say to a friend, "Hey, shouldn't we do something?" But your friend says, "Nah, this happens all the time." What do you do?

B4. Sara is a senior geology undergraduate who has applied for a part-time job with a well-known professor. You are a friend of Sara and one of two male graduate students already working with the professor. You are helping him review the candidates for the job and he tells you to ignore all the female students because they "simply will not cut it hiking all that way and carrying the heavy equipment needed." What are the consequences if you comply with the professor's instructions? What should you do?

B5. Kay is a high school senior who has earned a competitive position as an intern doing summer research in the Rocky Mountains with a group of graduate students and their advisors. She is one of three females on a 10-person research team, working long hours in the field and then staying overnight in the campground in trailers. There are two male professors working with them and the research is expected to take six weeks in the field. During the first week, the women find themselves being assigned the cooking and cleaning chores while the men are tasked with organizing and hauling the field equipment. In the field, all three women are asked to take notes rather than conduct the experiments. They also find themselves being referred to as "little ladies," and their appearance and the sexiness of their clothes are remarked upon daily, making them uncomfortable. What should Kay and the other women do?

B6. You are on an overnight field trip with a male and female professor and 10 undergraduate and graduate students. While camping that night, you head to the washrooms and see your female professor flirting with one of the undergraduate men, touching him on the shoulders, and notice that he backs away and appears uncomfortable. She grabs his hand and says, "Hey, you know you want it" and starts to kiss him. The student breaks it off, saying he has to get some sleep, and leaves. You go to the male professor's tent and tell him what happened. His response is "Gee, some guys get all the luck, forget about it, he can handle himself." What should you do?

B7. Carol is a new research scientist at an institution, working in the lab of a well-known geophysicist. She feels isolated by the fact that she is the only black woman in a lab of all white men. She finds herself often being told what to do by her peers, not given much of an opportunity to speak up at the weekly lab meeting, or being talked over at the meeting. She is asked to take notes whenever there is discussion that the professor wants recorded. Carol is also a very talented artist. The lab director asks Carol if she can help illustrate a paper he is doing with a large group of scientists in exchange for an acknowledgement. She agrees but finds it takes a significant portion of her time. While doing the illustrations, she discovers some important relationships in the parameters that the authors have missed, so she writes up several paragraphs about them and gives them to the lab director. He is very excited, tells her that they are important observations, and incorporates them into the paper. When she is given a final draft of the paper to review, she sees that she is only acknowledged for her illustrations. What should Carol do?

B8. Ben and Kevin are transfer students from a community college to a large public university. You share several classes and laboratories with them. You notice that a small group of students in your class has started talking about how Ben and Kevin don't belong and notice an altercation between Ben and one of the students regarding whose turn it is to use an instrument. The altercation ends in the student telling Ben to "go back where he belongs." Increasingly you see that Ben and Kevin tend to sit in the back of the class and leave quickly to avoid the other students. One evening, while working late on a laboratory assignment, you see the student who had the altercation with Ben appear to be tampering with Ben's experiment, then leave. What should you do?

C. Geoethics Case Studies

C1. You are conducting geologic mapping on a wildlife refuge when your trail is blocked by a "closed" sign because of nesting endangered birds. You can see a large outcrop about 10 feet away that would complete your transect and possibly provide important mapping and structural data. You were explicitly told by refuge personnel that the area is closed and that the nesting birds are very sensitive to the presence of humans and may abandon their nests. What do you do? What if you were mapping with your professor and they ignored the sign and went on in, urging you to do so too?

C2. You have been invited as part of a group of paleontologists and archaeologists to visit a Smithsonian research site near the Navajo Nation. The site is a possible source of clay for some of the ancient pottery of the Anasazi and other tribes. While on the site, you notice one of your colleagues has been pocketing the occasional specimen and pot shard. The site has strict regulations regarding sampling and you all have signed agreements to that fact. What do you do?

C3. Ellen is an engineering geologist with a specialty working in landslide areas. Her firm has been hired for a spectacular new development near the coast. This is a project where she could really make her name with the firm and also with the construction community. Her first day on the site is very promising; she believes this could be an ideal place for the kind of construction they want to build and reports this to her supervisor with enthusiasm. As the weeks pass, however, it is obvious that a portion of the development land has been affected by major landslides in the past and has a high potential for landslides in the future. She decides to cross-check her determination with a colleague at the firm before submitting her final report; he agrees but thinks the probability may be lower than she is estimating. She keeps her original high estimates and submits her report. Upon review, her supervisor also advises her to "dial down" her estimates. She goes on one last site visit, reaffirming her measurements and the hazard. Her supervisor however warns her that she must change her estimates or be fired. What should she do?

C4. Joe is a volcanologist on a prestigious team of government and university scientists that have spent six intensive weeks determining if a volcano in the Cascades is in the pre-eruptive stage. The seismic signal has been discontinuous, but the gas and thermal signals appear to be consistent with magma movement. Joe is convinced that the other scientists are being too cautious about issuing a warning, and is beginning to feel the odd man out because of his constant urging of the group to take action soon. Taking action would mean the mobilization of people and equipment, costing millions of dollars and requiring the evacuation of a nearby town that could potentially take a direct hit. Joe believes the caution is politically motivated both locally and federally and that large local industries and banks are also putting pressure on decision makers because of potential loss of business. Joe thinks there is a greater than 50% chance that the eruption is clearly underway and could happen within the next month. What should Joe do?

C5. Jill is a newly hired geochemist working with an oil company. One of her first projects is to sample and chemically analyze produced waters from a large oil and gas production field that is looking to expand and could impinge on a community's sole water source. In her report, she notes that several regulated contaminants are well over acceptable limits within the current oil and gas field. State regulations require companies to monitor the water quality in operating fields according to approved management plans and to report when there are problems (e.g., spills) and propose steps for mitigation or disposal. Environmental quality is also periodically audited. Further, her modeling of the fluid flow for the expansion of the field shows that the maximum projected concentrations of regulated contaminants are well over the health regulation and could easily reach the community reservoir. However, the median modeled concentrations are within acceptable limits. In the final public report for the expansion permit, her supervisor only includes the median concentrations calculated by the modeling. The proposed expansion passes public comment with few problems and will now move onto state and federal review. When Jill asks her supervisor if her whole report was passed on up to management, her supervisor tells her she has done her job and it is no longer her concern. What should she do?

C6. You are a member of an important government working group dealing with sea level rise estimates from present to 2100. Your group is struggling with uncertainty related to the rate of future sea level rise and some of the parameters that go into it. Several members feel strongly that your calculations of the acceleration of ice sheet melting are insufficient to include in the overall estimation and that only older observed rates should be used. You feel strongly that your calculations must be included, otherwise the projections will be woefully underestimated, but you acknowledge your calculations have high uncertainty. Also, your group has been directed to include only the most likely probabilities in the report, and not the whole distribution curve. You know that the results of this report are used by governments, emergency responders, and communities for preparing for the damaging effects of sea level rise and the spending of billions of dollars. What do you do?

C7. The state has commissioned your university to write a research report on the effects of fracking on groundwater, and you and your adviser have been asked to be on the team preparing the report. One of your jobs is researching the literature and extracting data related to fracking, groundwater, and wastewater. These results will be used by you and your advisor for a summary chapter on the impact of fracking. During your research you find out that your adviser sits on the board of a small oil and gas company that has fracking projects in your state. Your advisor prepares a first draft of the chapter using your summaries and data analyses. When you start reviewing the first draft, you feel that the advisor has put a positive spin on some of the results and excluded some of the more negative reports regarding fracking. When you mention the exclusion of the more negative results, the advisor calls them "gray literature" and says that he feels the science is unreliable. What do you do?

Resources and References

The following websites and references contain case studies, lesson plans, and/ or other resources that can help in designing and implementing integrity and ethics classes and workshops. The first three items have an emphasis on the geosciences while the last three items focus on scientific integrity.

1. Teaching Geoethics across the Geoscience Curriculum, David Mogk, Department of Earth Sciences, Montana State University, and Monica Bruckner, SERC, Carleton College, https://serc.carleton.edu/geoethics/index.html. (This website provides a primer on what geoethics is, how to teach it, and includes resources for both faculty and students including case studies and lesson plans.)

2. Wyss, M., and S. Peppoloni (2017), *Geoethics: Ethical Challenges and Case Studies in Earth Sciences*, 450 p. Elsevier, Amsterdam. (This book includes case studies in geoethics documenting "where experts have gone wrong and where key organizations have ignored facts, wanting assessments favorable to their agendas" that can provide the basis for discussion of a wide range of global issues in the geosciences.)

3. Peppoloni, S., and G. Di Capua (2015), Geoethics: The Role and Responsibility of Geoscientists. Geological Society, London, Special Publications, 419, http://sp.lyellcollection.org/online-first/419.(This special volume includes position papers, case studies, and discussion on numerous topics in geoethics including hazards, natural resources, uncertainty, and education.)

4. National Center for Case Study Teaching in Science, University at Buffalo http://sciencecases.lib.buffalo.edu/cs/. (The website contains case studies and lesson plans for a variety of scientific disciplines, including the geosciences, and topical areas such as ethics, legal issues, social justice, policy, pseudoscience, and others. The center's purpose is "to promote the nationwide application of active learning techniques to the teaching of science, with a particular emphasis on case studies and problem-based learning.")

5. Macrina, F. (2017), *Scientific Integrity: Text and Cases in Responsible Conduct of Research*, 530 p., American Society for Microbiology Press, Herndon, Virginia. (This book covers numerous scientific integrity subjects in depth, such as scientific record keeping, authorship, mentoring, conflict of interest,

and specific issues related to the biological and medical sciences. It includes numerous useful case studies that could be used by any discipline.)

6. National Academy of Sciences, National Academy of Engineering, and Institute of Medicine (2009), *On Being a Scientist: A Guide to Responsible Research Conduct*, National Academies Press, Washington, D.C., https://www.nap.edu/catalog/12192/on-being-a-scientist-a-guide-to-responsible-conduct-in. (This free downloadable book provides an overview of scientific integrity, the responsibilities of being a scientist, and case studies from various disciplines for discussion and practice.)

Appendix B

RESOURCES AND REFERENCES FOR SCIENTIFIC INTEGRITY, ETHICS, AND GEOETHICS

The chapters in this book can serve as valuable tools for students, faculty, instructors, and scientists, for understanding, teaching, and implementing the principles of scientific integrity and geoethics. It is also useful for professionals in industry and government, public policy, and decision makers who seek to understand the latest advances in scientific integrity and geoethics. Below is a list of selected resources and references for those who wish to delve deeper. Most of these resources and references were released in the last decade. The content focuses on scientific integrity and geoethics, and the list is divided into three kinds of resources: books and reports, journal articles, and websites. References were selected to include recent national and international codes of conduct, statements and codes of conduct from professional groups, and recent studies and reports on scientific integrity, geoethics, and misconduct.

Scientific Integrity

Books and Reports

Andrews, G.C. (2013), *Canadian professional engineering and geoscience: Practice and ethics*, Nelson Education Ltd., Toronto, Canada.

Council of Canadian Academies (2010), *Honesty, Accountability and Trust: Fostering Research Integrity in Canada*, Ottawa, ON, Canada.

European Science Foundation and All European Academies (2017), *The European code of conduct for research integrity* (revised ed.), Berlin, Germany.

Gardiner, S.M., S. Caney, D. Jamieson, and H. Shue (2010), *Climate ethics: Essential readings*, 386 p., Oxford University Press, New York.

Giorgio, L., A. Massimo, G. Marco, O. Ricardo, and S. Peppoloni (Eds.) (2014), *Engineering geology for society and territory: Volume 7. Education, professional ethics*

Scientific Integrity and Ethics in the Geosciences, Special Publications 73,
First Edition. Edited by Linda C. Gundersen.

and public recognition of engineering geology, 274 p., Springer, International Publishing, Switzerland, ISBN:978-3319093031.

Institute of Medicine and National Research Council (2002), *Integrity in scientific research: Creating an environment that promotes responsible conduct*, National Academies Press, Washington, D.C.

InterAcademy Partnership (2016), *Doing global science: A guide to responsible conduct in the global research enterprise*, Princeton University Press, Princeton, New Jersey.

Koepsell, D. (2015), *Scientific integrity and research ethics: An approach from the ethos of science*, Springer, Cham Switzerland. doi:10.1007/978-3-319-51277-8

Macrina, F. (2017), *Scientific integrity: Text and cases in responsible conduct of research*, American Society for Microbiology Press, Herndon, Virginia.

National Academies of Sciences, Engineering, and Medicine (2017), *Fostering integrity in research*, National Academies Press, Washington, D.C.

National Academy of Sciences, National Academy of Engineering, Institute of Medicine (2009), *Ensuring the integrity, accessibility and stewardship of research data*, National Academies Press, Washington, D.C.

National Academy of Sciences, National Academy of Engineering, and Institute of Medicine (2009), *On being a scientist: A guide to responsible research conduct*, National Academies Press, Washington, D.C.

National Academy of Sciences, National Academy of Engineering, and Institute of Medicine (1992), *Responsible science: Ensuring the integrity of the research process*, National Academies Press, Washington, D.C.

Oreskes, N., and E.M. Conway (2010), *The merchants of doubt*, Bloomsbury Press, New York.

Peppoloni, S., and G. Di Capua (Eds.) (2011), Geoethics and geological culture. Reflections from the Geoitalia Conference 2011, *Annals of Geophysics, 2012*, 55(3), 163.

Peppoloni, S., and G. Di Capua (Eds.) (2015), Geoethics: The role and responsibility of geoscientists, Geological Society, London, Special Publications, 419, http://sp.lyellcollection.org/online-first/419

Quinn, M.J. (2015), *Ethics for the information age*, (6th ed.), Pearson Education Inc., Upper Saddle River, New Jersey.

Resnik, D.B. (2005), *The ethics of science, an introduction*, Taylor and Francis, London, U.K.

Shamoo, A.E., and D.B. Resnick (2009), *Responsible conduct of research* (2nd ed.), Oxford University Press, New York.

Steneck, N.H. (2009), *Introduction to the Responsible Conduct of Research*, Office of Research Integrity, U.S. Government Printing Office, Washington D.C. https://ori.hhs.gov/sites/default/files/rcrintro.pdf

Steneck, N.H. (2013), Responsible Advocacy in Science: Standards, Benefits, and Risks, AAAS Report, Workshop on Advocacy in Science (http://www.aaas.org/report/report-responsible-advocacy-science-standards-benefits-and-risks).

Steneck, N.H., T. Mayer, M. Anderson, and S. Kleinert (Eds.) (2015), *Integrity in the global research arena*, World Scientific Publishing Ltd., Singapore.

Wessel, G.R., and J.K. Greenberg (Eds.) (2016), *Geoscience for the public good and global development: Toward a sustainable future*, Geological Society of America Special Paper 520.

Wyss, M., and S. Peppoloni (2014), *Geoethics: Ethical challenges and case studies in Earth sciences*, 450 p. Elsevier, Amsterdam.

Journal Articles

Bird, S. (2014), Social Responsibility and Research Ethics: Not Either/Or but Both, American Association for the Advancement of Science, Spring 2014 Professional Ethics Report. http://www.aaas.org/news/social-responsibility-and-research-ethics-not-eitheror-both

Bohle M. (2016), Handling of Human-Geosphere Intersections, *Geosciences*, 6(1), 3, doi:10.3390/geosciences6010003.

Clancy, K.B.H., R.G. Nelson, J.N. Rutherford, and K. Hinde (2014), Survey of Academic Field Experiences (SAFE): Trainees report harassment and assault, *PLoS ONE*, 9(7), e102172, doi:10.1371/journal.pone.0102172.

De Marco, P. (2015) Rachel Carson's environmental ethic: A guide for global systems decision making, *Journal of Cleaner Production*, 140, 127–133, Elsevier, Amsterdam, 10.1016/j.jclepro.2015.03.058.

Di Capua, G., and S. Peppoloni (2014), Geoethical aspects in natural hazards management, doi:10.1007/978-3-319-09303-1, in *Engineering geology for society and territory: Volume 7*, edited by G. Lollino et al., pp. 59–62, Springer International Publishing, Switzerland, doi:10.1007/s10584-015-1404-4.

DuBois, J.M., J.T. Chibnall, R. Tait, and J. Vander Wal (2016), Misconduct: Lessons from researcher rehab, *Nature Comment*, 534, 173–175, doi:10.1038/534173a.

Fanelli, D. (2009), How many scientists fabricate and falsify research? A systematic review and meta-analysis of survey data, *PLoS One*, 4(5), 10.1371/journal.pone.0005738.

Fang, F.C., R.G. Steen, and A. Casadevall (2012), Misconduct accounts for the majority of retracted scientific publications, *Proceedings of the National Academy of Sciences*, doi:10.1073/pnas.1212247109.

Guzzetti, F. (2015), Forecasting natural hazards, performance of scientists, ethics, and the need for transparency, *Toxicological & Environmental Chemistry*, doi:10.1080/02772248 .2015.1030664.

Hill C., C. Corbet, and A. St. Rose (2010), Why So Few? A Report on Women in STEM, American Association of University Women, Washington D.C. http://www.aauw.org/research/why-so-few/

Lerback, J., and B. Hanson (2017), Journals invite too few women to referee, *Nature*, 542, 455–457, doi:10.1038/541455a.

Markowitz, E.M., M. Grasso, and D. Jamieson (2015), Climate ethics at a multidisciplinary crossroads: Four directions for future scholarship, *Climatic Change*, 130, 465–474, doi:10.1007/s10584-015-1404-4.

Martinson, B.C., M.S. Anderson, and R. de Vries (2015), Scientists behaving badly, *Nature*, 435, 737–738, doi:10.1038/435737a.

Matteucci, R., G. Gosso, S. Peppoloni, S. Piacente, and J. Wasowski (2014), The "geoethical promise": A proposal. *Episodes*, 37(3), 190–191.

McNutt, M., M. Bradford, J. Drazen, B. Hanson, B. Howard, K. Hall Jamieson, V. Kiermer, M. Magoulias, E. Marcus, B. Kline Pope, R. Schekman, S. Swaminathan, P. Stang, and I. Verma (2017), Transparency in authors' contributions and responsibilities to promote integrity in scientific publication, *bioRxiv*, 140228, Cold Spring Harbor Laboratory, doi:10.1101/140228.

Peppoloni, S., P. Bobrowsky, and G. Di Capua (2015) Geoethics: A Challenge for Research Integrity in Geosciences pp. 287–294, doi:10.1142/9789814632393_0035.

Peppoloni, S., and G. Di Capua (2012), Geoethics and geological culture: Awareness, responsibility and challenges, *Annals of Geophysics*, 55(3), doi:10.4401/ag-6099.

Robinson, W. (2016), A departmental approach to addressing the problem of sexual harassment and assault in field experiences, *In the Trenches*, 6(2), http://nagt.org/nagt/publications/trenches/articles/v6n2-9.html.

Scoles, S. (2016), Month by month, 2016 cemented science's sexual harassment problem, *Wired*, https://www.wired.com/2016/12/can-build-calendar-sexual-harassment-stories-science/.

St. John, K., E. Riggs, and D. Mogk (2016), Sexual harassment in the sciences: A call to geoscience faculty and researchers to respond, *Journal of Geoscience Education*, 64, 255–257, http://nagt-jge.org/doi/pdf/10.5408/1089-9995-64.4.255.

Steen, R.G. (2011), Retractions in the scientific literature: Is the incidence of research fraud increasing? *Journal of Medical Ethics*, 37(4): 249–253, doi:10.1136/jme.2010.040923.

Stewart, I.S., and J.C. Gill (2017), Social geology: Integrating sustainability concepts into Earth sciences, *Proceedings of the Geologists' Association*, 128(2), 165–172, 10.1016/j.pgeola.2017.01.002.

Tollefson, J. (2015). Earth science wrestles with conflict-of-interest policies, *Nature*, 522, 403–404. doi:10.1038/522403a.

Websites

American Association for the Advancement of Science, Scientific Integrity, http://www.aaas.org/page/scientific-integrity

American Association of Petroleum Geologists:
  Code of Ethics, https://www.aapg.org/portals/0/docs/AAPGCodeOfEthics_102013.pdf
  Division of Professional Affairs, ethics videos, http://dpa.aapg.org/video/video.aspx

American Astronomical Society, Code of Ethics, http//aas.org/ethics

American Geophysical Union Scientific Integrity and Professional Ethics Policy, http://ethics.agu.org/policy/

American Geosciences Institute, Guidelines for Ethical Professional Conduct, http://www.americangeosciences.org/community/agi-guidelines-ethical-professional-conduct

American Institute for Professional Geologists, http://www.aipg.org

AIPG Code of Ethics, http://www.aipg.org/about/ethics.htm

AIPG Disciplinary Summary, updated annually, http://www.aipg.org/About/disciplinary.htm.

Association of American Geographers Statement on Professional Ethics, http://www.aag.org/cs/resolutions/ethics

Center for the Study of Ethics in the Professions, Illinois Institute of Technology, http://ethics.iit.edu

Codes of Ethics Collection, http://ethics.iit.edu/ecodes/

Modules in Applied Ethics, http://ethics.iit.edu/projects/modules-applied-ethics

Coalition on Publishing Data in the Earth and Space Sciences, www.copdess.org

Committee on Publication Ethics, https://publicationethics.org/

Ethics in Science and Engineering National Clearinghouse, http://www.umass.edu/sts/digitallibrary/

Ethics Unwrapped, University of Texas, McCombs School of Business, http://ethicsunwrapped.utexas.edu

Geological Society of America Code of Conduct, https://www.geosociety.org/GSA/ Membership/Code_of_Conduct/GSA/Membership/Code_of_Conduct

Impact CS Project, George Washington University, http://www.seas.gwu.edu/%7Eimpactcs// index.html

International Association for Promoting Geoethics, http://www.geoethics.org

IAPG Codes of ethics and codes of conduct, http://www.geoethics.org/#!codes/c24p2

IAPG Cape Town Statement, http://www.geoethics.org/ctsg

International Council for Science, Statute 5 Universality of Science http://www.icsu.org/ freedom-responsibility/cfrs/statute-5

Internet Encyclopedia of Philosophy—a peer-reviewed academic resource, http://www.iep. utm.edu/ethics/

Murdough Center for Engineering Professionalism, Texas Tech University, http://www. depts.ttu.edu/murdoughcenter/

National Association of State Boards of Geology, http://www.asbog.org

ASBOG Code of Ethics, http://asbog.org/documents/Code of Ethics--2014.pdf

National Center for Case Study Teaching in Science, University at Buffalo http://sciencecases. lib.buffalo.edu/cs/

National Center for Professional & Research Ethics, Ethics CORE, https://www. nationalethicscenter.org

National Science Foundation, Responsible Conduct of Research, http://www.nsf.gov/bfa/ dias/policy/rcr.jsp

National Society of Professional Engineers, Ethics, http://www.nspe.org/resources/ethics

Northeast Ethics Education Partnership (NEEP), http://brown.edu/research/research-ethics/neep

NEEP training materials can be accessed via http://www.brown.edu/research/research-ethics/ northeast-ethics-education-partnership/training-materials/training-materials

Office of Science and Technology Policy, Executive Office of the President (2000), *Federal Policy on Research Misconduct*, Federal Register 65:76260-76264. https://www.federalregister. gov/documents/2000/12/06/00-30852/executive-office-of-the-president-federal-policy-on-research-misconduct-preamble-for-research

Online Ethics Center for Engineering and Research, http://www.onlineethics.org/CMS/ about/UserGuide/18848.aspx

Poynter Center for the Study of Ethics and American Institutions, https://provost.indiana. edu/poynter-center/

Stanford Encyclopedia of Philosophy, http://plato.stanford.edu

Teaching Geoethics across the Geoscience Curriculum, David Mogk, Department of Earth Sciences, Montana State University and Monica Bruckner, SERC, Carleton College, https://serc.carleton.edu/geoethics/index.html

U.S. Department of Health & Human Services, Office of Research Integrity, http://ori.hhs. gov

Wikipedia Portal for Ethics, http://en.wikipedia.org/wiki/Portal:Ethics

World Conference on Research Integrity WCRI (2013), Montreal Statement on Research Integrity in Cross-Boundary Research Collaborations, http://www.researchintegrity.org/ Statements/Montreal%20Statement%20English.pdf

World Conference on Research Integrity WCRI (2010), Singapore Statement on Research Integrity, www.singaporestatement.org/statement.html

INDEX

Scientific Integrity and Ethics in the Geosciences, Special Publications 73,
First Edition. Edited by Linda C. Gundersen.
© 2018 American Geophysical Union. Published 2018 by John Wiley & Sons, Inc.